Soil Remediation and Improvement Through the Application of By-Product Amendments, 2nd Edition

Soil Remediation and Improvement Through the Application of By-Product Amendments, 2nd Edition

Guest Editors

Elias Afif Khouri

Rubén Forján Castro

MDPI

Basel • Beijing • Wuhan • Barcelona • Belgrade • Novi Sad • Cluj • Manchester

Guest Editors
Elias Afif Khouri
Department of Organisms and Systems Biology
University of Oviedo
Mieres
Spain

Rubén Forján Castro
Department of Soil Science and Agricultural Chemistry
University of Salamanca
Salamanca
Spain

Editorial Office
MDPI AG
Grosspeteranlage 5
4052 Basel, Switzerland

This is a reprint of the Special Issue, published open access by the journal *Environments* (ISSN 2076-3298), freely accessible at: https://www.mdpi.com/journal/environments/special_issues/8L7Z725SO6.

For citation purposes, cite each article independently as indicated on the article page online and as indicated below:

Lastname, A.A.; Lastname, B.B. Article Title. *Journal Name* **Year**, *Volume Number*, Page Range.

ISBN 978-3-7258-8459-9 (Hbk)
ISBN 978-3-7258-8460-5 (PDF)
https://doi.org/10.3390/books978-3-7258-8460-5

Cover image courtesy of Rubén Forján Castro

Contents

About the Editors

Elias Afif Khouri

Elías Afif Khouri is a Professor at the Universidad de Oviedo. His research focuses on nutrient dynamics in agricultural and forest soils, as well as on methods and techniques for soil and plant analysis. He is currently involved in several national and European research projects related to the structure and dynamics of forest ecosystems, and to the use of iron-based nanoparticles and carbon-based amendments for the sustainable remediation of polluted soils.

Rubén Forján Castro

Rubén Forján Castro is an Assistant Professor at the Universidad de Salamanca. His research focuses on soil remediation and restoration, with particular interest in contaminated and degraded soils, waste-derived amendments, phytoremediation, nature-based solutions, and circular economy approaches. His scientific activity is mainly related to soil science, environmental chemistry, geochemistry, water science, and nanotechnology for environmental applications. He has authored around 70 scientific publications, edited a scientific book, and participated in 10 research projects, including Horizon Europe, LIFE, national, and regional projects.

Preface

The management of degraded and contaminated soils remains one of the major environmental challenges associated with land use, industrial activity, mining, agriculture, and waste generation. In this context, the use of by-product amendments offers an opportunity to combine soil remediation and soil improvement with the valorization of residual materials. This Reprint gathers contributions that explore different approaches for restoring soil quality, reducing contaminant mobility, improving soil fertility, and promoting more sustainable management practices. We believe that the studies included here provide useful insights for researchers, practitioners, and decision-makers working on soil protection, circular economy, and environmental recovery.

Elias Afif Khouri and Rubén Forján Castro
Guest Editors

environments

MDPI

Editorial

Soil Remediation and Improvement Through the Application of By-Product Amendments, 2nd Edition

Elias Afif [1] and Rubén Forján [2,3,*]

[1] Agroforestry Engineering Area, Department of Organisms and Systems Biology, Polytechnic School of Mieres, University of Oviedo, 33600 Mieres, Spain; elias@uniovi.es
[2] Soil Sciences and Agricultural Chemistry Area, Faculty of Agricultural and Environmental Sciences, Universidad de Salamanca, 37007 Salamanca, Spain
[3] Environmental Biogeochemistry & Raw Materials Group, Institute of Natural Resources and Territorial Planning (INDUROT), University of Oviedo, 33600 Mieres, Spain
* Correspondence: forjanruben@usal.es

Modern society is currently facing two pressing environmental crises: the systemic loss of fertile, healthy soil due to various forms of degradation and the escalating generation of diverse waste streams. These phenomena are intrinsically linked to the intensification of global warming and the acceleration of anthropogenic climate change; however, ongoing research increasingly supports Nature-Based Solutions (NBSs) as an effective framework for addressing these challenges [1,2]. By adopting a circular economy approach [3], by-products that are traditionally viewed as environmental burdens in one sector can be repurposed as valuable resources in another, providing cost-effective strategies that not only restore soil health but also play a critical role in carbon sequestration and the global effort to mitigate climatic instability [4,5].

This second edition of the Special Issue brings together nine contributions that address these challenges from complementary perspectives, including residual nutrient use in crop rotations, vermicompost and vermicompost tea as organic fertilizers, biochar-based soil restoration, organic amendments in vineyards, valorization of sewage sludge and pruning residues, agro-industrial waste-derived circular fertilizers, and soil amendments for improving crop performance under water limitation. Collectively, these studies show that sustainable soil management cannot be based on a single amendment or agronomic practice. Rather, it requires the integration of organic residues, carbon-rich materials, biological processes, crop-specific responses, and rigorous monitoring of soil and plant indicators.

A coherent reading of the contributions included in this Special Issue shows that the value of circular amendments does not lie only in replacing synthetic fertilizers, but in reshaping the soil–plant system through residual fertility, biological activation, carbon stabilization, improved crop quality, and stress mitigation. The two studies by Oyege and Balaji Bhaskar provide an appropriate starting point because they demonstrate that vermicompost-derived nutrients may remain agronomically relevant beyond the crop to which they are initially applied. In a strawberry–corn double cropping system, moderate vermicompost application improved strawberry yield, biomass accumulation, and fruit quality without additional fertilization during the strawberry phase (Contribution 1). In a complementary corn–soybean rotation, residual vermicompost and vermicompost tea inputs enhanced soybean biomass, yield, chlorophyll status, stomatal conductance, and nutrient uptake (Contribution 2). These findings shift the interpretation of organic amendments from short-term nutrient sources to residual fertility tools capable of supporting rotational sustainability.

However, these same studies also introduce one of the central messages of this Special Issue: circular amendments are not automatically beneficial merely because they are organic. In the strawberry system, excessive vermicompost rates, particularly when associated with elevated sodium content, reduced plant vigor and marketable yield. Similarly, the soybean study showed that positive responses depended on balanced amendment combinations rather than maximal input application. This dose-dependent behavior connects directly with the broader evidence presented throughout the issue: the agronomic performance of waste-derived materials depends on feedstock composition, maturity, salinity, nutrient ratios, crop sensitivity, soil properties, and water availability. Therefore, circularity must be accompanied by characterization, calibration, and agronomic control.

This transition from nutrient supply to plant functional response is further developed by Rehman et al., who investigated the effect of vermicompost tea on the metabolic profile of *Diplotaxis muralis* using NMR spectroscopy (Contribution 3). Their work shows that liquid organic amendments can modify not only growth-related parameters but also the accumulation of amino acids, organic acids, flavonoids, glucosinolates, and phenolic compounds. This is a relevant conceptual advance because sustainable fertilization is frequently evaluated through yield alone, whereas crop nutritional quality and secondary metabolism are often treated as secondary outcomes. In this context, vermicompost tea emerges not only as a nutrient input, but also as a biological modulator of crop quality.

The same need to move beyond a purely fertilizing perspective is evident in the studies focused on carbon-rich amendments. Tang provides a mechanistic framework by reviewing how biochar can influence soil restoration through nutrient retention, reduced leaching, carbon, nitrogen and phosphorus cycling, enzymatic activity, and microbial community dynamics (Contribution 4). Nonetheless, this review also clarifies that biochar is not a uniform product: its behavior depends on feedstock, pyrolysis conditions, pH, surface chemistry, soil mineralogy, and microbial context. This point aligns with the evidence from the vermicompost studies: the effectiveness of a circular amendment is determined not only by its origin but by its physicochemical identity and interaction with the receiving soil.

The field relevance of this principle is illustrated by Pierini et al., who evaluated straw mulch and straw mulch combined with biochar in vineyards of the Alentejo wine region (Contribution 5). Their results show that organic amendments interacted with cultivar and topographic position to influence grape chemical composition, phenolic parameters, and color-related traits. This contribution is particularly valuable because it places soil amendment research in a high-value perennial system, where soil protection, water regulation, topography, and fruit quality are inseparable. In this manner, soil restoration is connected with not only edaphic recovery but also the quality of the final agricultural product.

The circular use of vineyard and agro-industrial residues is further expanded through two complementary technological routes. Nascimento-Gonçalves et al. assessed vermicompost and leachate produced from sewage sludge and vineyard pruning residues using cucumber germination and early seedling growth as bioassays (Contribution 6). Their results showed pathogen-free products, metal concentrations below regulatory limits, and high germination rates under the tested conditions, while also confirming that higher vermicompost doses did not necessarily improve seedling performance. Díaz-Rasero et al., in contrast, transformed vineyard stumps into hydrochars and pyrochars, with and without iron addition, showing that hydrothermal carbonization and pyrolysis generate materials with distinct structural and chemical properties (Contribution 7). Their preliminary grapevine tolerance screening did not reveal evident phytotoxic effects based on chlorophyll index measurements. Together, these studies show two complementary routes for residue valorization: biological stabilization through vermicomposting and thermochemi-

 https://doi.org/10.3390/environments13050260

cal conversion into carbonaceous materials. Both routes are promising, but both require safety assessment, phytotoxicity screening, and long-term soil–plant validation.

This functional differentiation among circular products is also central to the study by Maffia et al., who evaluated fertilizers derived from olive pomace, citrus residues, wood sawdust, straw, and sulfur-bentonite enrichment (Contribution 8). Their findings indicate that vermicompost-based formulations enhanced total organic carbon, total nitrogen, microbial biomass carbon, and enzymatic activities, whereas compost-based formulations contributed more strongly to organic matter accumulation and longer-term soil resilience. This distinction is important because it suggests that circular fertilizers should be classified not only by origin but also by function. Vermicompost may be more suitable when rapid biological activation and nutrient cycling are desired, compost may be more appropriate for building persistent organic matter and structural stability, and mineral-enriched formulations may target specific nutrient or soil constraints.

The final contribution, by Kruspe and Koyro, extends the discussion from fertility and soil restoration to climatic stress adaptation. By examining biochar and superabsorbent polymers under contrasting soil water contents in *Chenopodium quinoa*, they showed that amendments can modify plant water-use efficiency, photosynthetic regulation, oxidative stress responses, lipid peroxidation, and proline accumulation (Contribution 9). Their results are particularly relevant because future soil management will have to address not only nutrient depletion and organic matter decline but also drought frequency, irregular rainfall, and reduced irrigation reliability. In this regard, amendment performance must be interpreted through plant physiological responses, not only through soil chemical properties.

Taken together, these studies present a holistic view of circular soil management. Vermicompost and vermicompost tea contribute to residual fertility, microbial activation, nutrient uptake, and crop quality; biochar and related carbonaceous materials offer opportunities for nutrient retention, carbon stabilization, microbial habitat formation, and water regulation; and composted or vermicomposted agro-industrial residues provide practical pathways for recycling organic waste into agronomically useful products. Across these contributions, the same principle emerges repeatedly: circular amendments are most effective when they are selected and applied according to their function, chemical composition, receiving soil, target crop, and environmental constraint.

This integrated perspective also identifies the main knowledge gaps that remain. Long-term field studies are still needed to determine whether short-term improvements in plant performance, microbial activity, nutrient availability, and crop quality translate into durable soil carbon accumulation, stable productivity, and measurable climate mitigation. More attention should also be paid to feedstock variability, salinity, heavy metals, maturity indices, nutrient release kinetics, and potential organic contaminants. The evidence assembled in this Special Issue shows that Nature-Based Solutions and circular economy approaches are scientifically robust and agronomically promising, but their implementation must move from generalized application toward precision management. The future of circular soil restoration will depend not on applying more residues, but on applying better-characterized amendments in the right dose, at the right time, and for the right soil–crop system.

In conclusion, this Special Issue shows that Nature-Based Solutions and circular economy strategies are becoming operational tools for restoring degraded soils, recycling organic residues, reducing dependence on synthetic inputs, improving crop quality, and strengthening plant resilience under climatic stress. Organic residues and carbon-rich amendments can substantially contribute to sustainable agriculture, but their success depends on rigorous characterization, appropriate dosing, site-specific management, and long-term validation. By bringing together field trials, mechanistic review, physiological

https://doi.org/10.3390/environments13050260

analysis, metabolomics, residue valorization, and microbiological assessment, this second edition provides a coherent and timely contribution to the development of circular, climate-resilient, and biologically functional agricultural systems.

Conflicts of Interest: The authors declare no conflicts of interest.

List of Contributions:

1. Oyege, I.; Balaji Bhaskar, M.S. Residual Impacts of Vermicompost-Derived Nutrients on a Strawberry-Corn Double Cropping System Under Plasticulture in South Florida. *Environments* **2025**, *12*, 171.
2. Oyege, I.; Balaji Bhaskar, M.S. Sustainable Soybean Production Using Residual Vermicompost Inputs in Corn-Soybean Rotation. *Environments* **2025**, *12*, 333.
3. Rehman, S.u.; De Castro, F.; Aprile, A.; Benedetti, M.; Fanizzi, F.P. Influence of Vermicompost Tea on Metabolic Profile of *Diplotaxis muralis*: An NMR Spectroscopic Analysis. *Environments* **2025**, *12*, 366.
4. Tang, K.H.D. Biochar Amendments for Soil Restoration: Impacts on Nutrient Dynamics and Microbial Activity. *Environments* **2025**, *12*, 425.
5. Pierini, M.; Harjivan, S.G.; Sieli, N.; Cabrita, M.J.; Prats, S.; Catarino, S.; Ricardo-da-Silva, J.M. Effect of Organic Soil Amendments and Vineyard Topographic Position on the Chemical Composition of Syrah, Trincadeira, Alicante Bouschet, and Antão Vaz Grapes (*Vitis vinifera* L.) in the Alentejo Wine Region. *Environments* **2026**, *13*, 44.
6. Nascimento-Gonçalves, E.; Azevedo, T.; Lopes, H.; Medeiros, C.; Falco, V.; Coelho, A.C.; Sousa, J.R.; Coimbra, A.M.; Roboredo, M.; Oliveira, P.A.; Morais, M.C. Vermicompost and Leachate from Sewage Sludge and Vineyard Pruning Residues: An Exploratory Assessment of Cucumber Germination and Early Seedling Growth. *Environments* **2026**, *13*, 59.
7. Díaz-Rasero, J.M.; Sosa, T.; Ledesma, B.; Román, S. Physicochemical Characterization of Vineyard Stump-Derived Hydrochars and Pyrochars and Preliminary Grapevine Tolerance Screening. *Environments* **2026**, *13*, 83.
8. Maffia, A.; Marra, F.A.; Battaglia, S.; Mallamaci, C.; Muscolo, A. Enhancing Soil Functionality Through Circular Fertilizers Derived from Agro-Industrial Wastes: Insights into Microbiological and Biochemical Dynamics. *Environments* **2026**, *13*, 93.
9. Kruspe, N.; Koyro, H.-W. Impact of Biochar and Superabsorbent Polymer at High and Low Soil Water Content on Physiological and Biochemical Response of *Chenopodium quinoa* Willd. (cv. UDEC-5). *Environments* **2026**, *13*, 136.

References

1. Lal, R. Soil Carbon Sequestration Impacts on Global Climate Change and Food Security. *Science* **2004**, *304*, 1623–1627. [CrossRef] [PubMed]
2. European Commission. Nature-Based Solutions. Available online: https://research-and-innovation.ec.europa.eu/research-area/environment/nature-based-solutions_en (accessed on 27 April 2026).
3. European Commission. *Towards a Circular Economy: A Zero Waste Programme for Europe. Communication from the Commission to the European Parliament, the Council, the European Economic and Social Committee and the Committee of the Regions; COM (2014) 398 final*; European Commission: Brussels, Belgium, 2014.
4. Zabaniotou, A.; Rovas, D.; Libutti, A.; Monteleone, M. Boosting Circular Economy and Closing the Loop in Agriculture: Case Study of a Small-Scale Pyrolysis-Biochar Based System Integrated in an Olive Farm in Symbiosis with an Olive Mill. *Environ. Dev.* **2015**, *14*, 22–36. [CrossRef]
5. Yrjälä, K.; Ramakrishnan, M.; Salo, E. Agricultural Waste Streams as Resource in Circular Economy for Biochar Production towards Carbon Neutrality. *Curr. Opin. Environ. Sci. Health* **2022**, *26*, 100339. [CrossRef]

 https://doi.org/10.3390/environments13050260

environments

MDPI

Article

Impact of Biochar and Superabsorbent Polymer at High and Low Soil Water Content on Physiological and Biochemical Response of *Chenopodium quinoa* Willd. (cv. UDEC-5)

Natalie Kruspe * and Hans-Werner Koyro

Institute of Plant Ecology, Justus Liebig University Giessen, 35392 Giessen, Germany; hans-werner.koyro@uni-giessen.de
* Correspondence: n.kruspe@outlook.de

Abstract

In agriculture, soil amendments like compost, manure, superabsorbent polymers (SAP) and biochar (BC) are already in use to mitigate the effects of water shortage and to obtain a higher yield and survivability. The present study focuses on the impact of BC and SAP under moderate and reduced soil water content (SWC) on the physiological and biochemical response of *Chenopodium quinoa* Willd. (cv. UDEC-5), a naturally drought-resistant and strategic crop in arid regions, with the aim of further improving its resilience and biomass production. Plants were grown in the presence or absence (control) of SAP (1% or 0.1% g/100 g SAP) or BC (3% g/100 g BC) by taking into account the smallest possible amount of irrigation necessary for optimal growth of the control. Sixty-five days after sowing, the reduced watering approaches started. The irrigation amount was reduced slowly until plants without any amendment showed a significant reduction in CO_2/H_2O gas exchange and further significant changes in 23 morphological, physiological and biochemical symptoms of water shortage. Each amendment already caused individual plant response in wet conditions: The soil amendments of SAP (1% and 0.1%) and BC had no significant effect on biomass production but caused changes in PS I (portion of oxidized and open centers in PS I), the C/N ratio and N content. The addition of SAP (0.1% and 1%) led to a decrease in gH^+, ECStmAu $\times$ gH^+, R_D, R_L, the C_i/C_{atm} ratio and ETR/A_{gross} ratio and to an increase in water use efficiency (WUE), especially in the 0.1% SAP treatment. In moderate conditions, 0.1% SAP and 3% BC caused a significant increase in both the LOP and C/N ratio. In the moderate treatments, the application of 0.1% SAP promoted an increased A_{net}, while 3% BC promoted a significant reduction in malondialdehyde (MDA). The results of the present quinoa experiment indicate the drought avoidance mechanism of the control under low SWC. The reduced transpiration led to increased WUE due to the efficient use of the substomatal CO_2 reservoir under low C_s and low E. It could also be confirmed that quinoa plants balanced low soil water potential by the accumulation of compatible solutes to lower the LWP and LOP. Drought led, especially in leaves in the 1% SAP treatment, to significant reductions in CO_2/H_2O gas exchange (A_{net}, R_D), decreases in Y (II) and ETR in PS II, and an increase in the ETR/A ratio and over-reduced centers in PS I, pointing to an increased appearance of reactive oxygen species (ROS) in the chloroplasts. The latter change was indicated by higher levels of lipid peroxidation (MDA). It could be shown that the response of the test species *Chenopodium quinoa* to the addition of BC and SAP proved to be highly adaptable. The plant reacted in a very coordinated and specific way to both the danger of oversupply of SAP soil amendments under water shortage conditions and an effective adaptation to a limited water supply with 3% BC and 0.1% SAP by increasing WUE and proline content. However, BC also had a mitigating effect on the level of reactive oxygen species (ROS). It can be assumed that this effect is based on a more plant-compatible, less one-sided ion composition of BC. The results presented indicate that SAP and BC can have an impact on the water and nutrient accessibility for plants. Therefore, optimal biomass production and plant response can only be reached if plant soil interactions and competition between SAP, BC and the plant roots are taken into account when planning for climate-resilient, water-saving agriculture.

Keywords: biochar; superabsorbent polymers; quinoa; drought; photosynthesis; malondialdehyde; proline

1. Introduction

In recent decades, a continuous increase in the incidence of extreme weather events associated with climate change has been observed all around the world [1]. Rising global temperatures have led to rapid climate changes, together with more intense rainfall events and longer periods of drought [2], leading, among other things, to overall reduced levels of soil moisture [3–5]. The rise in temperature will increase evapotranspiration rates, which in turn will result in longer, harder and more frequent periods of drought [5–8]. In arid and semi-arid regions, in particular, water retention capacity will decline due to the sandy texture and lack of organic matter in the soil [9].

For vegetation, drought is a stressful environmental factor in which insufficient precipitation leads to a progressive reduction in the amount of water available in the soil [10], affecting plant development and other metabolic processes [11]. Plants have evolved defense mechanisms to cope with drought [12].

Different components, such as photosynthetic pigments and photosystems, the electron transport system and CO_2/H_2O gas exchange processes, can be threatened by water deficiency. In the first response, plants tend to decrease stomatal conductance to increase water use efficiency and photosynthetic productivity, finally leading, among other effects, to a restriction of growth [13]. This is a survival mechanism to conserve water, as plants reduce their transpiration rate to maintain their water balance and protect against wilting. Additionally, plants often increase ATP and NADPH consumption and reduce ROS production through increasing photorespiration in C_3 plants and high photosynthetic rates and efficiency in C_4 plants [13].

However, to prevent damage to photosynthesis, the supply and demand of electrons must be regulated to maintain a balance. Drought stress impairs the activity of PS I and PS II, which can lead, among other things, to an increase in over-reduction on the PS I acceptor side. One way to reduce PS I acceptor over-reduction while reducing electron demand would be to adjust the cyclic electron flows [14]. The reduction in the supply of electrons is achieved by increasing the level of non-photochemical quenching (NPQ), which dissipates light energy and decreases the efficiency of photochemical reactions in photosynthesis [15].

If the plant's response to drought stress is not strong enough to protect it, drought can cause severe damage to the photosystems through the generation of oxygen species (ROS) in chloroplasts, which can directly attack membrane lipids and increase lipid peroxidation [16]. Polyunsaturated fatty acids (PUFAs) are major components of the biomembrane and particularly susceptible to attacks by ROS [17]. The degradation of lipid hydroperoxides produces a wide variety of aldehydes, some of which are highly reactive, such as malondialdehyde (MDA), one of the most common reactive aldehydes [17]. As one of the final products of peroxidation of unsaturated fatty acids in phospholipids, MDA aggravates oxidative stress through the production of lipid radicals that can react them-

 https://doi.org/10.3390/environments13030136

selves and damage proteins [18,19]. The amount of free MDA, which causes damage to cell membranes under stressful conditions, has been widely used as a marker for oxidative stress [18,20].

Due to these associated changes, many climate models and emission scenarios assume negative effects of climate change on crop yields [21]. In densely populated regions where agriculture is intensively practiced, such as South-Central Asia, the Southeast of South America, the Southeast of the United States and Central Europe, there is an increased risk to global food security because these regions are exposed to a higher risk of drought [22,23]. Despite worsening conditions, the world population is anticipated to grow to nine billion people in the coming decade [24]. However, if droughts in agriculture increase in the future, it may make sense to use plants that survive and thrive under water shortage conditions such as arid and semi-arid regions. In this context, the cultivation of the resilient Andean crop quinoa (*Chenopodium quinoa* Willd.) is of particular interest to global agriculture due to its high genetic biodiversity and its adaptability to varying environmental conditions, including nutrient-poor and saline soils and water shortages [25–27]. Moreover, quinoa seeds show a superior nutritional profile compared to other, more commonly consumed grains [28]. In addition, due to evolutionary adaptions to dry weather conditions, quinoa plants are particularly efficient in coping with long periods of drought in terms of their morphology and physiology due to drought avoidance mechanisms [29]. Osmotic adjustment plays an important role in quinoa, as a low osmotic potential allows the turgor of the leaves to be maintained even when the soil water potential is extremely low [30]. In addition, quinoa limits the growth and closes stomata during soil drying, maintaining leaf water potential (LWP) and a lower rate of photosynthesis that results in an increased efficiency of water usage [31,32].

This can contain the accumulation of osmolytes such as amino acids (e.g., proline) and soluble sugars. These solutes reduce the osmotic potential of the cell, causing water to enter the cell initiating the major mechanism for maintaining cell turgor when the water potential decreases [33]. Additionally, these solutes in the cell cytoplasm are able to protect enzymes involved in photosynthesis and metabolism and therefore in growth and productivity [34]. Quinoa accumulates protective proteins and solutes such as proline to scavenge free radicals and thus prevents membrane protein denaturation during water shortage [35]. One reason to choose quinoa for the present study was that the cultivation of such pseudo-cereal might be necessary to ensure long-term food security worldwide despite climate change. However, to achieve successful cultivation with optimal yields for quinoa in regions around the world where water shortage is increasing, it is reasonable to optimize soil quality in which plants grow, thereby improving the abilities of this resilient species. It does not make any sense to look at the plant response in isolation and study only certain modes of water transfer. A general consideration of the plant water relations of the whole plant in nature needs to involve the soil–plant–atmosphere continuum (SPAC) [36,37] and especially the soil.

One opportunity to improve the quality of soils with low water retention capacity and high evaporation is to add biochar (BC) [38]. Several studies reported that a positive effect of BC addition on water holding capacity could be shown in a higher extent in coarse-textured and sandy soils, typical for arid regions, than in fine textured and fertile soils [38,39]. The addition of biochar significantly improves the physiochemical properties of the soil, such as organic matter content and water retention capacity [40,41]. Biochar addition also supported the accumulation of osmotic active substance in the plant tissues, which led to improved water uptake by the plants [42,43]. This had a positive effect on the interaction between soil-plant-water, resulting in improved photosynthetic performance through enhanced water use efficiency [44–46]. Zhang et al. [47] show that the addition of biochar

 https://doi.org/10.3390/environments13030136

even led to a significantly improved water use efficiency (WUE) due to leaf stomatal closure and the reduction in transpiration rate during drought stress by soybeans. Additionally, its porous structure and the related surface area enlargement enable increased microbial activity and improve the binding of essential anions (such as nitrate and phosphorous) and important nutritive cations (such as K, Ca and Mg), the latter one because of its high cation exchange capacity (CEC) [41,48]. Besides declining productivity, drought also declines the concentration of nitrogen and phosphorus in plants [49]. The influence of biochar addition on plant leaf nitrogen is discussed quite controversially. Xu et al. [45] showed that biochar increases leaf N accompanied with improved soil available N and biological N fixation. In contrast to that, some studies reported that biochar immobilizes mineral N in the soil due to an increased C/N ratio [50,51]. However, in drought conditions, the uptake of nitrogen can be of eminent importance to enhance oxidative stress response. It has been shown that nitrogen supply could alleviate the adverse effects of drought stress on plants by enhancing antioxidant defense and by inducing proline accumulation [52–54].

Another possible option for soil improvement is the application of superabsorbent polymers (SAP). The crosslinked hydrophilic structure of SAP enables them to absorb and store water in their three-dimensional networks, whose weight exceeds their own by a hundred to a thousand times, leading to an extremely high water absorption and retention capacity [55–59]. The correlation analyses of [60] could show that SAP improve the utilization rate of soil water (water use efficiency WUE) and nutrients (nutrient use efficiency NUE) by improving the soil structure, increasing soil organic carbon and microbial enzyme activity. As logical consequence, this led to an increase in yield of winter wheat. Several studies have shown, apart from the capability to increase yields, due to its water-saving and fertilizer-conserving properties, that the water-storage capacity of SAP supports survivability of plants during periods of drought in arid and semi-arid areas. Refs. [61–63] reported that ion exchange capacity of SAP in a saline soil hindered an excessive accumulation of toxic ions like Cl^- and Na^+ to enter plant organs. However, when used in soil, SAP can also influence the overall C/N ratio of the soil by holding water and nutrients. Therefore, SAP can improve plant production in times of water shortage by retaining more water and nutrients in the active rooting zone [64]. However, by retaining water and nitrogen compounds, they can also indirectly and adversely change the soil's C/N balance over time. Additionally, they can affect nutrient availability (such as Ca^{2+}, Zn, K and Mg^{2+}) for plants, which can negatively affect plant growth and nutrient cycling [65].

Although both BC and SAP share a high potential to improve soil parameter in dry regions. However, the addition of BC and SAP may also have disadvantages and limitations as they could compete with plant roots for scarce water resources. Based on these considerations, we developed the following hypotheses:

1. The increasing scarcity of water necessitates the most efficient possible use of irrigation water. However, BC and SAP effect a negative response on biomass production in case of the smallest possible amount of irrigation necessary for optimal growth of the control is applied.
2. The increase in WHC caused by the addition of additives such as BC or SAP will result in a higher water demand than in control plants and even at high SWC in symptoms of drought stress.
3. The soil additives SAP or BC may prime quinoa for an improved response to subsequent stress at low soil water content.
4. Harmful competition of water uptake between plants and the soil additives BC and SAP may cause adverse effects at low soil water content.
5. The plant's ability to handle oxidative stress is linked to its nitrogen (N) uptake and metabolism.

 https://doi.org/10.3390/environments13030136

6. When quinoa needs to decide because of limited resources at low SWC between two desirable but incompatible features being either growth (e.g., biomass production) or long survival (e.g., resistance to stress, risk avoidance), it will choose latter one.

Given its economic and industrial significance, enhancing the survivability and yield of *Chenopodium quinoa* Willd. is crucial in times of global climate changes. The research examines and compares the impact of super absorbent polymers or biochar on plant performance, biomass production, photosynthesis activity and ROS defense of quinoa at lower and higher soil water content. The goal is to develop efficient sustainable growth conditions that minimize water usage while optimizing survival and yields of quinoa in arid and semi-arid regions.

2. Materials and Methods

2.1. Plant Material and Growth Conditions

The experiment was carried out in a greenhouse at the University of Giessen, Germany (latitude 50.5873 and longitude 8.67554), under controlled conditions at an average temperature of 22–24 °C/17 °C (day/night), a relative humidity of 60–80% (±10) and a photoperiod of 16/8 h (day/night). The test species was *Chenopodium quinoa* (Willd.). Seeds of the variety UDEC-5 were kindly provided by the Institute of Plants Ecology, Giessen, Germany. For preparation, the seeds were germinated in a pot with vermiculite. Then, 14 days after germination, the plants were transferred in QuickPot plates with until soil type 0—zero soil (H. Nitsch & Sohn GmbH & Co. KG, Kreuztal, Germany). Fourty-one days after the start of the culture, the emerged seedlings were transferred to pots with a capacity of 1.5 kg (one plant per pot) filled with a mixture of topsoil (69%) and sand (29%) to obtain a low water-holding capacity and perlite (2%) to favor better aeration. We divided the 56 pots into 4 groups (14 pots per treatment): control without any amendment, two different treatments with superabsorber 1% (1% SAP = 15 g/pot) or 0.1% (0.1% SAP = 1.5 g/pot) superabsorber, and one treatment with 3% biochar (3% BC = 30 g/pot). The polymer used in the experiment was STOCKOSORB®, a granular potassium polyacrylate manufactured by Evonik Nutrition & Care GmbH, Krefeld, Germany [66–68]. The biochar (Schottdorf-type) used in this study, produced by Carbon Terra, Augsburg, Germany, consisted primarily of silicon (Si), calcium (Ca), potassium (K), iron (Fe), magnesium (Mg), sodium (Na), phosphorus (P) and sulfur (S) [67–69]. The treatments of 0.1%SAP and 3% BC were chosen because of a similar water holding capacity. The plants of all four treatments received exactly the same amount of water (7.72 mL H_2O/100 g dw soil) which was necessary for optimal growth of the control (=20% water holding capacity). Water-holding capacity was determined according to the methods of [68]. Although this resulted in a constant water content in the soil for all treatments, it also led to a decrease in the percentage water holding capacity at high SWC due to the addition of soil additives (1% SAP: 13.06; 0.1% SAP: 19.41; 3% BC 19.22% WHC). The division of the treatments in wet and dry treatments started at day 65 after sowing. Based on the plants' daily water usage, the dry treatments received 10 mL less water per day (4.6 mL H_2O/100 g dw soil) than the wet treatments. The slow reduction in the soil water content induced a continuous but slowly increase in the water shortage to stop reduction before crossing the permanent wilting point [69]. In this period, pots were weighed every day to control the water usage and to adapt the water supply to the increasing demand of the growing control plants. As a result, the plants of all four wet treatments received exactly the same amount of water (high SWC = 9.65 mL H_2O/100 g dw soil), which was necessary to reach 25% WHC of the control soil. Over the entire period, the treatments were fertilized once using a nutrient solution modified after [70]. After 78 days of treatment, a final harvest was carried out.

 https://doi.org/10.3390/environments13030136

2.2. *Growth Parameter*

At the day of harvest, we started to separate the root system from the soil by washing them before moisture was quickly removed from the surface of the plants' roots with tissue. Then, we measured the fresh weight (FW) of root, stem and leaves separately as well as 100 g soil of each sample. Dry weight (DW) was determined after drying the samples at 90 °C for 24 h in the oven. The water content (WC) was determined by calculating the difference between the FW and DW of plants.

2.3. *Water Relations*

The leaf water potential was determined with the Scholander pressure chamber [71,72]. For this purpose, the plants ($n = 3$ replicates) stayed in a darkroom of the greenhouse at night. The measurements were taken on the following day from the first fully developed leaf of the plant. In addition, the concentration of osmotically active particles was measured with an Osmomat 030, Gonotec^{TM}, Berlin, Germany. According to [73], the osmotic potential (in MPa) was calculated.

2.4. *Light Dependent Reactions of Photosynthesis*

The impact of the soil amendments and the water supply to the light dependent reactions of photosynthesis was determined with a MultispeQ V 2.0TM (PhotosynQ Inc., East Lansing, MI, USA) at the first fully developed leaf of each plant ($n = 4$ replicates). For the recording of the light saturation curves, we used a protocol provided by PhotosynQ "Photosynthesis RIDES Actinic Series 10 2x2000 of Uwe Grueters".

According to [74], for photosystem II, the potential maximum efficiency of PS II (F_v/F_m), the distribution of the photochemical quantum yield of PS II (Y (II)), the quantum yield of regulated non photochemical energy loss in PS II (Y (NPQ)), and the quantum yield of non-regulated non-photochemical energy loss in PS II, equivalent to Y (NO), were measured, and the electron transport rate (ETR) was calculated. Additionally, we indirectly studied photosystem I and the electrochromic shift generated across the thylakoid membrane during photosynthesis, (ECStmAu) caused by the buildup of a proton gradient and the membrane ATP-synthase activity with the parameters steady state proton flux (vH^+), proton conductivity (gH^+). By means of extinction-based parameters, we also determined the share of oxidized, open and over reduced centers of the PS I.

2.5. *Adjustment of the Stomatal Regulation During Light Independent Reactions of Photosynthesis (CO_2/H_2O Gas-Exchange)*

Parameters of the CO_2/H_2O gas exchange were determined by using a LI-6400/XT portable photosynthesis system (Li-Cor Inc., Lincoln, NE, USA) with a 6400-40(B) LED light source attached to the leaf chamber at light saturation conditions with a PPFD of 1500 μmol PAR m^{-2} s^{-1}. Net CO_2 assimilation rate (A_{net} in μmol m^{-2} s^{-1}), intercellular CO_2 concentration (C_i in ppm CO_2), stomatal conductance (C_s in mmol m^{-2} s^{-1}) and transpiration rate (E in mmol m^{-2} s^{-1}) were measured at the first fully developed leaf of the plant ($n = 4$ replicates). Water use efficiency (WUE) was calculated as the ratio of A_{net}/E. According to [75], a saturation curve was created as well as the dark respiration point (R_D in μmol CO_2 m^{-2} s^{-1} at 0 μmol PAR m^{-2} s^{-1}). Photorespiration (R_L in μmol PAR m^{-2} s^{-1}) was estimated as $1/12 \times (ETR - 4\,(A_{net} + R_D))$ according to [76]. The gross CO_2 assimilation rate (A_{gross} in μmol PAR m^{-2} s^{-1}) was calculated as the sum of net primary production (A_{net}) and respiratory loss (R_D and R_L).

2.6. *MDA Measurements—Biomarker for Lipid Peroxidation*

According to [77] the level of lipid peroxidation and oxidative stress was estimated indirectly by the formation of malondialdehyde (MDA). To determine the content of

https://doi.org/10.3390/environments13030136

MDA, leaf material (20 mg FW) was homogenized with a mortar and pestle in 2 mL of ice-cold trichloroacetic acid TCA (1%, w/v) and centrifuged at 15,000× g for 15 min at 4 °C. After that, 2mL of supernatant was mixed with 0.5% (w/v) of thiobarbituric acid (TBA) and heated at 95 °C for 30 min before quickly cooled in an ice bath. The samples were centrifuged (15,000× g for 10 min at 4 °C) and the absorbance was measured in the supernatant at 532 and 600 nm by using a spectrophotometer CAMSPEC M501 (CamSpec Ltd. Sawston, Cambridge, UK). The concentration of MDA was calculated with the help of the extinction coefficient 155 mM^{-1} cm^{-1}.

2.7. Proline Measurements

Proline concentration was determined according to the method of [78]. An amount of 0.2 g of plant dry material was homogenized with a mortar and pestle in 2 mL of sulphosalicylic acid (3% w/v). Then, the mixture was mixed with the acid ninhydrin solution (2 mL) and the glacial acetic acid (2 mL). The mixture was transferred at 90 °C for 1 h in a water bath and the reaction was stopped using ice. Proline was extracted by adding 4 mL of toluene to each tube, and the absorbance of toluene fraction (aspired from the liquid phase) was measured at 520 nm using a spectrophotometer CAMSPEC M501 (CamSpec Ltd. Sawston, Cambridge, UK). Proline concentration was determined by using calibration curve as µmol proline g^{-1} DW.

2.8. C/N-Analyse

For the determination of the carbon and nitrogen content an amount of 0.2 g of plant dry material was homogenized with a mortar and pestle and assayed with the Vario MAX CNS analyzer (Elementar Analysensystem GmbH, Hanau, Germany). As standard with known carbon and nitrogen amount, we used increasing quantities of hay and L-glutamic acid. The results of the percentage of carbon and total nitrogen amount were used to calculate the C/N ratio and to calculate the C and N content per g plant dry material.

2.9. Statistics

Between three and four replicates were used for data analyses. Statistical analyses were carried out by one-way analyses of variance using SigmaPlot software version 12.5 [79]. For the statistics of the data collected, the individual values were checked by using a one-way analysis of variance (One way ANOVA) according to Holm–Sidak and a normality test according to Shapiro–Wilk.

3. Results

3.1. Subsection

3.1.1. Growth

The fresh weight of control plants (control high SWC) was on average 23.1 g (n = 4 replicates) at the time of harvest. Reducing the water content led to a significant reduction in the fresh weight (Figure 1). Compared to the control plants, the addition of biochar and SAP did not result in any change in weight at a high soil water content. Drought also led to a decrease in fresh weight in the treatments with added SAP and BC. The dry treatments of 0.1% SAP (9.1 g) and 3% BC (9.2 g) showed the lowest value of the fresh weight.

The values of the dry weights show a similar result (Figure 1). The dry weight of the control plants was about 3.8 g. A reduced water content led to a tendency decrease to 3.1 g. As compared to the control plants, all treatments with additions (1% SAP%; 0.1% SAP; 3% BC) show a tendency to decrease in the dry weight at less water. Only the decrease in the treatments with 0.1% SAP and 3% BC at low SWC was significant.

 https://doi.org/10.3390/environments13030136

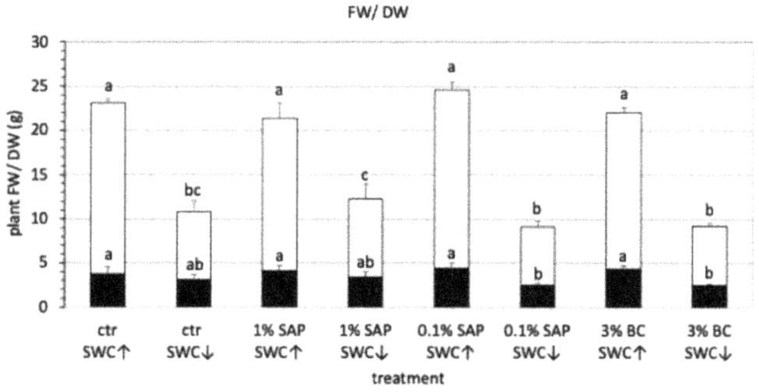

Figure 1. Effect of superabsorber or biochar amendment and additional water shortage on plant growth parameters of *Chenopodium quinoa* Willd. (UDEC-5) at day 78 after sowing. Fresh weight (sum of white and black bar), dry weight (black bar) of the whole plant without the roots. The difference between the FW and DW shows the water content in the plants. Values represent mean ± SE (n = 4 replicates) and the different letters a to c indicate significant differences between the treatments. High soil water content (SWC↑), low soil water content (SWC↓).

It was also shown that drought caused significant reduction in the length of the plant stems and the number of leaves in all treatments.

3.1.2. Water Relation

At low SWC, 0.1% SAP and biochar amendments led to a significant increase in the leaf osmotic potential (LOP) from −2.2 MPa (control) to −1.8 (0.1% SAP) and −1.9 MPa (3% BC), respectively (Figure 2). The plants of all treatments buffered the reduction in soil water potential by a reduction in the leaf water potential, which was mainly reached by a decrease in the leaf osmotic potential. All dry treatments responded to low SWC with a significant decrease in leaf water potential. The addition of 0.1% and 3% BC led a particularly strong decrease in the leaf water potential to less than −2.5 MPa in the mean under dry conditions. It is worth mentioning that the treatment with biochar showed in comparison with the remaining cultures the significantly lowest osmotic potential at low soil water content.

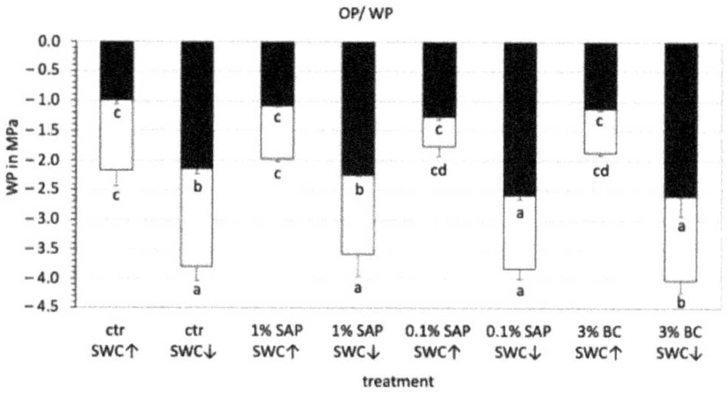

Figure 2. Effect of superabsorber or biochar amendment and additional water shortage on the plant leaf water potential in *Chenopodium quinoa* Willd. (UDEC-5) at day 78 after sowing. Leaf osmotic potential (sum of white and black bar), leaf water potential (black bar) and the difference between the leaf osmotic potential (LOP) and the leaf water potential (LWP) enable statements about the remaining turgor. Values represent mean ± SE (n = 4 replicates) and the different letters a to d indicate significant differences between the treatments. High soil water content (SWC↑), low soil water content (SWC↓).

https://doi.org/10.3390/environments13030136

3.1.3. Light Dependent Reactions of Photosynthesis (at PAR_{1200})

ETR was calculated on base of Y (II). The soil amendments had hardly any effect on the photochemical quantum yield Y (II) and ETR at high soil water content with the exception of 1% SAP (Figure 3A,D). The latter one showed a tendency towards a decrease in Y (II) and ETR. The low soil water content caused a significant decrease in Y (II) and ETR in the SAP treatments (1% SAP and 0.1% SAP) and a non-significant decrease in the control and BC treatments. The changes of Y (II) were balanced mainly by non-photochemical quenching Y (NPQ) (Figure 3B) with two exceptions. The addition of 0.1% SAP and 3% BC caused at low soil water content a significant decrease in non-regulated non-photochemical energy (Y (NO)) (Figure 3C). As shown in Figure 3D, the addition of SAP led to a significant decrease on the electron transport rate (ETR) of PS II at the low soil water content.

The addition of 1% and 0.1% SAP led to a significant increase in oxidized reaction centers at the PS I at high soil water content (Figure 4A). The portion of oxidized centers increased at low SWC in all treatments to the same level. This increase was mirrored by a decrease in the open reaction centers of the PS I (Figure 4B). As shown in Figure 4C, the addition of SAP (1% and 0.1%) caused a significant increase in over-reduced centers at low SWC.

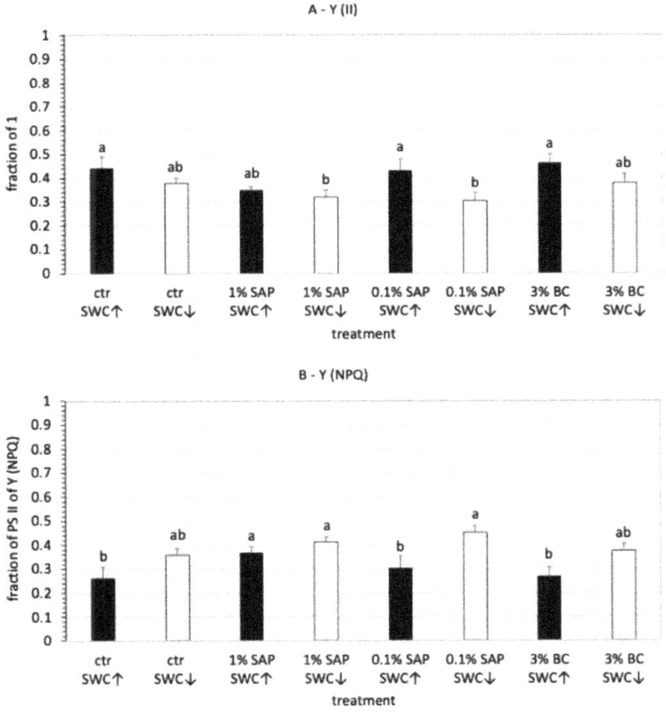

Figure 3. *Cont.*

https://doi.org/10.3390/environments13030136

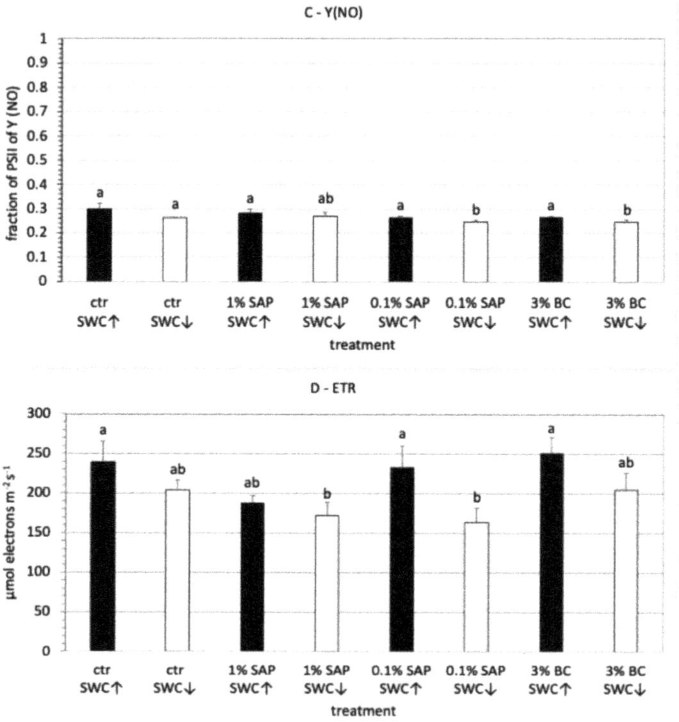

Figure 3. (**A**) Effect of superabsorber or biochar amendment on the photochemical quantum yield Y (II) in *Chenopodium quinoa* Willd. (UDEC-5) at day 78 after sowing. (**B**) Effect of superabsorber or biochar amendment on the regulated non photochemical quenching Y (NPQ) in *Chenopodium quinoa* Willd. (UDEC-5) at day 78 after sowing. (**C**) Effect of superabsorber or biochar amendment on the non-regulated non-photochemical quenching Y (NO) in *Chenopodium quinoa* Willd. (UDEC-5) at day 78 after sowing. (**D**) Effect of superabsorber or biochar amendment on the electron transport rate in *Chenopodium quinoa* Willd. (UDEC-5) at day 78 after sowing. Values represent mean ± SE (*n* = 4 replicates) and the different letters a to b indicate significant differences between the treatments. High soil water content (SWC↑), low soil water content (SWC↓).

Low SWC in the control, 0.1% SAP and 3% BC treatments led to a significant and at 1% SAP to a not significant decrease in vH^+. SAP (0.1% and 1%) and BC had hardly any effect on the steady-state proton flux (vH^+) at high and low SWC (Figure 5A). The addition of SAP (1%; 0.1%) led in contrast to BC (where it was similar to the control) already at high SWC to a significant reduction in steady-state proton conductivity (gH^+). Low SWC caused only in the control and BC treatments a decrease in gH^+ until both reached similar values as in the treatments with SAP (0.1% and 1%). In comparison to control is the decrease in vH^+ and gH^+ at high SWC in all treatments an indicator for a reduced potential for ATP synthase activity at low SWC. The addition of SAP or BC had no significant effect on electro-chromatic shift across the thylakoid membrane (ECStmAu) at high SWC. Low SWC led to a decrease in ECStmAu only in both SAP (1% and 0.1%) treatments. The product of ECStmAu $\times$ gH^+ was calculated as an indicator of the proton transport through the thylakoid membrane, which may be related to the rate of ATP synthesis. The addition of SAP (1%) led, in contrast to 0.1% SAP and BC (where it was similar to the control), to a significant reduction in ECStmAu $\times$ gH^+. Low SWC led only in the control, 0.1% SAP and

 https://doi.org/10.3390/environments13030136

BC treatments to a decrease in ECStmAu $^{\times}$ gH^+ until they reached values similar to the treatment with 1% SAP.

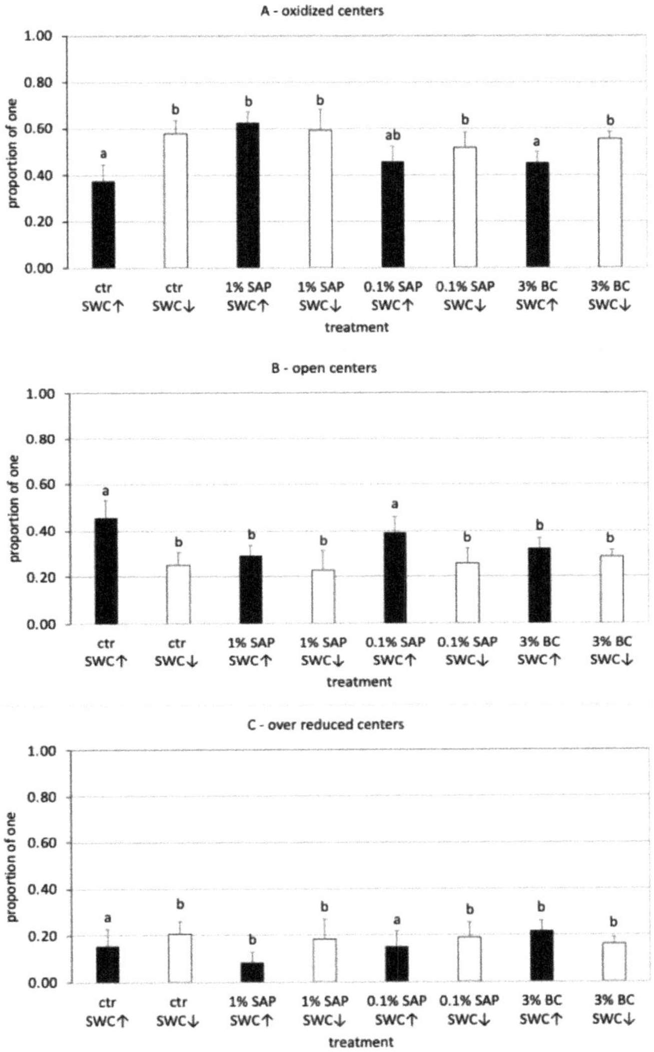

Figure 4. (**A**) Effect of superabsorber or biochar amendment on oxidized reaction centers at PS I in leaves of *Chenopodium quinoa* Willd. (UDEC-5) at day 78 after sowing. (**B**) Effect of superabsorber or biochar amendment on open reaction centers at PS I in leaves of *Chenopodium quinoa* Willd. (UDEC-5) at day 78 after sowing. (**C**) Effect of superabsorber or biochar amendment on over reduced centers at PS I in leaves of *Chenopodium quinoa* Willd. (UDEC-5) at day 78 after sowing. Values represent mean ± SE (n = 4 replicates) and the different letters a to b indicate significant differences between the treatments. High soil water content (SWC↑), low soil water content (SWC↓).

 https://doi.org/10.3390/environments13030136

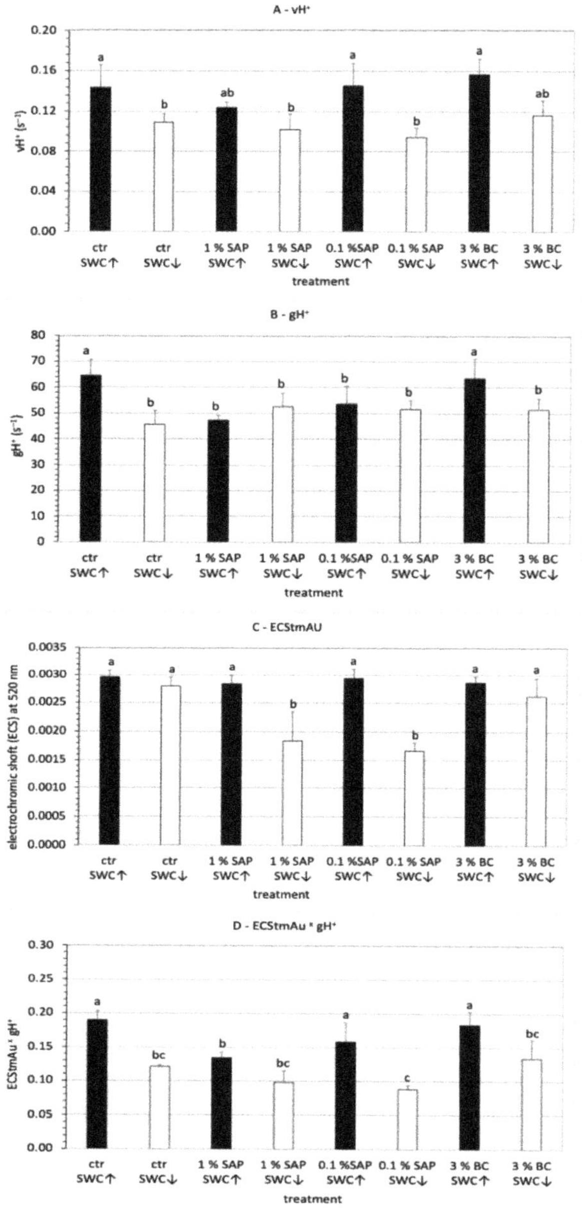

Figure 5. (**A**) Effect of superabsorber or biochar amendment on the steady-state rate of proton flux (vH^+) in leaves of *Chenopodium quinoa* Willd. (UDEC-5) at day 78 after sowing. (**B**) Effect of superabsorber or biochar amendment on the proton conductivity (gH^+) in leaves of *Chenopodium quinoa* Willd. (UDEC-5) at day 78 after sowing. (**C**) Effect of superabsorber or biochar amendment on the electrochromic shift (ECStmAu) in leaves of *Chenopodium quinoa* Willd. (UDEC-5) at day 78 after sowing. (**D**) Effect of superabsorber or biochar amendment on the product of ECStmAu $^\times$ gH^+ in leaves of *Chenopodium quinoa* Willd. (UDEC-5) at day 78 after sowing. Values represent mean ± SE (n = 4 replicates) and the different letters a to c indicate significant differences between the treatments. High soil water content (SWC↑), low soil water content (SWC↓).

https://doi.org/10.3390/environments13030136

3.1.4. Leaf CO_2/H_2O Gas Exchange

Plant leaves reached in the control treatment a net CO_2 assimilation rate (A_{net}) of 10.08 µmol CO_2 m^{-2} s^{-1} (Table 1). Treatment with 0.1% SAP with high soil water content led to an increase in A_{net} by 46.7% (14.79 µmol CO_2 m^{-2} s^{-1}) (Table 1). In all treatments, drought induced a significant decrease in the max. A_{net}.

Table 1. Effect of superabsorber or biochar amendment on leaf gas-exchange and chlorophyll fluorescence parameters of *Chenopodium quinoa* Willd. (UDEC-5) at light saturation (PPFD of 1000 µmol PAR m^{-2} s^{-1}) at day 78 after sowing. Values represent mean ± SE (n = 4 replicates) and the different letters a to f indicate significant differences between the treatments.

Treatment	ctr.		1% SAP		0.1% SAP		3% BC	
Parameter	**SWC↑**	**SWC↓**	**SWC↑**	**SWC↓**	**SWC↑**	**SWC↓**	**SWC↑**	**SWC↓**
A_{net} (µmol CO_2 m^{-2} s^{-1})	10.08 ±1.28 (b)	7.46 ±1.37 (c)	10.93 ±0.57 (b)	4.02 ±0.77 (d)	14.79 ±1.48 (a)	6.12 ±0.75 (c)	11.65 ±0.84 (b)	5.78 ±1.49 (c)
R_D (µmol CO_2 m^{-2} s^{-1})	2.23 ±0.20 (a)	1.60 ±0.23 (ab)	1.47 ±0.22 (b)	1.15 ±0.36 (b)	1.41 ±0.08 (b)	1.79 ±0.52 (ab)	2.27 ±0.22 (a)	1.32 ±0.29 (b)
R_L (µmol CO_2 m^{-2} s^{-1})	13.26 ±0.88 (a)	11.91 ±0.48 (ab)	9.79 ±0.56 (b)	10.62 ±0.47 (b)	10.66 ±1.24 (b)	9.88 ±0.68 (b)	12.88 ±1.15 (a)	12.70 ±0.58 (a)
WUE (A_{net}/E)	5.01 ±0.55 (b)	6.99 ±0.40 (c)	6.09 ±0.86 (b)	6.53 ±0.89 (bc)	8.24 ±0.72 (d)	10.45 ±0.19 (e)	5.35 ±0.31 (a)	6.65 ±1.06 (bc)
ETR/A_{tot} A_{gross} (µmol CO_2 m^{-2} s^{-1})	8.04 ±0.13 (b)	8.73 ±0.07 (d)	7.34 ±0.04 (a)	9.89 ±0.39 (f)	7.09 ±0.11 (a)	8.45 ±0.08 (c)	7.89 ±0.12 (b)	9.04 ±0.25 (e)
C_s (µmol CO_2 m^{-2} s^{-1})	0.13 ±0.04 (a)	0.06 ±0.01 (b)	0.12 ±0.04 (a)	0.04 ±0.02 (b)	0.16 ±0.05 (a)	0.03 ±0.01 (b)	0.15 ±0.01 (a)	0.05 ±0.02 (b)
E (µmol CO_2 m^{-2} s^{-1})	1.92 ±0.45 (a)	0.95 ±0.09 (b)	1.79 ±0.49 (a)	0.70 ±0.36 (b)	1.95 ±0.57 (a)	0.48 ±0.08 (b)	2.14 ±0.06 (a)	0.83 ±0.20 (b)
C_i/C_{atm} ppm/ppn	0.67 ±0.04 (a)	0.50 ±0.12 (b)	0.62 ±0.06 (b)	0.57 ±0.16 (b)	0.57 ±0,09 (b)	0.36 ±0.04 (c)	0.67 ±0.03 (a)	0.51 ±0.08 (bc)

Abbreviations: High soil water content (SWC↑), low soil water content (SWC↓); Net CO_2 assimilation (A_{net}); stomatal conductance C_s; transpiration rate (E); dark respiration (R_D); light respiration (R_L); water use efficiency (WUE = A_{net}/E), ETR/A_{gross} and the C_i/C_{atm} ratio.

There was a significant correlation between the values of A_{net}, stomatal conductivity (C_s) and transpiration (E). Reduced soil water content led to a significant decrease in C_s in all treatments, consequently leading to a significant decrease in A_{net} and E at lowSWC. In addition, the reduced SWC caused a decrease in C_i/C_{atm} ratio in all treatments independent of the amendment. Compared to the control, treatment with the addition of 1% SAP showed the lowest A_{net} at low SWC. Consequently, the control treatment and the treatments with the addition of 0.1% SAP and 3% BC were able to improve water use efficiency (WUE) at low SWC. This increase in WUE was highest at high SWC and resulted from a particularly sharp decline in C_s and E. Furthermore, both SAP treatments showed the lowest R_L and R_D

https://doi.org/10.3390/environments13030136

values at high and low SWC. Reduced SWC led to a decrease in R_D and R_L in the control and BC treatments until they reached the values of the SAP treatments.

3.1.5. ETR/A_{gross} Ratio

The increase in the ETR/A_{gross} ratio can be used as an indicator of the risk of oxidative stress. The addition of SAP or BC shows an increase in the ETR/A_{gross} ratio at low SWC (Table 1). The highest increase is observed in the treatment with the addition of 1% SAP. This result correlated with the increase in over-reduced centers at low soil water content.

3.1.6. MDA Content

The MDA contents in the control treatment were on average 15.9 nm/g FW (Figure 6). Only the addition of biochar led to a significant reduction in the MDA content at high SWC. The MDA content in leaves stabilized in the control and SAP treatments at low SWC in the range of 20 nm/g FW. However, the MDA content in the BC treatment stayed significantly lower than in all other treatments, as well as at low SWC.

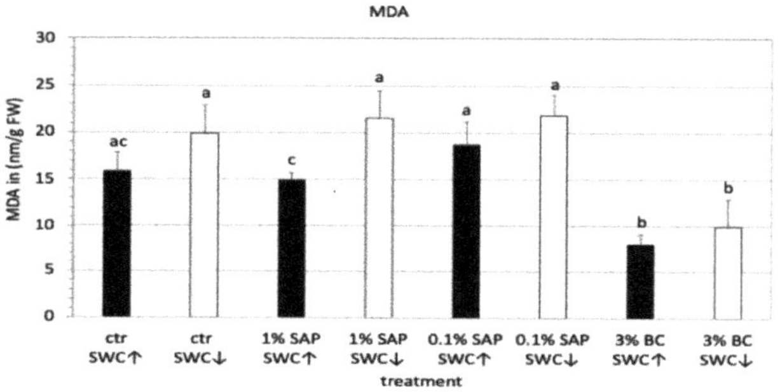

Figure 6. Effect of superabsorber or biochar amendment on MDA content in leaves of *Chenopodium quinoa* Willd. (UDEC-5) at day 78 after sowing. Values represent mean ± SE (n = 4 replicates) and the different letters a to c indicate significant differences between the treatments. High soil water content (SWC↑), low soil water content (SWC↓).

3.1.7. C/N Ratio

The N content (Figure 7B) had a major impact on the C/N ratio (Figure 7A). The control plants presented a high N content in line with the low C/N ratio at high and low SWC. The addition of 1% SAP resulted in a low N content at high and low SWC, and consequently in a high C/N ratio. Reduced SWC caused in the treatments with 0.1% SAP and 3% BC increase in N content up to the level of the control plants.

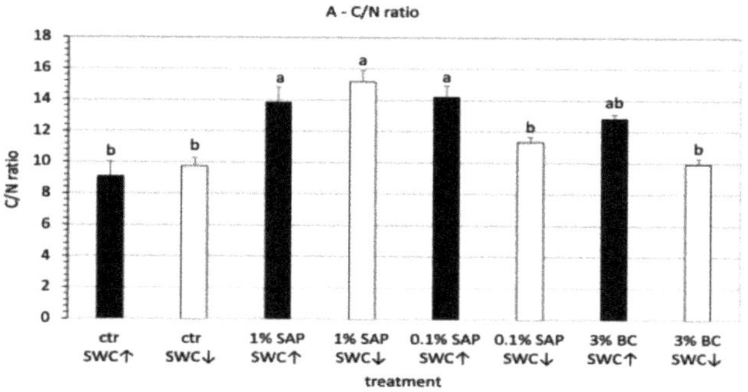

Figure 7. *Cont.*

https://doi.org/10.3390/environments13030136

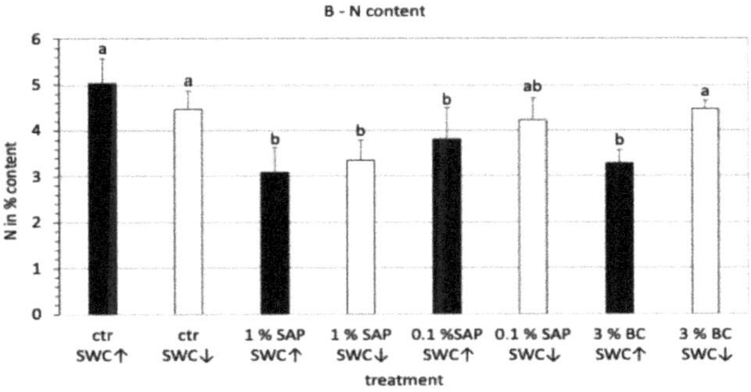

Figure 7. (**A**) Effect of superabsorber or biochar amendment on C/N ratio in leaf tissues of *Chenopodium quinoa* Willd. (UDEC-5) at day 78 after sowing. (**B**) Effect of superabsorber or biochar amendment on N in % in leaf tissues of *Chenopodium quinoa* Willd. (UDEC-5) at day 78 after sowing. Values represent mean ± SE (*n* = 4 replicates) and the different letters a to b indicate significant differences between the treatments. High soil water content (SWC↑), low soil water content (SWC↓).

3.1.8. Proline Content

The proline content was at high SWC not significantly different in the control (3.85 mmol/kg) and all other treatments (Figure 8). Reduced SWC caused a significant increase in proline content only in treatments with 0.1% SAP and BC.

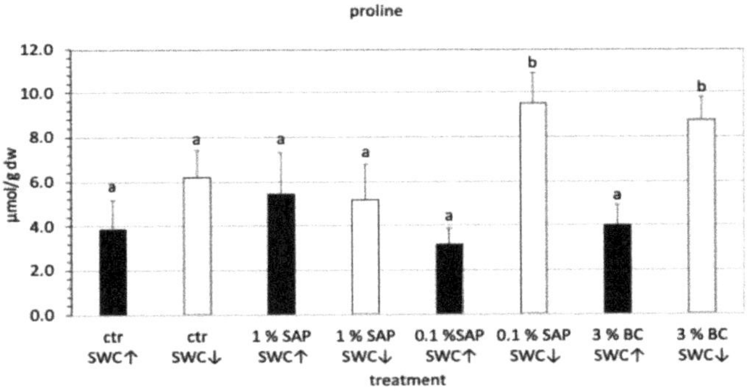

Figure 8. Effect of superabsorber or biochar amendment on proline content in leaves of *Chenopodium quinoa* Willd. (UDEC-5) at day 78 after sowing. Values represent mean ± SE (*n* = 4 replicates) and the different letters a to b indicate significant differences between the treatments. High soil water content (SWC↑), low soil water content (SWC↓).

4. Discussion

In this study we analyzed the impact of various soil amendments on growth and physiological responses of *Chenopodium quinoa* Willd. (cv. UDEC-5) in response to a persistent moderate water stress period. Abiotic stress resistance of plants has proven to be complex. It is shown in this study that the response of the test species to drought and the addition of BC and SAP proved to be very precisely tailored to the respective conditions. However, the plant response patterns were partially surprising and did not match all initial hypotheses and expectations made at the beginning, as will be shown in the following chapters.

https://doi.org/10.3390/environments13030136

Impact of soil additives at high SWC

First, we looked at the effect of soil additives on plant growth at high SWC. In accordance with Hypothesis 4 the soil amendment of 0.1% SAP and 3% BC at high SWC caused a significant decrease in plant dry weight (Figure 1) and changes in PS1 (portion of oxidized and open centers in PS1) (Figure 4), C/N ratio and N content (Figure 7). The addition of SAP (0.1% and 1%) led to a decrease in the gH^+ and ECStmAu $\times$ gH^+ (indicator of electrical potential across the membrane) (Figure 5), R_D, R_L, C_i/C_{atm} ratio and ETR/A ratio and to an increase in the WUE (Table 1), especially in the 0.1% SAP treatment. Treatment with 0.1% SAP and 3% BC, both at high SWC, caused a significant increase in the leaf osmotic potential (Figure 2) and no significant increase in C/N ratio (Figure 7A), decrease in N content (Figure 7B) and opposing effects on MDA content (Figure 6). The C/N ratio increased with increasing SAP amendment.

This impact of soil additives is not uncommon. In agreement with the present results, [60] shows that the application of SAP can already, at high SWC, lead to increases in soil organic carbon, soil microbial biomass carbon, soil sucrase and cellulase activities, soil water consumption, water consumption, A_{net}, leaf WUE, nitrogen use efficiency (NUE) and yield of winter wheat. The correlation analyses of [60] could show that SAP improve the utilization rate of soil water and nutrients by improving the soil structure and increasing soil organic carbon and microbial enzyme activity. However, plants grown with 1% SAP also showed a significant reduction in PS II activity (Y (II) (Figure 3A) and ETR (Figure 3D) in favor of a high Y (NPQ) (Figure 3B) as signs of negative effect from SAP addition.

However, BC treated plants showed significantly lower MDA contents than all other treatments. Several studies have shown that biochar can lead to an improved ion balance, increased proline accumulations and decreases in antioxidant enzyme activity, H_2O_2 and MDA contents and, thus, improved membrane stability [54,80]. Derbali et al. [81] interpreted this coordinated response to the BC addition as an accompanying priming effect or preparation for possible upcoming stress.

Impact of reduced SWC on biomass production and water relation of the control plants

In agreement with several other studies [31,32,82], low SWC in the control treatment had a negative effect on net rate of CO_2 assimilation (Table 1), Y (II), and ETR (Figure 3D), and consequently the production of fresh and dry matter of the plant (Figure 1). The results of the present quinoa experiment indicate, in accordance with [83], a drought-avoidance mechanism of the control: the reduced transpiration led to increased water use efficiency due to the decrease in the C_i/C_{atm} ratio, stomatal conductance, transpiration (Table 1) and leaf area development. We also confirm that quinoa plants improved water uptake at low SWC by the accumulation of compatible solutes to lower the leaf water potential and the leaf osmotic potential (Figure 2) [35]. There was no significant impact of low SWC on C/N ratio and N content (Figure 7), but surprisingly also not on the proline content (Figure 8) of control plants, as shown by [84].

Impact of soil additives on biomass production and water relation at reduced SWC

Contrary to expectations, the addition of 0.1% SAP and 3% BC resulted, at low SWC, in a significantly higher reduction in fresh and dry weight (Figure 1) and water potential (Figure 2) compared to the control and 1% SAP treatment. It can be assumed that plants change, under both treatments (0.1% SAP low SWC and 3% BC low SWC), their metabolic activities from growth to the physiological and biochemical responses of the plant to perceived stress [85]. However, in accordance with hypotheses 1.4 and 6, plants grown at low SWC at 0.1% SAP and 3% BC switched from preferred growth (e.g., biomass production) to long survival (e.g., enhanced resilience). The strategy behind this plant response might at least partially be a compensation of the reduced biomass production by an extension in

https://doi.org/10.3390/environments13030136

plant development [86]. This compensation phenomenon after temporary abiotic stress has previously been described in maize and quinoa leaves [81,87–89] and is thought to maintain the capacity to resume biomass production for a longer period in anticipation of water becoming available again.

The energy reserves saved by the reduced biomass production were used in the 0.1% SAP and 3% BC treatments, among other things, for the accumulation of compatible solutes to lower the osmotic potential and, consequently, also the leaf water potential (Figure 2). The addition of 0.1% SAP (−2.58 WP in MPa) and 3% BC (−2.60 WP in MPa) caused a higher decrease in the water potential than the control and of the 1% SAP treatments. This decrease in the water potential was achieved in the 3% BC treatment with a significantly higher decrease in the osmotic potential than at 0.1% SAP at low SWC and consequently also higher turgor potential. Wang et al. [90] show that with a decrease in soil moisture content, the content of malondialdehyde (MDA) increased significantly in soybean leaves. The BC-mediated higher investigation in the release of soil water resources could be one reason for the lower MDA contents than in other treatments.

Light dependent reactions of photosynthesis

Drought stress has a significant effect on the light-dependent reactions of photosynthesis, because they are highly dependent on water [13]. Light reactions utilize light energy absorbed by chloroplast for the breakdown of water (photolysis) and to generate ATP via a proton gradient across the thylakoid membrane and NADPH via photosystem II (PS II) and I (PS I), respectively. If the absorbed light energy exceeds the capacity of the photosynthetic machinery because of H_2O deficiency, this can lead to light-induced damage in the photosystems (PS I and PS II) [91]. To estimate the response of the reaction centers of PS II, we measured Y (II), Y (NPQ) and Y (NO) among other things.

Impact of soil additives at high SWC on PS (II) and proton flow of photosynthesis

Non-photochemical quenching (NPQ) mechanisms are responsible for the dissipation of excess excitation energy to generate heat and to prevent overexcitation of chlorophyll molecules, leading to the photodamage of PS II [13,92,93]. Unexpectedly, but in accordance with Hypothesis 2, the treatment with 1% SAP, even at high SWC, led to a decrease in Y (II) and an increase in Y (NPQ) (Figure 3). Dissipation of excess absorbed light energy through zeaxanthin- and ΔpH-dependent photosystem II antenna quenching is considered the major mechanism for non-photochemical quenching and photoprotection [94]. However, the significant increase in Y (NPQ) (Figure 3B) at 1% SAP at high SWC was accompanied by a significant decrease in ECStmAu (electrochromic shift across the thylakoid membrane) (Figure 3C) and ECStmAu × gH^+ (indicator of electrical potential across the membrane) (Figure 3D), gH^+ (indicator of proton conductivity of the thylakoid membrane) (Figure 5B) and vH^+ (steady state proton flux) (Figure 5A). Ivanov et al. [94] interpreted this phenomenon as evidence of a zeaxanthin-independent pathway for dissipation of excess light energy, based within the PSII reaction center, which may also play a significant role in photoprotection. They assumed that PSII reaction centers can be reversibly interconverted from photochemical energy transducers that convert light into ATP and NADPH to efficient, non-photochemical energy quenchers that protect the photosynthetic apparatus from photodamage. It is not clear why this unusual form of energy transduction is carried out at 1% SAP at high SWC but it supports photoprotection by the shown reduction in the ETR in PS (II) (Figure 3D) and the otherwise possible increase in the ETR/A_{gross} (Table 1). It can be assumed that this response, in agreement with Hypothesis 2, is an indicator of a higher water requirement at high SWC of 1% SAP than of control plants and an incipient undersupply of the former.

https://doi.org/10.3390/environments13030136

Impact of soil additives at reduced SWC on PS (II) and proton flow of photosynthesis

In agreement with the expectations, the reduction in soil water availability in all low SWC treatments, with the exception of 1% SAP, led to a decrease in the quantum yield of PS II (Y (II) (Figure 3A) and to an increase in non-photochemical quenching Y (NPQ) (Figure 3B). The neglectable increase in 1% SAP can be attributed to the early response of the 1% SAP treatment at high SWC (see above).

To prevent damage from photosynthesis, the supply and demand of electrons must be regulated to maintain a balance. However, drought stress impairs the activity of PS I and PS II, which can lead, among other things, to an increase in over-reduction on the PS I acceptor side. One way to reduce PS I acceptor over-reduction while reducing electron demand would be the adjustment of the cyclic electron flows [14]. The electrons in the cyclic electron flow are not passed on to NADP reductase but return to oxidized PS I and absorb light energy again. This energy is used to transport protons to the thylakoid lumen and increase the proton motive force. Huang et al. [95] show that the use of cyclic electron flow during drought stress provides effective protection against over-reduction on the PS I acceptor side.

However, as shown for the 1% SAP treatment at high SWC before, the significant increase in Y(NPQ) at all other low SWC treatments was accompanied by a significant decrease in ECStmAu (Indicator of pH gradient), ECStmAu $^{\times}$ gH^+ (indicator of electrical potential across the membrane) gH^+ (indicator of proton conductivity of the thylakoid membrane) and vH^+ (steady state proton flux) (Figure 5). Thus, it appears, after [94], that any environmental condition that increases the reduction state of QA will improve the probability of photoprotection in Quinoa through reaction center quenching. Ivanov et al. [94] argued that the accumulation of reduced QA increases the probability for non-radiative charge recombination within the PS II reaction center with a concomitant decrease in the yield of thermoluminescence.

However, the dissipation of excess excitation to prevent oxidative stress was highest in both SAP treatments. In contrast, at low SWC, the 3% BC treatment showed Y (II) and Y (NPQ) values similar to the control at low SWC but higher gH^+, vH^+, ECStmAu and ECStmAu $^{\times}$ gH^+ values, indicating a higher transfer of excess light energy on ATP synthesis. This result indicates that PS II was less affected under these conditions by the addition of BC than both SAP (1%; 0.1%) treatments.

The decreases in quantum yield of PS II (Y (II) (Figure 5A), leaf water potential (Figure 2) and ETR (Figure 3D) are typical responses of plants to water deficiency. However, Y (II) and ETR were more reduced in water deficiency in plants treated with SAP (1%; 0.1%) than in those treated without amendment or with biochar (3%). Derbali et al. [91] showed that a reduced photosynthetic efficiency and ETR are not necessarily a sign of a destructive effect. In fact, this reaction is even important when the energy demand of the non-light-depended reaction is also reduced.

Impact of soil additives on PS (I) at high SWC

Plants use photosystems I and II (PSI/II) to transfer electrons along the linear electron transfer (LET) chain to $NADP^+$ and to produce a proton gradient to drive ATP synthesis [96]. Under (stromal redox) pressure: $NADP^+$ synthesis controls photosystem I. To measure the fraction of active reaction centers of PS I, we used a saturation pulse followed by light irradiation [97]. Treatment with 0.1% SAP and 3% BC, in comparison to the wet control treatment (SWC↑), had no significant impact on the sunlight-driven transmembrane electron transport: the oxidation of plastocyanin in the inner side of the thylakoid membrane and the reduction in ferredoxin in the stroma [98]. The present results showed that plants of the control, 0.1% SAP and 3% BC treatments had significantly higher fractions

https://doi.org/10.3390/environments13030136

of open reaction centers than plants of the 1% SAP treatment (Figure 4B). This one was balanced by an increased fraction of oxidized reaction centers (Figure 4A). This recurring peculiarity of 1% SAP fits well with Hypothesis 3 that non-photochemical energy quenchers in 1% SAP reduce the LET from PS (II) to PS (I) in preparation for a further reduced water supply [94–96] and effective protection against over-reduction on the PS I acceptor side.

Impact of soil additives on PS (I) at reduced SWC

Water deficiency (SWC↓) led to a reduction in electron uptake of PS I via open reaction centers in all treatments with the exception of 1% SAP. This one was already low at 1% SAP. We interpret the already low fraction of open centers of the 1% SAP treatment at high SWC at a preadaptation to reduced water availability (see Hypothesis 3). This decreasing ability to absorb electrons from PS II correlated with the decrease in quantum yields (Y (II)) and ETR. Lima-Melo et al. [99] reported that an imbalance between the electrons from the ETR and acceptor sides of the photosystem I can lead to a PS I photoinhibition. To determine the state of PS I, it is useful to examine the oxidized reaction centers (Figure 4A). We could show the in the mean highest fraction of over reduced centers in plants treated with SAP (1%; 0.1%). A reduction in water content led only in these both treatments to a decrease in open centers (Figure 4B), followed by an increase in over reduced centers (Figure 4C). It can be assumed that an increase in the fraction in over-reduced center can provide evidence for possible oxidative stress (Figure 4C). The so-called PS I photoinhibition is a lethal event for oxygenic phototrophs, and it is prevented by keeping the reaction center chlorophyll (P700) oxidized in excess light conditions [100,101]. Therefore, it can be assumed that the highest danger for a disproportionate supply of electrons at the PS I and oxidative stress occurs at 0.1% SAP and 1% SAP. Additionally, the non-photochemical energy quenchers [94] seem not to be effective enough to avoid the above discussed protection at 1% SAP against over-reduction on the PS I acceptor side.

Light independent reactions of photosynthesis

Impact of soil additives at high SWC

The addition of 3% BC at high SWC contrary to Hypothesis 2 had hardly any impact on the CO_2/H_2O gas exchange (Table 1). However, SAP amendment in both treatments (0.1% SAP and 1% SAP), in accordance with Hypothesis 2, caused a significant reduction in R_D, R_L and increase in ETR/A_{gross} and WUE. The last one was reached in the 1% SAP treatment by a low ETR and in the 0.1% SAP treatment by a significant increase in C_s and A_{net} (control high SWC: 10.1 and 0.1% SAP high SWC: 14.8 CO_2 μmol m^{-2} s^{-1}). The combination of these changes according to Hypothesis 3 can be interpreted as a priming effect or a positive impact of eustress on CO_2/H_2O gas exchange. Effective application of eustress (positive stress), such as mild water deficit or nutritional stress, can provoke tailored plant responses including the activation of physiological and molecular mechanisms and the strategic accumulation of bioactive compounds such as proline necessary for adaptation to suboptimal environments [102]. A study on *Arabidopsis thaliana* showed a hormetic response of the effective quantum yield of PS II to mild water deficit and the activation of ROS defense responses [103]. The results of quinoa at 0.1% SAP and 1% SAP confirm the assumption of eustress in the present study. A significant increase in the WUE together with a decrease in ETR/A_{gross} point to an adjustment of the CO_2/H_2O gas exchange. Sepehri et al. [104] suggest that uncertainty exists regarding water accessibility for plants when SAP are utilized. This may lead to competition for water absorption between plants and SAP, potentially reducing the amount of water available for plants [105]. Unlike SAP, there was no proof of any eustress at high SWC when BC was added to the soil, not even a change in the proline content (Figure 8).

https://doi.org/10.3390/environments13030136

Impact of reduced SWC on light independent reactions of photosynthesis of the control plants

Previous studies have shown that the rate of photosynthetic CO_2 assimilation (A_{net}) is influenced by a decrease in relative water content and leaf water potential [106]. Environmental stress such as water shortage decreased the CO_2 assimilation rate in barley leaves due to reduced stomatal conductance [107,108]. In line with these results, the present study also showed a decrease in CO_2 assimilation of the control of Quinoa at low SWC condition. This reduction was accompanied by a decreased ETR and a significant increase in the ETR/A_{gross} ratio and a higher risk of oxidative stress [99]. Drought affects both the H_2O and the CO_2 uptake of the plant. As a primary resource for plant growth and productivity, H_2O and CO_2 are directly or indirectly involved in photosynthesis [109]. Plant leaf CO_2/H_2O exchange and water use efficiency (WUE), the ratio of net photosynthesis (A_{net}) to transpiration (E) are main parameters for estimating plant response, survival and productivity as well as for planning water use in arid and semi-arid area [109]. In the present study, Quinoa responded in with a significant increase in the WUE because of a greater reduction in E than in A_{net} (Table 1). The latter was supported by a reduction in the C_i/C_{atm} ratio and a more efficient CO_2 assimilation at closed stomata. There was a correlation between the decrease in C_s and a decrease in transpiration (E) and a reduction in A_{net} at low SWC (Table 1).

Impact of soil additives at reduced SWC on light independent reactions of photosynthesis

In contrast to previous studies, the addition of BC and 1% SAP, in comparison to the control, caused no significant increase in stomatal conductance (C_s), transpiration (E) and CO_2 fixation at low SWC [91,110]. The 1% SAP treatment even had a significant negative impact on A_{net} and ETR/A_{gross} ratio (Table 1). However, the treatment at 0.1% SAP (low SWC) caused a significant increase in the WUE (control low SWC: 7.90 and 0.1% SAP low SWC: 10.45 µmol CO_2/mmol H_2O) due to a disproportionate reduction in stomatal conductivity C_s and a significantly higher reduction in E (control low SWC: 0.95 and 0.1% SAP low SWC: 0.48 mmol H_2O m^{-2} s^{-1}). In accordance with [111], the decrease in stomatal conductance during water shortage (Table 1) correlates with the decrease in leaf water potential (Figure 2), which significantly decreased in the present study in quinoa plants treated with 0.1% SAP. In addition, plants of this treatment also showed a significantly higher increase in the water use efficiency during drought than the control (Table 1). It can be assumed that 0.1% SAP had an additional impact on the active and effective adaption of the CO_2/H_2O gas exchange of quinoa plants at reduced SWC. Previous studies have already shown that the addition of SAP can improve the WUE, stomatal conductance and photosynthetic activity during water shortage [62].

The different responses of plants to 0.1% and 1% SAP are presumably caused by an increasing competition for water between plant and soil with increasing SAP concentration. In agreement with [25], the decrease in A_{net} at 1% SAP at low SWC correlated well with the decrease in photochemical quantum yield of PS II (Y (II) and ETR at low SWC (Figure 3D).

Indicator of oxidative stress

Finally, plants of all treatments in response to reduced water supply showed a significant increase in ETR/A_{gross} and MDA at low SWC. These results of both parameters are indicators for higher risk of oxidative stress [99] and show that the drought avoidance mechanisms were not sufficient to prevent oxidative stress. Increased ROS generations at water shortage condition according to [91] are indicated by higher levels of lipid peroxidation, significantly higher levels of antioxidant activities, and high proline levels.

https://doi.org/10.3390/environments13030136

Impact of soil additives on MDA content

Increased lipid peroxidation in the form of MDA (malondialdehyde) accumulation serves as an indicator of oxidative damage that can be caused by the synthesis and accumulation of reactive oxygen species (ROS) under drought stress [112]. It can be assumed that only the addition of biochar mitigated oxidative stress levels under drought conditions (Figure 6). Previous studies have shown that biochar amendment significantly decreased the content of MDA, $O_2^{\bullet -}$ and H_2O_2 in the leaves of *Malus hupehensis* Rehd. [113]. According to the present study (Figure 6), the MDA concentration in plants treated with biochar (3%) was remarkably lower (~50%) than in the control group, even when water supply was sufficient. Furthermore, there was no significant increase in MDA at low SWC. The presented results agree with the results of [114], who reported that the addition of BC reduced the MDA concentration both under normal irrigation and during drought stress. These results confirm the assumption that the addition of biochar mitigated oxidative stress levels under drought conditions by regulating photosynthetic systems, thereby preventing over reduction on the acceptor side of photosystem I.

Cao et al. [115] showed that the addition of SAP (0.1% until 0.4% SAP) could also improve the drought resistance of *Cynodon dactylon* (L.) seedlings. The MDA concentration decreased significantly in *Cynodon dactylon* (L.) at concentrations between 0.1% until 0.4% SAP either at sufficient or limited water supply. However, we were not able to confirm this positive impact of SAP in the present study. The MDA content of quinoa leaves was in all SAP treatments under both irrigation regimes at least as high as in the control plants (Figure 6). It can be assumed that the offered soil water contents led to a competition between plants and SAP for water absorption, which potentially decreases the amount of water available for plants [105]. Sepehri et al. [104] suggested that uncertainty exists regarding water accessibility for plants when SAP are utilized. This may lead to competition between plants and SAP for water absorption, potentially reducing the amount of water available for plants [105]. Therefore, the ability of SAP to store various amounts of water within itself, and thus in the soil, could not be used to improve but to reduce the water availability for plants [116]. This assumption also fits with the response of SAP-treated plants (0.1% and 1%) at low SWC where quinoa was neither able to hinder over reduction in PS I nor to prevent a higher ROS accumulation at low SWC. It seems that optimal plant growth on soil with SAP amendment depends on higher soil water content than control plants and that the application of an appropriate SAP concentration is crucial to ensure optimal growth conditions [117].

Another reason for the missing improvement of the drought resistance of quinoa may be the interaction of soil and SAP due to the physiochemical properties of the soil as well as the chemical properties of SAP. The last one is not only its variation in water absorbing capacity (WAC) but also the possibly negative impact on ion composition in soil water since SAP are cation-absorbing polymers [65,118].

Furthermore, factors such as soil structure, temperature, pH and mineral composition in the water may have impact on the efficiency of SAP, their water retention capacity and overall performance [119].

Impact of soil additives on Carbon, nitrogen (C/N ratio)

The SAP Stockosorb® is a potassium-based nutrient-free co-polymer [120]. SAP' primary function is to improve water and nutrient retention in the soil. However, when used in soil, hydrophilic polymers like Stockosorb® can influence the overall C/N ratio of the soil by holding water and nutrients. By retaining water and nutrients, it can indirectly influence the soil's C/N balance over time and affect nutrient availability for plants which can affect plant growth and nutrient cycling. Biochar is known as soil improver but with the

https://doi.org/10.3390/environments13030136

side effect to reduce the plant-available nitrogen (N) owing to their high carbon (C) content, which usually results in N-immobilization in soil [51]. Several recent analyses [121–123] showed that reduction in soil N is a widespread response to biochar application. Nguyen et al. [121] found biochar application decreased NH_4–N by an average of 11% and NO_3–N by 10%. In agreement with these studies and Hypothesis 5, we could show that mainly the addition of SAP (0.1% and 1% SAP) but also of 3% BC led to a significant decrease in the leaf N content and leaf C/N ratio at high SWC (Figure 7). There also seems to be a clear negative correlation between a higher SAP content in the soil and C/N ratio and N content in the leaf (Figure 7) at low SWC. It was found out that low N-concentrations in plant tissues as shown for 1% SAP (Figure 7B) can contribute to changes in leaf photosynthetic capacity, in transpiration fluxes and in WUE, which has consequences for C sequestration capacity [124]. However, besides no significant reduction in the N-content in the control and the 1% SAP at reduced SWC, in the present study, we also found a non-significant increase in the leaf N-content in the 0.1% SAP and a significant increase in the 3% BC treatment at reduced SWC. It can be assumed that the increase in N-content and the related decrease in the C/N ratio caused in these two treatments the upregulation of the adaptation mechanisms of plants against oxidative stress but on cost of biomass production (see Hypothesis 6). One indicator for this N-dependent upregulation was an increased proline accumulation in both treatments. Furthermore, it is clear that plants treated with 1% SAP have difficulties to absorb and accumulate nitrogen (Figure 7), which confirms the assumption of a plant-SAP competition for nitrogen. Together with the antioxidant enzymes, the non-enzymatic antioxidant defense represents a major ROS-scavenging force and eminent importance [125].

Impact of soil additives on Proline content

Plants of all high SWC treatments showed equally low proline values. There was no clear correlation between proline content and N-content or C/N ratio in leaves of quinoa plants at high SWC. It is well known that quinoa plants respond to drought stress with various stress avoidance mechanisms, including ROS detoxification and the accumulation of protective proteins and solutes such as proline [35]. This is also confirmed by the present study. High SWC especially in plants treated with 3% BC or 0.1% SAP led to more than twice or three times higher proline contents (Figure 8) as part of the upregulation of the adaption mechanisms of plants und oxidative stress. It can be assumed that the addition of 0.1% SAP and 3% BC enables plants to synthesize proline to protect cell functions by scavenging reactive oxygen species [126]. Consequently, it can be assumed that the low increase in the proline content in leaf tissues of the water deficient control or 1% SAP treatments can be explained by disturbances in the regulation of ROS defense. This problem might be enhanced at 1% SAP (high SWC) by insufficient nitrogen supply (Figure 7). Moreover, studies of [127,128] have indicated that SAP can absorb various cations, which may potentially affect plant growth dynamics [104]. Situ et al. [129] demonstrated that cation exchange between potassium-based SAP can result in excessive accumulation of K^+, reducing plant development, root biomass, length, and area, while also causing deficiencies in Ca^{2+} and Mg^{2+}. This interaction may have also affected the efficiency of SAP in the present study under water shortage conditions. In contrast to 1% SAP, in this study, it is shown that the amendment of biochar significantly reduced oxidative stress caused by non-enzymatic ROS response of quinoa during water shortage. It can be assumed that plants treated with biochar were protected from harmful conditions through an improved soil environment (see MDA content). Ref. [130] reported that soil conditions may prime the organism for an improved response to subsequent stress. For example, exposing a plant to mild abiotic stress can improve its resistance to stress that occurs later, such as severe drought. This optimized environment of plants by biochar (3%) may have

https://doi.org/10.3390/environments13030136

supported an increased resistance of quinoa to further and more severe reductions in water supply. Compared to both SAP treatments (1%; 0.1%), the lower ROS damage (Figure 6) in the biochar treatment at reduced SWC can in part be explained by the support of N-accumulation and the therefore also possible increase in proline content (Figure 8) as a carbon and nitrogen reservoir to protect the plant from free radical damage and to maintain osmotic balance between the cell vacuole and cytoplasm in the plant cell in combination with an avoidance of over-reduction in PS I centers (Figure 4C) and successful dissipation of light energy.

5. Summary and Conclusions

The purpose of this study was to determine the impact of BC and SAP on the biomass production, extent and intensity of plants' efforts to water deficit. At the beginning of the study, six hypotheses were formulated to prove which properties the two soil additives BC and SAP share besides a high potential to improve soil parameter in dry regions, as well as their disadvantages and limitations. In summary, the examination of the six hypotheses yielded the following statements:

- Hypothesis (1): BC and SAP effect a negative growth response when the smallest possible amount of irrigation necessary for optimal growth of the control is applied: It is shown, in accordance with hypotheses 1 and 6, that plants grown at low SWC switched at 0.1% SAP and 3% BC from preferred growth (e.g., biomass production) to long survival (e.g., enhanced resilience). The strategy behind this plant response of the plant could be, at least partly, to compensate for this reduction in growth rate by extending the duration of growth.
- Hypothesis (2): The increase in WHC caused by the addition of additives such as BC or SAP will result in a higher water demand as in control plants and even at high SWC to symptoms of drought stress: Unexpectedly, but in accordance with Hypothesis 2, the treatment with 1% SAP even at high SWC led to a decrease in Y (II) and an increase in Y (NPQ) (Figure 3). Dissipation of excess absorbed light energy through zeaxanthin- and ΔpH-dependent photo-system II antenna quenching is considered the major mechanism for non-photochemical quenching and photoprotection. However, the significant increase in Y (NPQ) was accompanied at 1% SAP (high SWC) by a significant decrease in ECStmAu, ECStmAU*H^+, gH^+ and vH^+. This phenomenon can be explained by mounting evidence of a zeaxanthin-independent pathway for dissipation of excess light energy based within the PS II reaction center that may also play a significant role in photoprotection. It can be assumed that this response, in agreement with Hypothesis 2, is an indicator of a higher water requirement at high SWC of 1% SAP than of control plants and an incipient undersupply of the former. The addition of 3% BC at high SWC, contrary to Hypothesis 2, had hardly any impact on the CO_2/H_2O gas exchange. However, SAP amendment in both treatments (0.1% SAP and 1% SAP), in accordance with Hypothesis 2, caused a significant reduction in RD, RL and increase in ETR/A_{gross} and WUE.
- Hypothesis (3): The soil additives SAP or BC may prime quinoa for an improved response to subsequent stress at low soil water content: The explanations for Hypothesis 2 can also be interpreted, according to Hypothesis 3, as a priming effect or a positive impact of eustress on CO_2/H_2O gas exchange. The present results showed that plants of the control, 0.1% SAP and 3% BC treatments had significantly higher fractions of open reaction centers than plants of the 1% SAP treatment. The last one was balanced by an increased fraction of oxidized reaction centers. This recurring peculiarity of 1% SAP fits well with Hypothesis 3 that non-photochemical energy quenchers reduce in 1% SAP the LET from PS (II) to PS (I) in preparation for a further reduced water

 https://doi.org/10.3390/environments13030136

supply and effective protection against over-reduction on the PS I acceptor side. Water deficiency led to a reduction in electron uptake of PS I via open reaction centers in all treatments, with the exception of 1% SAP. This one was already low at 1% SAP. We interpret the already low fraction of open centers at high SWC at 1% SAP as a preadaptation to reduced water availability. This decreasing ability to absorb electrons from PS II correlated with the decrease in quantum yields (Y (II)) and ETR. It can be assumed that an imbalance between the electrons from the ETR and acceptor sides of the photosystem I can lead to a PS I photoinhibition.

- Hypothesis (4): Harmful competition of water absorption between plants and the soil additives BC and SAP may bring about adverse effects at low soil water content: It could be shown that 3% BC and 0.1% SAP can cause a reduction in plant dry weight at high and reduced SWC. This effect can at least partially be explained by the enhanced capacity of soils to bind water in the soil and to reduce soil water potential.
- Hypothesis (5): The plant's ability to handle oxidative stress is linked to its nitrogen (N) uptake and metabolism: Several recent analyses have shown that reduction in soil N is a widespread response to biochar application. In agreement with Hypothesis 5, it is shown that mainly the addition of SAP (0.1% and 1% SAP), but also of 3% BC, led to a significant decrease in the leaf N content and leaf C/N ratio at high SWC (Figure 7). There seems to be a clear negative correlation between a higher SAP content in the soil and C/N ratio and N content in the leaf at low SWC.
- Hypothesis (6): When quinoa needs to decide at low SWC between either growth or long survival it, will choose the latter: Besides no significant reduction in the N-content in the control and the 1% SAP at low SWC in the present study, we could also observe a tendency for increasing leaf N-content in the 0.1% SAP and a significant increase in the 3% BC treatment at reduced SWC. It can be assumed that the increase in N-content and the related decrease in the C/N ratio in the last two treatments ensure the upregulation of the adaptation mechanisms of plants under oxidative stress but on cost of growth. One indicator for this N-dependent upregulation was an increased proline accumulation in both treatments.

It is shown that the response of the test species *Chenopodium quinoa* to the addition of BC and SAP proved to be extraordinarily flexible, including fine-tuning to the specific soil conditions. The plant reacted with highly coordinated measures to both the danger of oversupply of SAP soil amendments and water shortage. The targeted adaptation to limited water supply became particularly evident in the 3% BC and 0.1% SAP treatments, among other things, by increased WUE and proline contents. However, BC also had a mitigating effect on the level of reactive oxygen species (ROS). It can be assumed that this effect is based on a more plant-compatible, less one-sided ion composition of BC. Due to the positive effects of biochar on plant adaptation and survivability at water shortage, the use of biochar is recommended. However, the application of SAP in conjunction with biochar may further improve soil properties and the physiological status of quinoa at water shortage. Management practices such as the presented ones play an essential role in soil water and nutrient storage and, by taking the soil conditions into consideration, there is an essential need to combine them with other soil amendments like hydrochar, compost or fertilizer and study their impact on plant response to water shortage.

Author Contributions: Conceptualization, N.K. and H.-W.K.; methodology, N.K. and H.-W.K.; software, N.K. and H.-W.K.; validation, N.K.; formal analysis, N.K. and H.-W.K.; investigation, N.K. and H.-W.K.; resources, H.-W.K.; data curation, N.K.; writing—original draft preparation, N.K.; writing—review and editing, H.-W.K.; visualization, N.K.; supervision, H.-W.K.; project administration, H.-W.K.;. All authors have read and agreed to the published version of the manuscript.

 https://doi.org/10.3390/environments13030136

Funding: This research received no external funding.

Data Availability Statement: The original contributions presented in this study are included in the article. Further inquiries can be directed to the corresponding author.

Acknowledgments: The authors thank Pieter Andreas Lennard Claas for reviewing the English language and style of the manuscript and for editing our manuscript.

Conflicts of Interest: The authors declare no conflicts of interest.

Abbreviations

A_{gross} (gross photosynthesis); A_{net} (net photosynthesis); BC (biochar); CEC (cation exchange capacity); C_i/C_{atm} ratio (carbon intercellular/carbon atmospheric ratio); C/N ratio (carbon/nitrogen ratio); C_s (stomatal conductivity); ctr (control); dw (dry weight); E (transpiration); ECStmAu (electrochromic shift); ECStmAu $\times$ gH^+ (indicator of the proton transport through the thylakoid membrane); ETR (electron transport rate); fw (fresh weight); gH^+ (proton conductivity); LOP (leaf osmotic potential); LWP (leaf water potential); MDA (malondialdehyde); NUE (nutrient use efficiency); PS I/II (photosystem I/II); QA (quinon); RD (dark respiration); RL (light respiration); ROS (reactive oxygen species); SAP (superabsorbent polymers); SPAC (soil-plant-atmosphere-continuum); SWC (soil water content); TBA (Thiobarbituric acid); TCA (Trichloracetic acid); vH^+ (steady-state rate of proton flux); WC (water content); WUE (water use efficiency), Y (II) photochemical quantum yield of PS II; Y (NPQ) (non-photochemical energy loss in PS II); Y (NO) (non-regulated non-photochemical energy loss in PS II).

References

1. IPCC. Summary for Policymakers. In *Climate Change 2023: Synthesis Report. Contribution of Working Groups I, II and III to the Sixth Assessment Report of the Intergovernmental Panel on Climate Change*; Core Writing Team, Lee, H., Romero, J., Eds.; IPCC: Geneva, Switzerland, 2023; pp. 1–34. [CrossRef]
2. Trenberth, K.E. Changes in precipitation with climate change. *Clim. Res.* **2011**, *47*, 123–138. [CrossRef]
3. Dai, A.; Zhao, T.; Chen, J. Climate Change and Drought: A Precipitation and Evaporation Perspective. *Curr. Clim. Change Rep.* **2018**, *4*, 301–312. [CrossRef]
4. Greve, P.; Seneviratne, S.I. Assessment of future changes in water availability and aridity. *Geophys. Res. Lett.* **2015**, *42*, 5493–5499. [CrossRef]
5. Samaniego, L.; Thober, S.; Kumar, R.; Wanders, N.; Rakovec, O.; Pan, M.; Zink, M.; Sheffield, J.; Wood, E.F.; Marx, A. Anthropogenic warming exacerbates European soil moisture droughts. *Nat. Clim. Change* **2018**, *8*, 421–426. [CrossRef]
6. Ruiz, K.; Biondi, S.; Oses, R.; Acuña-Rodríguez, I.; Antognoni, F.; Martinez-Mosqueira, E.; Coulibaly, A.; Canahua-Murillo, A.; Pinto, M.; Zurita-Silva, A.; et al. Quinoa biodiversity and sustainability for food security under climate change. A review. *Agron. Sustain. Dev.* **2014**, *34*, 349–359. [CrossRef]
7. Grillakis, M.G. Increase in severe and extreme soil moisture droughts for Europe under climate change. *Sci. Total Environ.* **2019**, *660*, 1245–1255. [CrossRef]
8. Corwin, D.L. Climate change impacts on soil salinity in agricultural areas. *Eur. J. Soil. Sci.* **2021**, *72*, 842–862. [CrossRef]
9. Huang, J.; Hartemink, A.E. Soil and environmental issues in sandy soils. *Earth-Sci. Rev.* **2020**, *208*, 103295. [CrossRef]
10. Fita, A.; Rodríguez-Burruezo, A.; Boscaiu, M.; Prohens, J.; Vicente, O. Breeding and Domesticating Crops Adapted to Drought and Salinity: A New Paradigm for Increasing Food Production. *Front. Plant Sci.* **2015**, *6*, 978. [CrossRef] [PubMed]
11. Begna, T. Effects of drought stress on crop production and productivity. *Int. J. Res. Stud. Agric. Sci.* **2020**, *6*, 34–43. [CrossRef]
12. Farooq, M.; Wahid, A.; Lee, D.J. Exogenously applied polyamines increase drought tolerance of rice by improving leaf water status, photosynthesis and membrane properties. *Acta Physiol. Plant.* **2009**, *31*, 937–945. [CrossRef]
13. Qiao, M.; Hong, C.; Jiao, Y.; Hou, S.; Gao, H. Impacts of Drought on Photosynthesis in Major Food Crops and the Related Mechanisms of Plant Responses to Drought. *Plants* **2024**, *13*, 1808. [CrossRef]
14. Zhou, R.; Kan, X.; Chen, J.; Hua, H.; Li, Y.; Ren, J.; Feng, K.; Liu, H.; Deng, D.; Yin, Z. Drought-induced changes in photosynthetic electron transport in maize probed by prompt fluorescence, delayed fluorescence, P700 and cyclic electron flow signals. *Environ. Exp. Bot.* **2019**, *158*, 51–62. [CrossRef]
15. Graßes, T.; Pesaresi, P.; Schiavon, F.; Varotto, C.; Salamini, F.; Jahns, P.; Leister, D. The role of ΔpH-dependent dissipation of excitation energy in protecting photosystem II against light-induced damage in Arabidopsis thaliana. *Plant Physiol. Biochem.* **2002**, *40*, 41–49. [CrossRef]

https://doi.org/10.3390/environments13030136

16. Anjum, S.A.; Wang, L.; Farooq, M.; Khan, I.; Xue, L. Methyl Jasmonate-Induced Alteration in Lipid Peroxidation, Antioxidative Defence System and Yield in Soybean Under Drought. *J. Agron. Crop Sci.* **2011**, *197*, 296–301. [CrossRef]
17. Esterbauer, H.; Schaur, R.J.; Zollner, H. Chemistry and biochemistry of 4-hydroxynonenal, malonaldehyde and related aldehydes. *Free Radic. Biol. Med.* **1991**, *11*, 81–128. [CrossRef] [PubMed]
18. Sharma, P.; Jha, A.B.; Dubey, R.S.; Pessarakli, M. Reactive Oxygen Species, Oxidative Damage, and Antioxidative Defense Mechanism in Plants under Stressful Conditions. *J. Bot.* **2012**, *2012*, 217037. [CrossRef]
19. Halliwell, B.; Gutteridge, J.M.C. *Free Radicals in Biology and Medicine*; Oxford University Press: Oxford, UK, 1989.
20. Yamauchi, Y.; Furutera, A.; Seki, K.; Toyoda, Y.; Tanaka, K.; Sugimoto, Y. Malondialdehyde generated from peroxidized linolenic acid causes protein modification in heat-stressed plants. *Plant Physiol. Biochem.* **2008**, *46*, 786–793. [CrossRef] [PubMed]
21. Gammans, M.; Mérel, P.; Ortiz-Bobea, A. Negative impacts of climate change on cereal yields: Statistical evidence from France. *Environ. Res. Lett.* **2017**, *12*, S239–S286. [CrossRef]
22. Carrão, H.; Naumann, G.; Barbosa, P. Mapping global patterns of drought risk: An empirical framework based on sub-national estimates of hazard, exposure and vulnerability. *Glob. Environ. Change* **2016**, *39*, 108–124. [CrossRef]
23. Wen, W.; Timmermans, J.; Chen, Q.; van Bodegom, P.M. A Review of Remote Sensing Challenges for Food Security with Respect to Salinity and Drought Threats. *Remote Sens.* **2021**, *13*, 6. [CrossRef]
24. Zeifman, L.; Hertog, S.; Kantorova, V.; Wilmoth, J. A World of 8 Billion. United Nations Department of Economic and Social Affairs. 2022. Available online: https://www.un.org/development/desa/dpad/wp-content/uploads/sites/45/publication/PB_140.pdf (accessed on 29 January 2026).
25. Hinojosa, L.; González, J.A.; Barrios-Masias, F.H.; Fuentes, F.; Murphy, K.M. Quinoa Abiotic Stress Responses: A Review. *Plants* **2018**, *7*, 106. [CrossRef] [PubMed]
26. FAO; CIRAD. *State of the Art Report on Quinoa Around the World in 2013*; Bazile, D., Bertero, H.D., Nieto, C., Eds.; Food and Agriculture Organization of the United Nations (FAO): Santiago, Chile, 2015. Available online: https://agritrop.cirad.fr/575489/1/document_575489.pdf (accessed on 29 January 2026).
27. Vega-Gálvez, A.; Miranda, M.; Vergara, J.; Uribe, E.; Puente, L.; Martínez, E.A. Nutrition facts and functional potential of quinoa (*Chenopodium quinoa* Willd.), an ancient Andean grain: A review. *J. Sci. Food Agric.* **2010**, *90*, 2541–2547. [CrossRef]
28. Pereira, E.; Encina-Zelada, C.; Barros, L.; Gonzales-Barron, U.; Cadavez, V.; Ferreira, I.C.F.R. Chemical and nutritional characterization of *Chenopodium quinoa* Willd (quinoa) grains: A good alternative to nutritious food. *Food Chem.* **2019**, *280*, 110–114. [CrossRef] [PubMed]
29. Ain, Q.T.; Siddique, K.; Bawazeer, S.; Ali, I.; Mazhar, M.; Rasool, R.; Mubeen, B.; Ullah, F.; Unar, A.; Jafar, T.H. Adaptive mechanisms in quinoa for coping in stressful environments: An update. *PeerJ* **2023**, *11*, e14832. [CrossRef] [PubMed]
30. Bhargava, A.; Shukla, S.; Ohri, D. *Chenopodium quinoa*—An Indian perspective. *Ind. Crops Prod.* **2006**, *23*, 73–87. [CrossRef]
31. Jacobsen, S.E.; Liu, F.; Jensen, C.R. Does root-sourced ABA play a role for regulation of stomata under drought in quinoa (*Chenopodium quinoa* Willd.). *Sci. Hortic.* **2009**, *122*, 281–287. [CrossRef]
32. Sun, Y.; Liu, F.; Bendevis, M.; Shabala, S.; Jacobsen, S.-E. Sensitivity of Two Quinoa (*Chenopodium quinoa* Willd.) Varieties to Progressive Drought Stress. *J. Agron. Crop Sci.* **2013**, *200*, 12–23. [CrossRef]
33. Pérez-López, U.; Robredo, A.; Lacuesta, M.; Mena-Petite, A.; Muñoz-Rueda, A. The impact of salt stress on the water status of barley plants is partially mitigated by elevated CO_2. *Environ. Exp. Bot.* **2009**, *66*, 463–470. [CrossRef]
34. Chen, T.; Chen, Y.; Wang, X.; Zhang, L. Osmotic adjustment in roots and leaves of two cotton cultivars with different tolerance to soil salinity. *Biomed. J. Sci. Tech. Res.* **2019**, *15*, 1–11. [CrossRef]
35. Saddiq, M.S.; Wang, X.; Iqbal, S.; Hafeez, M.B.; Khan, S.; Raza, A.; Iqbal, J.; Maqbool, M.M.; Fiaz, S.; Qazi, M.A.; et al. Effect of Water Stress on Grain Yield and Physiological Characters of Quinoa Genotypes. *Agronomy* **2021**, *11*, 1934. [CrossRef]
36. Schweiger, A.H.; Zimmermann, T.; Poll, C.; Marhan, S.; Leyrer, V.; Berauer, B.J. The need to decipher plant drought stress along the soil–plant–atmosphere continuum. *Oikos* **2023**, *2023*, e10136. [CrossRef]
37. Philip, J.R. Plant water relations: Some physical aspects. *Annu. Rev. Plant Physiol.* **1966**, *17*, 245–268. [CrossRef]
38. Alkharabsheh, H.M.; Seleiman, M.F.; Battaglia, M.L.; Shami, A.; Jalal, R.S.; Alhammad, B.A.; Almutairi, K.F.; Al-Saif, A.M. Biochar and Its Broad Impacts in Soil Quality and Fertility, Nutrient Leaching and Crop Productivity: A Review. *Agronomy* **2021**, *11*, 993. [CrossRef]
39. Dubrovina, I. Biochar has Different Effects on Soil Fertility and Spring Barley Productivity in Two Boreal Soils with Contrasting Textures. *Agric. Res.* **2025**, 1–13. [CrossRef]
40. Hossain, M.Z.; Bahar, M.M.; Sarkar, B.; Donne, S.W.; Ok, Y.S.; Palansooriya, K.N.; Kirkham, M.B.; Chowdhury, S.; Bolan, N. Biochar and its importance on nutrient dynamics in soil and plant. *Biochar* **2020**, *2*, 379–420. [CrossRef]
41. Naz, Z.; Jahan, S.; Rashid, A.; Kamran, A.; Abd-Allah, E.F.; Avila-Quezada, G.D.; Hashem, A. Efficacy of modified biochar for enhancing maize photosynthesis and yield under various irrigation regimes. *Sci. Rep.* **2025**, *15*, 32216. [CrossRef]
42. Kammann, C.I.; Linsel, S.; Gößling, J.W.; Koyro, H.-W. Influence of biochar on drought tolerance of *Chenopodium quinoa* Willd and on soil–plant relations. *Plant Soil* **2011**, *345*, 195–210. [CrossRef]

43. Paneque, M.; De la Rosa, J.M.; Franco-Navarro, J.D.; Colmenero-Flores, J.M.; Knicker, H. Effect of biochar amendment on morphology, productivity and water relations of sunflower plants under non-irrigation conditions. *Catena* **2016**, *147*, 280–287. [CrossRef]
44. Uzoma, K.C.; Inoue, M.; Andry, H.; Fujimaki, H.; Zahoor, A.; Nishihara, E. Effect of cow manure biochar on maize productivity under sandy soil condition. *Soil Use Manag.* **2011**, *27*, 205–212. [CrossRef]
45. Xu, C.Y.; Hosseini-Bai, S.; Hao, Y.; Rachaputi, R.C.N.; Wang, H.; Xu, Z.; Wallace, H. Effect of biochar amendment on yield and photosynthesis of peanut on two types of soils. *Environ. Sci. Pollut. Res.* **2015**, *22*, 6112–6125. [CrossRef]
46. Rawat, J.; Saxena, J.; Sanwal, P. *Biochar: A Sustainable Approach for Improving Plant Growth and Soil Properties*; IntechOpen: London, UK, 2019. [CrossRef]
47. Zhang, Y.; Ding, J.; Wang, H.; Su, L.; Zhao, C. Biochar addition alleviate the negative effects of drought and salinity stress on soybean productivity and water use efficiency. *BMC Plant Biol.* **2020**, *20*, 288. [CrossRef]
48. Atkinson, C.J.; Fitzgerald, J.D.; Hipps, N.A. Potential mechanisms for achieving agricultural benefits from biochar application to temperate soils: A review. *Plant Soil* **2010**, *337*, 1–18. [CrossRef]
49. Bista, D.R.; Heckathorn, S.A.; Jayawardena, D.M.; Mishra, S.; Boldt, J.K. Effects of Drought on Nutrient Uptake and the Levels of Nutrient-Uptake Proteins in Roots of Drought-Sensitive and -Tolerant Grasses. *Plants* **2018**, *7*, 28. [CrossRef]
50. Ippolito, J.A.; Novak, J.M.; Busscher, W.J.; Ahmedna, M.; Rehrah, D.; Watts, D.W. Switchgrass biochar affects two Aridisols. *J. Environ. Qual.* **2012**, *41*, 1123–1130. [CrossRef]
51. Phillips, C.L.; Meyer, K.M.; Garcia-Jaramillo, M.; Weidman, C.S.; Stewart, C.E.; Wanzek, T.; Grusak, M.A.; Watts, D.W.; Novak, J.; Trippe, K.M. Towards predicting biochar impacts on plant-available soil nitrogen content. *Biochar* **2022**, *4*, 9. [CrossRef]
52. Li, S.; Zhou, L.; Addo-Danso, S.D.; Ding, G.; Sun, M.; Wu, S.; Lin, S. Nitrogen supply enhances the physiological resistance of Chinese fir plantlets under polyethylene glycol (PEG)-induced drought stress. *Sci. Rep.* **2020**, *10*, 7509. [CrossRef]
53. Chen, X.; He, H.-Z.; Chen, G.-K.; Li, H.-S. Effects of biochar and crop straws on the bioavailability of cadmium in contaminated soil. *Sci. Rep.* **2020**, *10*, 9528. [CrossRef] [PubMed]
54. Alam, M.Z.; McGee, R.; Hoque, M.A.; Ahammed, G.J.; Carpenter-Boggs, L. Effect of arbuscular mycorrhizal fungi, selenium and biochar on photosynthetic pigments and antioxidant enzyme activity under arsenic stress in mung bean (*Vigna radiata*). *Front. Physiol.* **2019**, *10*, 193. [CrossRef]
55. Chang, L.; Xu, L.; Liu, Y.; Qiu, D. Superabsorbent polymers used for agricultural water retention. *Polym. Test.* **2021**, *94*, 107021. [CrossRef]
56. Ai, F.; Yin, X.; Hu, R.; Ma, H.; Liu, W. Research into the super-absorbent polymers on agricultural water. *Agric. Water Manag.* **2021**, *245*, 106513. [CrossRef]
57. Oladosu, Y.; Rafii, M.Y.; Arolu, F.; Chukwu, S.C.; Salisu, M.A.; Fagbohun, I.K.; Muftaudeen, T.K.; Swaray, S.; Haliru, B.S. Superabsorbent Polymer Hydrogels for Sustainable Agriculture: A Review. *Horticulturae* **2022**, *8*, 605. [CrossRef]
58. Yang, Y.; Liang, Z.; Zhang, R.; Zhou, S.; Yang, H.; Chen, Y.; Zhang, J.; Yin, H.; Yu, D. Research Advances in Superabsorbent Polymers. *Polymers* **2024**, *16*, 501. [CrossRef]
59. Zohuriaan-Mehr, M.J.; Omidian, H.; Doroudiani, S.; Kabiri, K. Advances in non-hygienic applications of superabsorbent hydrogel materials. *J. Mater. Sci.* **2010**, *45*, 5711–5735. [CrossRef]
60. Yang, Y.; Zhang, S.; Wu, J.; Gao, C.; Lu, D.; Tang, D.W.S. Effect of long term application of super absorbent polymer on soil structure, soil enzyme activity, photosynthetic characteristics, water and nitrogen use of winter wheat. *Front. Plant Sci.* **2022**, *13*, 998494. [CrossRef] [PubMed]
61. Saha, A.; Sekharan, S.; Manna, U. Superabsorbent hydrogel (SAH) as a soil amendment for drought management: A review. *Soil Tillage Res.* **2020**, *204*, 104736. [CrossRef]
62. Yang, F.; Cen, R.; Feng, W.; Liu, J.; Qu, Z.; Miao, Q. Effects of Super-Absorbent Polymer on Soil Remediation and Crop Growth in Arid and Semi-Arid Areas. *Sustainability* **2020**, *12*, 7825. [CrossRef]
63. Shi, Y.; Li, J.; Shao, J.; Deng, S.; Wang, R.; Li, N.; Sun, J.; Zhang, H.; Zhu, H.; Zhang, Y.; et al. Effects of Stockosorb and Luquasorb polymers on salt and drought tolerance of *Populus popularis*. *Sci. Hortic.* **2010**, *124*, 268–273. [CrossRef]
64. Islam, M.R.; Xue, X.; Mao, S.; Ren, C.; Eneji, A.E.; Hu, Y. Effects of water-saving superabsorbent polymer on antioxidant enzyme activities and lipid peroxidation in oat (*Avena sativa* L.) under drought stress. *J. Sci. Food Agric.* **2011**, *91*, 680–686. [CrossRef]
65. Martin, C.A.; Ruter, J.M.; Roberson, R.W.; Sharp, W.P. Element absorption and hydration potential of polyacrylamide gels. *Commun. Soil Sci. Plant Anal.* **1993**, *24*, 539–548. [CrossRef]
66. *STOCKOSORB 660 Safety Data Sheet*; Version 1.0, Revision 11/07/2019; [PDF]; Evonik Industries: Essen, Germany, 2019. Available online: http://apofl.com/wp-content/uploads/2021/08/STOCKOSORB-660-SAFETY-DATA-SHEET.pdf (accessed on 29 January 2026).

 https://doi.org/10.3390/environments13030136

67. Friede, K. Untersuchungen und Bewertung der Biologischen Qualität von Biokohlekomposten und Deren Wirkungen auf die Lebenraumfunktion von Böden mit dem Ziel der Ableitung von Biologischen Qualittätskriterien. Doctoral Dissertation, Freie Universität Berlin, Berlin, Germany, 2021. Available online: https://refubium.fu-berlin.de/bitstream/handle/fub188/30975/Dissertation_Karin_Friede.pdf?sequence=3 (accessed on 29 January 2026).
68. Kammann, C.; Schmidt, H.P.; Messerschmidt, N.; Linsel, S.; Steffens, D.; Müller, C.; Koyro, H.-W.; Conte, P.; Joseph, S. Plant growth improvement mediated by nitrate capture in composted biochar. *Sci. Rep.* **2015**, *5*, 11080. [CrossRef]
69. Akram, M.Z.; Libutti, A.; Rivelli, A.R. Evaluation of Vegetative Development of Quinoa under Water Stress by Applying Different Organic Amendments. *Agronomy* **2023**, *13*, 1412. [CrossRef]
70. Hewitt, E.J. *Sand and Water Culture Methods Used in Study of Plant Nutrition*, 2nd ed.; Commonwealth Agricultural Bureaux: Farnham Royal, UK, 1966.
71. Scholander, P.F. The role of solvent pressure in osmotic systems. *Proc. Natl. Acad. Sci. USA* **1966**, *55*, 1407–1414. [CrossRef]
72. Matyssek, R.; Herppich, W.B. *Experimentelle Pflanzenökologie*; Springer Spektrum: Berlin, Germany, 2020; pp. 425–462.
73. Bowler, M.G. The physics of osmotic pressure. *Eur. J. Phys.* **2017**, *38*, 055102. [CrossRef]
74. Kuhlgert, S.; Austic, G.; Zegarac, R.; Osei-Bonsu, I.; Hoh, D.; Chilvers, M.I.; Roth, M.G.; Bi, K.; TerAvest, D.; Weebadde, P.; et al. MulDspeQ Beta: A tool for large-scale plant phenotyping connected to the open PhotosynQ network. *R. Soc. Open Sci.* **2016**, *3*, 160592. [CrossRef]
75. Schulte, M.; Offer, C.; Hansen, U. Induction of CO_2-gas exchange and electron transport: Comparison of dynamic and steady-state responses in Fagus sylvatica leaves. *Trees* **2003**, *17*, 153–163. [CrossRef]
76. Valentini, R.; Epron, D.; De Angelis, P.; Matteucci, G.; Dreyer, E. In situ estimation of net CO_2 assimilation, photosynthetic electron flow and photorespiration in Turkey oak (*Q. cerris* L.) leaves: Diurnal cycles under different levels of water supply. *Plant Cell Environ.* **1995**, *18*, 631–640. [CrossRef]
77. Heath, R.L.; Packer, L. Photoperoxidation in isolated chloroplasts: I. Kinetics and stoichiometry of fatty acid peroxidation. *Arch. Biochem. Biophys.* **1968**, *125*, 189–198. [CrossRef] [PubMed]
78. Bates, L.S.; Waldren, R.P.; Teare, I.D. Rapid determination of free proline for water-stress studies. *Plant Soil* **1937**, *39*, 205–207. [CrossRef]
79. *SigmaPlot for Windows*, Version 12; Systat Software Inc.: Bangalore, India, 2012. Available online: https://www.systatsoftware.com (accessed on 29 January 2026).
80. Ambaye, T.G.; Vaccari, M.; van Hullebusch, E.D.; Amrane, A.; Rtimi, S. Mechanisms and adsorption capacities of biochar for the removal of organic and inorganic pollutants from industrial wastewater. *Int. J. Environ. Sci. Technol.* **2021**, *18*, 3273–3294. [CrossRef]
81. Derbali, W.; Manaa, A.; Goussi, R.; Derbali, I.; Abdelly, C.; Koyro, H.-W. Post-stress restorative response of two quinoa genotypes differing in their salt resistance after salinity release. *Plant Physiol. Biochem.* **2021**, *164*, 222–236. [CrossRef]
82. Yang, A.; Akhtar, S.S.; Amjad, M.; Iqbal, S.; Jacobsen, S.E. Growth and physiological responses of quinoa to drought and temperature stress. *J. Agron. Crop Sci.* **2016**, *202*, 445–453. [CrossRef]
83. Stikic, R.; Jovanovic, Z.; Marjanovic, M.; Djordjevic, S. The effect of drought on water regime and growth of quinoa (*Chenopodium quinoa* Willd.). *Ratar. I Povrt.* **2015**, *52*, 80–84. [CrossRef]
84. Elewa, T.A.; Sadak, M.S.; Saad, A.M. Proline treatment improves physiological responses in quinoa plants under drought stress. *Biosci. Res.* **2017**, *14*, 21–33.
85. Lichtenthaler, H.K. The stress concept in plants: An introduction. *Ann. N. Y. Acad. Sci.* **1998**, *851*, 187–198. [CrossRef]
86. Verbraeken, L.; Wuyts, N.; Mertens, S.; Cannoot, B.; Maleux, K.; Demuynck, K.; De Block, J.; Merchie, J.; Dhondt, S.; Bonaventure, G.; et al. Drought affects the rate and duration of organ growth but not inter-organ growth coordination. *Plant Physiol.* **2021**, *186*, 1336–1353. [CrossRef] [PubMed]
87. Avramova, V.; AbdElgawad, H.; Vasileva, I.; Petrova, A.S.; Holek, A.; Mariën, J.; Asard, H.; Beemster, G.T.S. High antioxidant activity facilitates maintenance of cell division in leaves of drought tolerant maize hybrids. *Front. Plant Sci.* **2017**, *8*, 84. [CrossRef]
88. Avramova, V.; AbdElgawad, H.; Zhang, Z.; Fotschki, B.; Casadevall, R.; Vergauwen, L.; Knapen, D.; Taleisnik, E.; Guisez, Y.; Asard, H.; et al. Drought Induces Distinct Growth Response, Protection, and Recovery Mechanisms in the Maize Leaf Growth Zone. *Plant Physiol.* **2015**, *169*, 1382–1396. [CrossRef]
89. Nelissen, H.; Sun, X.-H.; Rymen, B.; Jikumaru, Y.; Kojima, M.; Takebayashi, Y.; Abbeloos, R.; Demuynck, K.; Storme, V.; Vuylsteke, M.; et al. The reduction in maize leaf growth under mild drought affects the transition between cell division and cell expansion and cannot be restored by elevated gibberellic acid levels. *Plant Biotechnol. J.* **2017**, *16*, 615–627. [CrossRef]
90. Wang, X.; Wu, Z.; Zhou, Q.; Wang, X.; Song, S.; Dong, S. Physiological Response of Soybean Plants to Water Deficit. *Front. Plant Sci.* **2022**, *12*, 809692. [CrossRef]
91. Derbali, I.; Derbali, W.; Gharred, J.; Manaa, A.; Slama, I.; Koyro, H.-W. Mitigating Salinity Stress in Quinoa (*Chenopodium quinoa* Willd.) with Biochar and Superabsorber Polymer Amendments. *Plants* **2024**, *13*, 92. [CrossRef]

92. Falouti, M.; Ellouzi, H.; Bounaouara, F.; Farhat, N.; Aggag, A.M.; Debez, A.; Rabhi, M.; Abdelly, C.; Slama, I.; Zorrig, W. Higher activity of PSI compared to PSII accounts for the beneficial effect of silicon on barley (*Hordeum vulgare* L.) plants challenged with salinity. *Photosynthetica* **2022**, *60*, 508–520. [CrossRef]
93. Moradi, F.; Ismail, A.M. Responses of Photosynthesis, Chlorophyll Fluorescence and ROS-Scavenging Systems to Salt Stress During Seedling and Reproductive Stages in Rice. *Ann. Bot.* **2007**, *99*, 1161–1173. [CrossRef]
94. Ivanov, A.G.; Sane, P.V.; Hurry, V.; Öquist, G.; Huner, N.P.A. Photosystem II reaction centre quenching: Mechanisms and physiological role. *Photosynth. Res.* **2008**, *98*, 565–574. [CrossRef] [PubMed]
95. Huang, W.; Yang, S.J.; Zhang, S.B.; Zhang, J.-L.; Cao, K.-F. Cyclic electron flow plays an important role in photoprotection for the resurrection plant Paraboearufescens under drought stress. *Planta* **2012**, *235*, 819–828. [CrossRef]
96. Degen, G.E. Under (stromal redox) pressure: NADP+ synthesis regulates photosystem I biogenesis. *Plant Physiol.* **2022**, *189*, 1890–1892. [CrossRef]
97. Kanazawa, A.; Ostendorf, E.; Kohzuma, K.; Hoh, D.; Strand, D.D.; Sato-Cruz, M.; Savage, L.; Cruz, J.A.; Fisher, N.; Froehlich, J.E.; et al. Chloroplast ATP Synthase Modulation of the Thylakoid Proton Motive Force: Implications for Photosystem I and Photosystem II Photoprotection. *Front. Plant Sci.* **2017**, *8*, 719. [CrossRef] [PubMed]
98. Yang, H.; Liu, J.; Wen, X.; Lu, C. Molecular mechanism of photosystem I assembly in oxygenic organisms. *Biochim. Biophys. Acta (BBA)-Bioenerg.* **2015**, *1847*, 838–848. [CrossRef]
99. Lima-Melo, Y.; Alencar, V.T.C.B.; Lobo, A.K.M.; Sousa, R.H.V.; Tikkanen, M.; Aro, E.-M.; Silveira, J.A.G.; Gollan, P.J. Photoinhibition of Photosystem I Provides Oxidative Protection During Imbalanced Photosynthetic Electron Transport in Arabidopsis thaliana. *Front. Plant Sci.* **2019**, *10*, 916. [CrossRef] [PubMed]
100. Zivcak, M.; Kalaji, H.M.; Shao, H.-B.; Olsovska, K.; Brestic, M. Photosynthetic proton and electron transport in wheat leaves under prolonged moderate drought stress. *J. Photochem. Photobiol. B Biol.* **2014**, *137*, 107–115. [CrossRef]
101. Shimakawa, G.; Müller, P.; Miyake, C.; Krieger-Liszkay, A.; Sétif, P. Photo-oxidative damage of photosystem I by repetitive flashes and chilling stress in cucumber leaves. *Biochim. Biophys. Acta (BBA)-Bioenerg.* **2024**, *1865*, 149490. [CrossRef] [PubMed]
102. Rouphael, Y.; Kyriacou, M.C. Enhancing Quality of Fresh Vegetables Through Salinity Eustress and Biofortification Applications Facilitated by Soilless Cultivation. *Front. Plant Sci.* **2018**, *9*, 1254. [CrossRef]
103. Sperdouli, I.; Ouzounidou, G.; Moustakas, M. Hormesis Responses of Photosystem II in Arabidopsis thaliana under Water Deficit Stress. *Int. J. Mol. Sci.* **2023**, *24*, 9573. [CrossRef]
104. Sepehri, S.; Abdoli, S.; Asgari Lajayer, B.; Astatkie, T.; Price, G.W. Changes in phytochemical properties and water use efficiency of peppermint (*Mentha piperita* L.) using superabsorbent polymer under drought stress. *Sci. Rep.* **2023**, *13*, 21989. [CrossRef]
105. Chen, X.; Mao, X.; Lu, Q.; Liao, Z.; He, Z. Characteristics and mechanisms of acrylate polymer damage to maize seedlings. *Ecotoxicol. Env. Saf.* **2016**, *129*, 228–234. [CrossRef] [PubMed]
106. Lawlor, D.W.; Cornic, G. Photosynthetic carbon assimilation and associated metabolism in relation to water deficits in higher plants. *Plant Cell Environ.* **2002**, *25*, 275–294. [CrossRef] [PubMed]
107. Reddy, A.R.; Chaitanya, K.V.; Vivekanandan, M. Drought-induced responses of photosynthesis and antioxidant metabolism in higher plants. *J. Plant Physiol.* **2004**, *161*, 1189–1202. [CrossRef]
108. Ghotbi-Ravandi, A.A.; Shahbazi, M.; Shariati, M.; Mulo, P. Effects of Mild and Severe Drought Stress on Photosynthetic Efficiency in Tolerant and Susceptible Barley (*Hordeum vulgare* L.) Genotypes. *J. Agron. Crop Sci.* **2014**, *200*, 403–415. [CrossRef]
109. Zhang, J.; Jiang, H.; Song, X.; Jin, J.; Zhang, X. The Responses of Plant Leaf CO_2/H_2O Exchange and Water Use Efficiency to Drought: A Meta-Analysis. *Sustainability* **2018**, *10*, 551. [CrossRef]
110. Sarma, B.; Gogoi, N.; Bharali, M.; Mali, P. Field evaluation of soil and wheat responses to combined application of hardwood biochar and inorganic fertilizers in aridic sandy loam soil. *Exp. Agric.* **2017**, *54*, 507–519. [CrossRef]
111. Fghire, R.; Anaya, F.; Ali, O.I.; Benlhabib, O.; Ragab, R.; Wahbi, S. Physiological and photosynthetic response of quinoa to drought stress. *Chil. J. Agric. Res.* **2015**, *75*, 174–183. [CrossRef]
112. Anjum, S.A.; Ashraf, U.; Tanveer, M.; Khan, I.; Hussain, S.; Shahzad, B.; Zohaib, A.; Abbas, F.; Saleem, M.F.; Ali, I.; et al. Drought Induced Changes in Growth, Osmolyte Accumulation and Antioxidant Metabolism of Three Maize Hybrids. *Front. Plant Sci.* **2017**, *8*, 69. [CrossRef]
113. Wang, Y.; Pan, F.; Wang, G.; Zhang, G.; Wang, Y.; Chen, X.; Mao, Z. Effects of biochar on photosynthesis and antioxidative system of Malus hupehensis Rehd. seedlings under replant conditions. *Sci. Hortic.* **2014**, *175*, 9–15. [CrossRef]
114. Khan, Z.; Khan, M.N.; Zhang, K.; Luo, T.; Zhu, K.; Hu, L. The application of biochar alleviated the adverse effects of drought on the growth, physiology, yield and quality of rapeseed through regulation of soil status and nutrients availability. *Ind. Crops Prod.* **2021**, *171*, 113878. [CrossRef]
115. Cao, Y.; Xie, Q.; Wang, J.; Yang, J.; Xie, Z. Effects of super absorbent polymer on the physiological characteristics and drought resistance of bermudagrass [*Cynodon dactylon* (L.) Pers.] seedlings. *Appl. Ecol. Environ. Res.* **2020**, *18*, 4075–4085. [CrossRef]

 https://doi.org/10.3390/environments13030136

116. Tohidi-Moghadam, H.R.; Shirani-Rad, A.H.; Nour-Mohammadi, G.; Habibi, D.; Mashhadi-Akbar-Boojar, M. Effect of Super Absorbent Application on Antioxidant Enzyme Activities in Canola (*Brassica napus* L.) Cultivars under Water Stress Conditions. *Am. J. Agri. Biol. Sci.* **2009**, *4*, 215–223.
117. Li, L.; Zhang, H.; Zhou, X.; Chen, M.; Lu, L.; Cheng, X. Effects of super absorbent polymer on scouring resistance and water retention performance of soil for growing plants in ecological concrete. *Ecol. Eng.* **2019**, *138*, 237–247. [CrossRef]
118. Abrisham, E.S.; Jafari, M.; Tavili, A.; Rabii, A.; Zare Chahoki, M.A.; Zare, S.; Egan, T.; Yazdanshenas, H.; Ghasemian, D.; Tahmoures, M. Effects of a super absorbent polymer on soil properties and plant growth for use in land reclamation. *Arid. Land Res. Manag.* **2018**, *32*, 407–420. [CrossRef]
119. Abdolmaleki, A.; Dapprich, P.; Bertram, H.; Hollensteiner, J.; Schmitz, M.; Schmitt, A.O.; Gültas, M. Impact of Super Absorbent Polymers (SAP) on the Morphological, Physiological, and Biochemical Responses of Rapeseed and Maize Under Drought Stress. *Crop Breed. Genet. Genom.* **2025**, *7*, e250009. [CrossRef]
120. Ghebru, M.G.; du Toit, E.S.; Steyn, J.M. Water and nutrient retention by Aquasoil® and Stockosorb® polymers. *S. Afr. J. Plant Soil* **2007**, *24*, 32–36. [CrossRef]
121. Nguyen, T.T.N.; Xu, C.Y.; Tahmasbian, I.; Che, R.; Xu, Z.; Wallace, H.M.; Bai, S.H. Effects of biochar on soil available inorganic nitrogen: A review and meta-analysis. *Geoderma* **2017**, *288*, 79–96. [CrossRef]
122. Liu, Q.; Zhang, Y.; Liu, B.; Amonette, J.E.; Lin, Z.; Liu, G.; Ambus, P.; Xie, Z. How does biochar influence soil N cycle? A meta-analysis. *Plant Soil* **2018**, *426*, 211–225. [CrossRef]
123. Gao, S.; DeLuca, T.H.; Cleveland, C.C. Biochar additions alter phosphorus and nitrogen availability in agricultural ecosystems: A meta-analysis. *Sci. Total Environ.* **2019**, *654*, 463–472. [CrossRef] [PubMed]
124. Sardans, J.; Penuuelas, J.; Estiarte, M.; Prieto, P. Warming and drought alter C and N concentration, allocation and accumulation in a Mediterranean shrubland. *Glob. Change Biol.* **2008**, *14*, 2304–2316. [CrossRef]
125. Noctor, G.; Reichheld, J.P.; Foyer, C.H. ROS-related redox regulation and signaling in plants. *Semin. Cell Dev. Biol.* **2018**, *80*, 3–12. [CrossRef] [PubMed]
126. Kaur, G.; Asthir, B. Proline: A key player in plant abiotic stress tolerance. *Biol. Plant.* **2015**, *59*, 609–619. [CrossRef]
127. Zhang, M.; Cheng, Z.; Zhao, T.; Liu, M.; Hu, M.; Li, J. Synthesis, characterization, and swelling behaviors of salt-sensitive maize bran–poly (acrylic acid) superabsorbent hydrogel. *J. Agric. Food Chem.* **2014**, *62*, 8867–8874. [CrossRef]
128. Liu, J.; Li, Q.; Su, Y.; Yue, Q.; Gao, B. Characterization and swelling-deswelling properties of wheat straw cellulose-based semi-IPNs hydrogel. *Carbohydr. Polym.* **2014**, *107*, 232–240. [CrossRef]
129. Situ, Y.; Yang, Y.; Huang, C.; Liang, S.; Mao, X.; Chen, X. Effects of several superabsorbent polymers on soil exchangeable cations and crop growth. *Environ. Technol. Innov.* **2023**, *30*, 103126. [CrossRef]
130. Hilker, M.; Schmülling, T. Stress priming, memory, and signalling in plants. *Plant Cell Environ.* **2019**, *42*, 753–761. [CrossRef] [PubMed]

 https://doi.org/10.3390/environments13030136

environments

MDPI

Article

Enhancing Soil Functionality Through Circular Fertilizers Derived from Agro-Industrial Wastes: Insights into Microbiological and Biochemical Dynamics

Angela Maffia, Federica Alessia Marra, Santo Battaglia, Carmelo Mallamaci and Adele Muscolo *

Department of AGRARIA, 'Mediterranea' University of Reggio Calabria, Feo di Vito, 89122 Reggio Calabria, Italy; angela.maffia@unirc.it (A.M.); federica.marra@unirc.it (F.A.M.); santo.battaglia@unirc.it (S.B.); carmelo.mallamaci@unirc.it (C.M.)

* Correspondence: amuscolo@unirc.it

Abstract

The growing demand for sustainable fertilization practices has stimulated interest in circular fertilizers derived from agro-industrial and agricultural wastes. This study assessed the agronomic and biological performance of several waste-based fertilizers—produced through composting, vermicomposting, and sulfur–bentonite enrichment—on chemical and microbiological soil properties. Composts and vermicomposts were prepared from olive pomace, citrus residues, wood sawdust, and straw, with or without elemental sulfur obtained from petroleum gas desulfurization. Field trials were conducted on a sandy loam soil (Motta San Giovanni, Italy) to compare the different formulations. After six months, soils amended with waste-based fertilizers exhibited significant improvements in key parameters relative to both the control and mineral fertilizer treatment. Vermicompost applications (SV1, SV2) increased total organic carbon by 20–30% (up to 2.1%), total nitrogen by 35–45% (0.22–0.23%), microbial biomass carbon by ~25% (≈1090 μg C g^{-1}), and dehydrogenase and fluorescein diacetate activities by 10–20% compared with compost or sulfur–bentonite treatments. Compost amendments (SC1, SC2) raised soil pH (8.2–8.3) and organic matter content (≈3.3–3.6%), while sulfur–bentonite formulations lowered pH to 7.1–7.3 and increased water-soluble phenols (up to 40 μg TAE g^{-1} d.s). The highest cation exchange capacity (22–23 cmol (+) kg^{-1}) was observed in vermicompost-amended soils. Microbial community analysis revealed greater fungal abundance under sulfur–bentonite treatments, whereas bacteria and actinomycetes predominated in compost-amended soils. Principal Component Analysis (explaining 76% of variance) identified two main functional pathways: vermicompost treatments clustered with indicators of high biological activity (TOC, TN, MBC, and enzyme activities), while compost and sulfur–bentonite treatments were associated with pH, phenolic compounds, and fungal biomass, reflecting slower but more stable organic matter turnover. Overall, vermicompost-based fertilizers proved most effective in enhancing short-term nutrient availability and microbial activation, whereas composts favored long-term soil carbon accumulation and stability. These results highlight the potential of circular fertilizers derived from agro-industrial wastes to restore soil health, close nutrient cycles, and reduce dependence on synthetic fertilizers—thereby advancing sustainable and circular agriculture.

Keywords: circular economy; circular fertilizers; soil microbial activity; nutrient cycling; soil health; sustainable agriculture

1. Introduction

Modern agriculture faces two interconnected challenges: sustaining crop productivity while preserving soil functionality and environmental quality. Intensive reliance on synthetic fertilizers has substantially increased yields but has also led to nutrient imbalances, soil degradation, and greenhouse gas emissions [1,2]. In contrast, circular fertilizers—produced by recycling agro-industrial and agricultural wastes—represent a sustainable strategy to restore soil fertility, close nutrient loops, and reduce dependency on finite mineral resources [3,4].

Among waste-recycling strategies, composting and vermicomposting are particularly effective in transforming organic residues such as food waste, wood sawdust, straw, olive pomace, and citrus residues into nutrient-rich soil amendments [5,6]. These processes enhance organic matter stability, improve nutrient availability, and stimulate beneficial microbial communities that sustain key soil biogeochemical cycles [7]. The integration of such waste-derived fertilizers into agricultural systems is consistent with the circular economy paradigm and promotes both agronomic efficiency and environmental sustainability [8].

Elemental sulfur was included in this study as a complementary mineral amendment, rather than a primary focus. Its use is particularly relevant in Mediterranean and calcareous soils, where sulfur deficiency and high pH frequently limit nutrient availability and microbial-mediated oxidation processes [9,10]. Sulfur enrichment can increase the solubility of phosphorus and micronutrients (Fe, Zn, and Mn) and stimulate specific microbial groups involved in sulfur oxidation, thereby enhancing soil biochemical functionality [11,12]. This rationale justified the inclusion of sulfur–bentonite formulations alongside composts and vermicomposts in our comparative assessment.

In this study, soil functionality was assessed through a comprehensive set of chemical, biochemical, and microbiological indicators. Chemical parameters included total organic carbon (TOC), total nitrogen (TN), cation exchange capacity (CEC), and pH (chemical indicators); biochemical indicators encompassed enzymatic activities (dehydrogenase and fluorescein diacetate hydrolysis) and water-soluble phenols; while microbiological indicators comprised microbial biomass carbon (MBC) and the abundance of fungi, bacteria, and actinomycetes. Together, these parameters provided an integrative assessment of soil health rather than isolated "single-index" measurements.

The objectives of this study were to: (i) evaluate the effects of compost-, vermicompost-, and sulfur-enriched fertilizers derived from mixed agro-industrial wastes on soil functionality under field conditions; (ii) compare their performance with that of conventional mineral (NPK) and traditional organic (manure) fertilizers; and (iii) identify the interrelationships among soil chemical, biochemical, and microbial properties that underpin improvements in soil health.

We hypothesized that circular fertilizers, particularly those obtained by vermicomposting and sulfur–bentonite enrichment, would enhance soil nutrient availability, microbial biomass, and enzymatic activity more effectively than conventional fertilizers, while simultaneously promoting long-term carbon sequestration and soil resilience.

The innovation of this work lies in the integration of multiple elements: (i) the use of circular waste streams derived from diverse agro-industrial biomass sources; (ii) mineral enrichment through elemental sulfur applied synergistically with organic inputs; (iii) a direct comparison with both mineral and traditional organic fertilizers; and (iv) a comprehensive evaluation of soil biological, biochemical, and chemical indicators of ecosystem functionality under real field conditions.

By combining these components, the study provides an integrated framework for evaluating whether waste-derived and sulfur-enriched fertilizers can effectively close nutrient cycles, restore soil health, and deliver agronomic performance comparable to—or

 https://doi.org/10.3390/environments13020093

surpassing—that of conventional or organic fertilizers. This approach advances the transition toward a more sustainable, circular model of agriculture.

2. Materials and Methods

2.1. Feeding Materials

Two distinct organic mixtures were prepared as feedstocks for composting:

- Compost 1 (C1): This mixture was composed of 90% olive pomace obtained from the traditional three-phase olive oil extraction process and 10% wheat straw as a bulking agent. The olive pomace contained approximately 43% lignin, 11.29% hemicellulose, and 9.55% cellulose.
- Compost 2 (C2): This mixture consisted of 90% orange-processing residues from the citrus industry, combined with 8% straw and 2% buffalo manure as a structural and nutrient supplement. Orange waste was characterized by 19% lignin, 7% hemicellulose, and 35% cellulose.

2.2. Composting Process Setup

Composting was conducted in electrically operated, temperature-controlled composters engineered to optimize aerobic decomposition. Each composter featured separate chambers to prevent the mixing of fresh and maturing material, allowing independent temperature regulation and promoting optimal microbial activity. The composting process was carried out using community composters (model "10 BASIC" Satriani snc design–manufacturing–startup–maintenance, Potenza, Italy) with an average treatment capacity of approximately 8 tons per year. Each unit was constructed from AISI 304 stainless steel and had approximate dimensions of 2 × 1 × 1 m. The system was equipped with mechanical mixing arms to ensure proper aeration and homogenization of the material, and an integrated heater with forced aeration to maintain optimal thermophilic conditions. Process control was managed through an electromechanical control panel, allowing for automated operation and temperature regulation. Each composter included a daily inspection door for monitoring and sample collection. The equipment was CE-certified in accordance with the EU Machinery Directive, ensuring compliance with European safety and performance standards. Each compost mixture was processed in triplicate following a defined thermal regime: an initial mesophilic phase of 8 days at 29 °C, a thermophilic phase of 20 days at 50 °C, and a secondary mesophilic phase of 92 days at 27 °C [9]. This was followed by a 30-day stabilization period at 20 °C to ensure compost maturity. Moisture content was maintained at approximately 50%, oxygen levels remained above 15%, and pH was 6.8. Temperature, moisture, and oxygen were continuously monitored using a multi-parameter probe (models EXP-421, EXP-422, EXP-831, and EXP-486; LSI Lastem, Settala, Milan, Italy), with data recorded via the manufacturer's RS485 or wireless logging system. Water was added as needed to maintain moisture, and daily turning ensured adequate aeration and uniform decomposition [10].

Upon completion, the composts were air-dried, sieved (<2 mm), and homogenized to obtain uniform samples. Both compost formulations reached full maturity within approximately six months [10].

2.3. Vermicomposting Process Setup

Vermicomposting was performed in 50 L bins (Vevor 5-Tray Worm Composter, model WB25101, Shanghai Vevor Industrial Co., Ltd., Shanghai, China). Two formulations were prepared:

- Vermicompost 1 (V1): 45% olive waste, 45% pine sawdust, and 10% wheat straw, with 20% earthworm inoculum.

https://doi.org/10.3390/environments13020093

- Vermicompost 2 (V2): 45% orange waste, 45% pine sawdust, and 10% wheat straw, with 20% earthworm inoculum.

Each bin was initially loaded with 20 kg of feedstock mixture, distributed in three uniform layers of approximately equal thickness to promote aeration and uniform decomposition. The bedding material was kept loose to facilitate oxygen diffusion and maintained in a slightly moist condition (50–60% humidity), avoiding waterlogging [11].

Red wigglers (*Eisenia fetida*) were introduced at a density of approximately 1000 individuals (≈1 lb) per square foot of surface area. During the four-month vermicomposting period, the feed was added weekly in thin layers (≈3–4 cm) to maintain a continuous supply of degradable material and stimulate worm activity. Moisture and temperature were monitored throughout the process to maintain optimal conditions for earthworm metabolism. At the end of the cycle, the worms had fully processed the substrate into mature vermicompost, characterized by a dark color, earthy odor, and fine granular texture [11].

2.4. Production of Sulfur–Bentonite Pellets with Composted Orange Waste and Olive Pomace

Pellet production (3–4 mm diameter) was carried out by Steel Belt System s.r.l. (Venegono Inferiore, Italy) using a proprietary pastillation technology. Formulation stage: The base mixture was prepared by blending 80% sulfur, 10% bentonite, and 10% organic additive derived from either composted orange-processing residues SBOWC2 or olive pomace SBOPC1.

Pelletizing stage: The molten mixtures were fed into a patented rotary pastillator system that distributed the material as droplets onto a continuously moving, cooled steel belt. This process enabled rapid solidification of the droplets into uniform spherical or lentil-shaped pellets.

Finishing and packaging: Once solidified, the pellets were collected via conveyor and elevator systems, transferred to storage silos, and subsequently weighed and packaged for use. The bentonite proportion—maintained at approximately 10% of the sulfur mass—was selected empirically to ensure optimal binding, structural integrity, and nutrient dispersion [12,13].

Before its application in the field, the SBOPC1 and SBOWC2 products were analyzed to ensure compliance with environmental and agricultural safety standards. Tests were performed to detect possible pathogenic microorganisms (total and fecal coliforms, *Salmonella* spp., and *Escherichia coli*) and heavy metals. Pathogen and heavy metal analysis was performed by a specialized and accredited laboratory, GEOLAB S.r.l., located in Via Trieste, 38–87036 RENDE (CS). The analytical results confirmed that the product was free of pathogens and heavy metal contamination, confirming its suitability for use as a soil amendment [14].

Prior to use, the SBOPC1 and SBOWC2 products were subjected to quality and safety analyses to ensure their suitability for agricultural application. Tests for pathogenic microorganisms (total coliforms, fecal coliforms, *Salmonella* spp., and *Escherichia coli*) and heavy metals were carried out. Pathogen and heavy metal analysis was performed by a specialized and accredited laboratory, B S.r.l., located in Via Trieste, 38–87036 RENDE (CS).

2.5. Sulfur–Bentonite Enriched with Vermicompost Produced with Orange Waste and Olive Pomace

The sulfur–bentonite, supplied as 3–4 mm pads by Steel Belt System s.r.l. (Venegono Inferiore, Italy), was mixed with orange waste-based vermicompost (SBOWV2) or olive waste-based vermicompost (SBOPV1). The mixture consisted predominantly of sulfur (80%), with bentonite (10%) and vermicompost (10%). Prior to use, the mixture was tested for potential pathogen and heavy metal contamination by the specialized and accredited laboratory GEOLAB S.r.l., located at Via Trieste 38, 87036 Rende (CS).

https://doi.org/10.3390/environments13020093

2.6. Soil Experiments

A sandy loam soil (11.85% clay, 23.21% silt, and 64.94% sand) located in Orfei farm, Motta San Giovanni, Loc. Liso, Italy (37.9991° N, 15.6999° E), was used for the trial, which was conducted from December to June. The area is characterized by a Mediterranean climate, with mild and rainy winters and warm, dry springs. During the experimental period, the average air temperature ranged from approximately 10–14 °C in winter to 18–24 °C in late spring, while total rainfall was mainly concentrated in the winter months, with limited precipitation occurring from April to June. The experimental field was divided into plots of 0.250 hectares each for the application of different fertilizer treatments. For each treatment, three replicates were used. The experiment was replicated for two consecutive years (Figure 1). Organic amendments (compost and vermicompost) and sulfur–bentonite formulations were surface-applied uniformly to each plot and subsequently incorporated into the top 0–20 cm of soil by mechanical tillage prior to crop establishment. Fertilizers were applied at the beginning of the growing season (December).

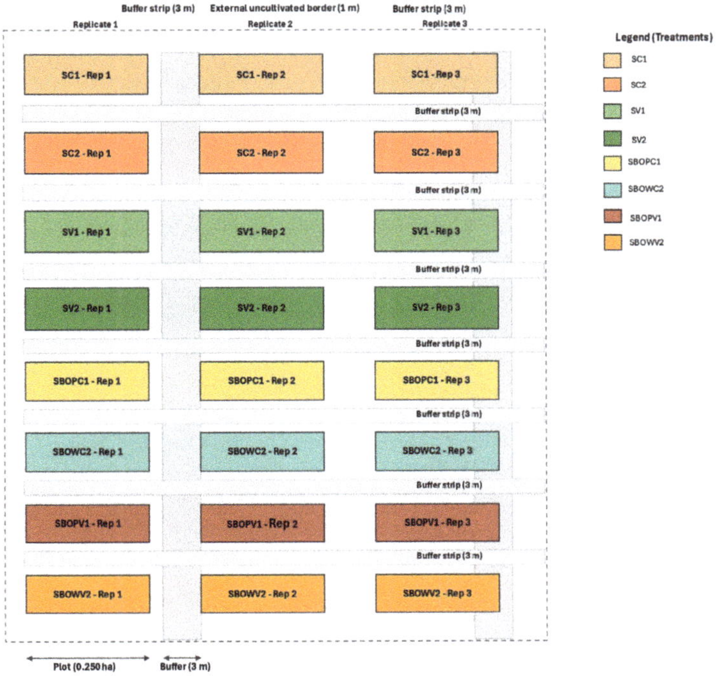

Figure 1. Schematic representation of the experimental field layout showing the spatial arrangement of fertilizer treatments. Each rectangle represents a plot of 0.250 ha. Eight fertilization treatments (SC1, SC2, SV1, SV2, SBOPC1, SBOWC2, SBOPV1, and SBOWV2) were arranged with three spatial replicates per treatment. Plots were separated by uncultivated buffer strips (≈3 m) to minimize cross-treatment interference, while an external uncultivated border surrounded the experimental area. The layout is representative of each experimental year.

Soil sampling was carried out at the end of the growing season (June), approximately six months after amendment application, to evaluate the effects of the different fertilization treatments on soil chemical, biochemical, and microbiological properties. Fertilizer application rates were selected according to common agricultural practices adopted by local farmers in Mediterranean environments of southern Italy (Calabria). The applied doses accounted for the different nutrient concentrations, physical properties, and agronomic roles

https://doi.org/10.3390/environments13020093

of composts, vermicomposts, and sulfur–bentonite formulations, in order to reproduce realistic field conditions.

The following fertilization treatments were applied:

- Sulfur–bentonite plus olive pomace compost (SBOPC1) or vermicompost (SBOPV1); or sulfur–bentonite orange compost or vermicompost was used at 200 kg per 0.250 ha;
- Olive or orange compost was used at 2123 kg per 0.250 ha;
- Olive or orange vermicompost was used at 1769 kg per 0.250 ha.

Soil samples were analyzed for their physical and biological properties. Various soil parameters were measured, including pH, electrical conductivity (EC) [15], organic carbon (OC) content [16], total nitrogen (TN) content [17], water-soluble phenols (WSPs) [18], and cation exchange capacity (CEC) [19]. Cations and anions were detected as reported in Muscolo et al. [13].

2.7. *Chemical and Biological Properties of Soils*

Enzyme activities, specifically fluorescein diacetate (FDA) hydrolysis [20] and dehydrogenase (DH) [21], have been detected. Microbial biomass carbon (MBC) was assessed using the chloroform-fumigation extraction procedure [22]. Fumigated and unfumigated soil extracts were analyzed for soluble organic carbon [16], with MBC calculated from the difference in organic C between the fumigated and unfumigated samples, applying an extraction efficiency coefficient of 0.38 [22]. To detect bacteria, fungi, and actinomycetes, 10 g of each soil sample was extracted with 95 mL of 0.1% (w/v) solution of sodium pyrophosphate. Soil extract solutions were diluted (10^{-1} to 10^{-7}), and the shares were plated on agarized culture media, each specific for bacteria, fungi, or actinomycetes [13]. Colony-forming units (CFU) for each microorganism were counted as reported in Maffia et al. [10]. The relative abundance of each microbial group was calculated as:

$$\text{Percentage } (\%) = \text{CFU of the group} \div \text{Total CFU} \times 100$$

where:

- CFU of the group = colony-forming units of fungi, bacteria, or actinomycetes.
- Total CFU = UFC of each group/UFC total × 100.

The biomass of each microbial group (MBC) (in $\mu g\ C\ g^{-1}$) was calculated using the fraction of that group within the total microbial population:

$$\text{MBC } \left(\mu\text{g C g}^{-1}\right) = \text{fraction of each group} \times \text{total}$$

where:

$$\text{Fraction of group} = \text{CFU of the group/Total CFU}$$

2.8. *Statistical Analysis*

Data are expressed as means of three analyses for each treatment. To analyze the effect of different fertilizers on various parameters measured, a two-way analysis of variance (ANOVA) with Tukey's Honestly Significant Difference (HSD) test was conducted with IBM SPSS Statistics (version 24, IBM Corp., Armonk, NY, USA). This statistical approach allows for comparison across treatments while accounting for potential interaction between the types of organic waste and the transformation processes. The analyses were carried out using JMP software (version 14, SAS Institute Inc., Cary, NC, USA) with significance differences at $p \leq 0.05$. To explore relationships among the different fertilizer formulations and their effects on soil chemical parameters, Principal Component Analysis (PCA) was performed using JMP software.

https://doi.org/10.3390/environments13020093

3. Results

3.1. Characteristics of Composts, Vermicomposts, and Sulfur–Bentonite-Enriched with Composts and Vermicomposts

The results showed significant differences in pH values among the various fertilizer formulations. The pH ranged from 6.5 and 6.8 in C1 and SBOPC1, or C2 and SBOPC2, respectively, to 7.4 and 7.6 in V1 and SBOPV1 or V2 and SBOWV2, respectively (Table 1).

Table 1. Chemical characteristics of fertilizers: C1 (90% olive pomace + 10% straw); C2 (90% orange waste and 10% of straw); SBOPC1 (sulfur–bentonite plus composted olive pomace); SBOWC2 (sulfur–bentonite plus composted orange waste); V1 (vermicomposted olive pomace: sawdust: straw (45:45:10) earthworm 20%); V2 (vermicomposted orange waste: sawdust: straw (45:45:10) earthworm 20%); SBOPV1 (sulfur–bentonite plus vermicomposted olive pomace); and SBOWV2 (sulfur–bentonite plus vermicomposted orange waste).

Chemical Characteristics	C1	C2	V1	V2	SBOPC1	SBOWC2	SBOPV1	SBOWV2
pH	$6.5^{b,*} \pm 0.5$	$6.8^{b} \pm 0.8$	$7.4^{a} \pm 0.3$	$7.6^{a} \pm 0.2$	$6.5^{b} \pm 0.2$	$6.8^{b} \pm 0.4$	$7.4^{a} \pm 0.3$	$7.5^{a} \pm 0.2$
EC	$1.3^{b} \pm 0.25$	$1.8^{b} \pm 0.2$	$3.18^{a} \pm 0.2$	$2.5^{ab} \pm 0.4$	$1.4^{b} \pm 0.2$	$1.5^{b} \pm 0.4$	$3.0^{a} \pm 0.2$	$2.6^{ab} \pm 0.4$
WC	$48^{a} \pm 3.2$	$42^{a} \pm 3$	$43^{a} \pm 2$	$42^{a} \pm 1$	$6.4^{b} \pm 0.4$	$6.9^{b} \pm 0.6$	$5.2^{b} \pm 0.5$	$4.8^{b} \pm 0.7$
TC	$44^{b} \pm 2.40$	$49^{b} \pm 2.4$	$51.2^{b} \pm 1.2$	$59.5^{a} \pm 2$	$5.5^{b} \pm 1.4$	$6.7^{b} \pm 0.4$	$6.8^{b} \pm 0.7$	$7.9^{b} \pm 1$
TN	$2.5^{a} \pm 0.12$	$2.7^{a} \pm 0.14$	$2.33^{a} \pm 0.09$	$2.66^{a} \pm 0.1$	$0.7^{b} \pm 0.02$	$0.5^{b} \pm 0.04$	$0.7^{b} \pm 0.04$	$0.9^{b} \pm 0.07$
C/N	$17.6^{b} \pm 1.1$	$18.1^{b} \pm 1.2$	$21.9^{a} \pm 0.6$	$22.3^{a} \pm 0.8$	$7.85^{c} \pm 0.5$	$13.4^{b} \pm 1$	$9.71^{c} \pm 0.9$	$8.77^{c} \pm 0.5$
Na^+	$1.1^{c} \pm 0.06$	$0.9^{c} \pm 0.02$	$4.6^{a} \pm 0.09$	$2.4^{b} \pm 0.07$	$0.9^{c} \pm 0.08$	$0.8^{c} \pm 0.1$	$0.7^{c} \pm 0.08$	$0.6^{c} \pm 0.1$
NH_4^+	$0.7^{a} \pm 0.02$	$0.6^{a} \pm 0.01$	$0.5^{a} \pm 0.04$	$0.33^{b} \pm 0.04$	$0.05^{a} \pm 0.02$	$0.05^{a} \pm 0.01$	$0.04^{a} \pm 0.02$	$0.06^{a} \pm 0.02$
K^+	$17^{a} \pm 1.50$	$18^{a} \pm 1.3$	$7.57^{b} \pm 0.2$	$9.65^{b} \pm 0.3$	$3.7^{c} \pm 0.02$	$3.1^{c} \pm 0.5$	$1.5^{d} \pm 0.02$	$0.19^{d} \pm 0.02$
Mg^{2+}	$1.1^{b} \pm 0.1$	$1.8^{a} \pm 0.2$	$1.3^{b} \pm 0.02$	$1.42^{b} \pm 0.3$	$1.53^{b} \pm 0.08$	$1.25^{b} \pm 0.06$	$0.5^{c} \pm 0.02$	$0.7^{c} \pm 0.02$
Ca^{2+}	$2.4^{a} \pm 0.3$	$2.9^{a} \pm 0.2$	$1.5^{b} \pm 0.03$	$2.3^{a} \pm 0.06$	$1.7^{b} \pm 0.2$	$2.6^{a} \pm 0.1$	$0.9^{c} \pm 0.1$	$1.1^{c} \pm 0.1$
Cl^-	nd	nd	$11.12^{a} \pm 0.9$	$9.23^{a} \pm 1.1$	nd	$0.48^{b} \pm 0.05$	nd	nd
NO_2^-	nd	nd	nd	0.33 ± 1.1	nd	nd	nd	nd
NO_3^-	$0.40^{b} \pm 0.002$	$0.49^{a} \pm 0.01$	nd	$0.84^{a} \pm 1.1$	$0.25^{b} \pm 0.03$	$0.45^{a} \pm 0.02$	nd	nd
PO_4^{3-}	$0.40^{b} \pm 0.03$	$0.93^{ab} \pm 0.03$	$1.27^{a} \pm 0.01$	$1.40^{a} \pm 0.02$	$0.46^{b} \pm 0.06$	$0.63^{b} \pm 0.04$	$0.57^{b} \pm 0.06$	$0.73^{b} \pm 0.04$
SO_4^{2-}	$0.39^{b} \pm 0.02$	$0.37^{b} \pm 0.02$	$0.38^{b} \pm 0.02$	$0.33^{b} \pm 0.02$	$0.55^{a} \pm 0.01$	$0.64^{a} \pm 0.02$	$0.68^{a} \pm 0.05$	$0.65^{a} \pm 0.05$
WSPs	$2.1^{a} \pm 0.06$	$1.7^{b} \pm 0.6$	$2.2^{a} \pm 0.03$	$1.8^{b} \pm 0.05$	$2.6^{a} \pm 0.01$	$1.9^{b} \pm 0.05$	$2.4^{a} \pm 0.5$	$1.6^{b} \pm 0.5$

pH; EC = electrical conductivity (dS/m); water content (WC %); total carbon (TC %); total nitrogen (TN %); C/N = carbon–nitrogen ratio; WC = soil moisture (%); cations and anions (mg * g^{-1} dw); and WSPs = water-soluble phenols (μg TAE g^{-1} d.s). Data are the means of three replicates ± standard deviation. nd: not determined. * Different letters in the same row indicate statistically significant differences among the treatments according to Tukey's test ($p < 0.05$).

The electrical conductivity (EC) showed the greatest values in fertilizers containing both vermicomposts (ranging from 2.4 to 3.18).

In general, composts and vermicomposts exhibited higher moisture content (averaging 45%) compared with sulfur-based fertilizer (*in media* 6.2). They also contained higher total carbon (averaging 50%) and total nitrogen (averaging 2.5), whereas the sulfur-based formulations had only 6.5% total carbon and 0.7% total nitrogen. Both composts were richer in potassium and calcium, while vermicomposts were comparatively richer in sodium and phosphate. As expected, sulfate concentrations were markedly higher in the sulfur-based fertilizers (Table 1). Moreover, water-soluble phenolic compounds were lower in compost and vermicompost derived from orange waste, including the sulfur-enriched variants. All produced fertilizers were free of pathogens and heavy metals, validating their environmental safety and suitability for soil application [14].

The Principal Component Analysis (PCA) of the fertilizer chemical parameters explained 77.9% of the total variance, with PC1 accounting for 50.4% and PC2 for 27.5%. The score plot (Figure 1) revealed a clear separation among fertilizer groups according to their chemical composition. Along PC1, composts and vermicomposts (C1, C2, V1, and V2) were positioned on the right side of the biplot and positively associated with total carbon (TC), total nitrogen (TN), C/N ratio, electrical conductivity (EC), NO_3^-, and PO_4^{3-}, reflecting higher nutrient content and organic matter quality. In contrast, sulfur–bentonite-enriched formulations (SBOPC1, SBOWC2, SBOPV1, and SBOWV2) clustered on the left,

 https://doi.org/10.3390/environments13020093

correlating with higher SO_4^{2-} and water-soluble phenols (WSPs) and slightly lower pH, indicative of the acidifying effect of elemental sulfur oxidation. PC2 further differentiated the sulfur–bentonite formulations according to the organic feedstock used. Those enriched with vermicompost (SBOPV1, SBOWV2) were characterized by higher levels of SO_4^{2-} and WSPs, while formulations containing compost (SBOPC1, SBOWC2) showed lower values of these parameters and a more balanced composition. The distribution pattern indicates that the addition of sulfur–bentonite modifies the chemical signature of the organic materials, shifting them toward a more mineralized and sulfur-rich profile compared to pure compost or vermicompost (Figure 2).

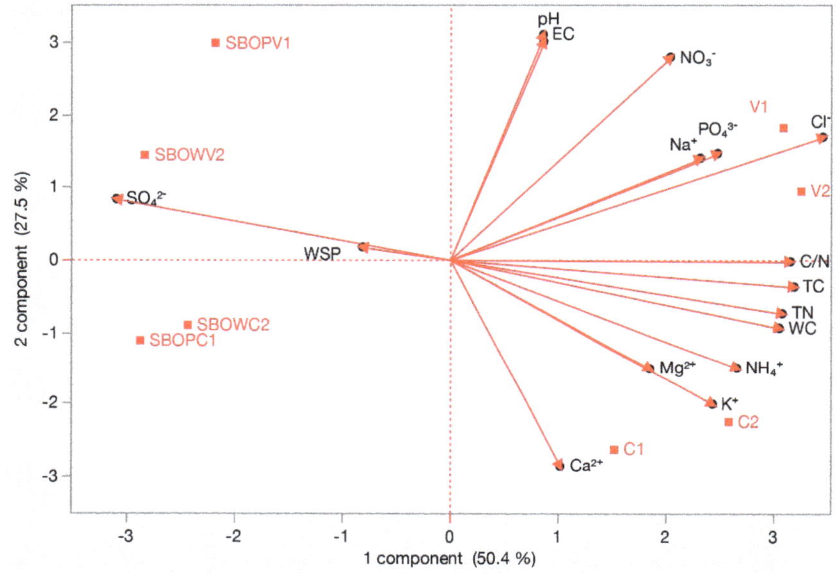

Figure 2. Principal Component Analysis of chemical characteristic of fertilizers: C1 (90% olive pomace + 10% straw); C2 (90% orange waste and 10% of straw); SBOPC1 (sulfur–bentonite plus composted olive pomace); SBOWC2 (sulfur–bentonite plus composted orange waste); V1 (vermicomposted olive pomace: sawdust: straw (45:45:10) earthworm 20%); V2 (vermicomposted orange waste: sawdust: straw (45:45:10) earthworm 20%); SBOPV1 (sulfur–bentonite plus vermicomposted olive pomace); and SBOWV2 (sulfur–bentonite plus vermicomposted orange waste).

3.2. Chemical and Biological Soil Properties of Fertilized Soils

Soil pH was the highest in both compost-treated soils (8.2 in SC1 and 8.3 in SC2). The electrical conductivity (EC) values indicated that, following fertilizer application, the soils were non-saline, ranging from 220 to 335 $\mu S\ cm^{-1}$ (Table 2). No significant differences in soil moisture content were observed among the treatments. The total organic carbon (TOC) content was the highest in SC2, as well as in SV1 and SV2, while the total nitrogen (TN) content reached its maximum in SV1 and SV2. FDA was the lowest in soil amended with both composts. The highest DH activity was found in soil treated with V2. The highest microbial biomass carbon (MBC) was recorded in soils treated with both vermicomposts, which also exhibited the greatest fluorescein diacetate (FDA) hydrolysis and dehydrogenase (DH) enzymatic activities. Cation exchange capacity was the highest in soil amended with fertilizers that contained both vermicomposts. The distribution of soil microbial groups relative to total microbial biomass carbon (MBC) revealed the highest fungal abundance in soils treated with sulfur–bentonite (SB) combined with either compost or vermicompost (Table 2).

https://doi.org/10.3390/environments13020093

Table 2. Chemical properties of soils treated with different composts and vermicomposts. Treatments: C1—compost from 90% olive pomace + 10% straw; C2—compost from 90% orange waste + 10% straw; V1—vermicompost from olive pomace, sawdust, and straw (45:45:10) with 20% earthworm biomass; V2—vermicompost from orange waste, sawdust, and straw (45:45:10) with 20% earthworm biomass; SBOPC1—sulfur–bentonite with composted olive pomace; SBOWC2—sulfur–bentonite with composted orange waste; SBOPV1—sulfur–bentonite with vermicomposted olive pomace; and SBOWV2—sulfur–bentonite with vermicomposted orange waste. Data are means of three replicates ± standard deviation. Different letters in the same column indicate significant differences (Tukey's test, $p \leq 0.05$).

	SC1	SC2	SV1	SV2	SBOPC1	SBOWC2	SBOPV1	SBOWV2
pH	$8.3^{a} \pm 0.55$	$8.2^{a} \pm 0.52$	$7.6^{b} \pm 0.80$	$7.6^{b} \pm 0.40$	$7.2^{c} \pm 0.40$	$7.1^{c} \pm 0.40$	$7.3^{c} \pm 0.40$	$7.2^{c} \pm 0.40$
EC	$310^{a} \pm 10$	$320^{a} \pm 12$	$333^{a} \pm 9$	$335^{a} \pm 12$	$22^{c} \pm 12$	$220^{c} \pm 11$	$251^{b} \pm 12$	$250^{b} \pm 10$
WC	$29^{a} \pm 1.6$	$22^{b} \pm 1.1$	$27^{a} \pm 1.4$	$28^{a} \pm 1.4$	$25^{a} \pm 1.7$	$26^{a} \pm 1.7$	$25^{a} \pm 1.5$	$26^{a} \pm 1.2$
WSPs	$33^{a} \pm 2$	$28^{b} \pm 1.4$	$33^{a} \pm 2.1$	$26^{b} \pm 1.1$	$40^{a} \pm 3.2$	$28^{b} \pm 2.8$	$35^{a} \pm 2.5$	$29^{b} \pm 1.3$
TOC	$1.7^{b} \pm 0.16$	$1.9^{a} \pm 0.16$	$2.0^{a} \pm 0.15$	$2.1^{a} \pm 0.25$	$1.3^{c} \pm 0.25$	$1.5^{c} \pm 0.19$	$1.6^{c} \pm 0.25$	$1.8^{bc} \pm 0.19$
TN	$0.16^{c} \pm 0.01$	$0.17^{c} \pm 0.01$	$0.23^{a} \pm 0.02$	$0.22^{b} \pm 0.04$	$0.11^{b} \pm 0.03$	$0.12^{c} \pm 0.02$	$0.14^{b} \pm 0.03$	$0.16^{c} \pm 0.02$
C/N	$10.6^{ab} \pm 0.5$	$11.7^{a} \pm 0.8$	$8.69^{b} \pm 0.4$	$9.54^{b} \pm 0.7$	$11.8^{a} \pm 0.9$	$12.5^{a} \pm 0.9$	$11.43^{a} \pm 0.8$	$11.25^{a} \pm 0.6$
SOM	$2.92^{c} \pm 0.3$	$3.27^{c} \pm 0.27$	$3.44^{b} \pm 0.25$	$3.6^{ab} \pm 0.13$	$2.24^{cb} \pm 0.13$	$2.58^{b} \pm 0.23$	$2.75^{b} \pm 0.24$	$3.1^{b} \pm 0.35$
FDA	$40^{b} \pm 1.6$	$41^{b} \pm 1.3$	$45^{a} \pm 1.6$	$44^{a} \pm 1.5$	$43^{a} \pm 1.2$	$45^{a} \pm 1.3$	$44^{a} \pm 1.5$	$46^{a} \pm 1.3$
DH	$50^{c} \pm 1.6$	$52^{bc} \pm 1.4$	$55^{b} \pm 1.5$	$60^{a} \pm 1.8$	$55^{b} \pm 1.6$	$51^{c} \pm 1.6$	$50^{c} \pm 1.5$	$56^{b} \pm 1.3$
MBC	$888^{d} \pm 5$	$989^{b} \pm 7$	$1087^{a} \pm 12$	$1099^{a} \pm 11$	$937^{c} \pm 13$	$945^{c} \pm 14$	$922^{c} \pm 13$	$977^{b} \pm 14$
CEC	$18.9^{b} \pm 1$	$18.7^{b} \pm 0.9$	$22^{a} \pm 1.1$	$23^{a} \pm 1.2$	$22^{a} \pm 1$	$19.9^{b} \pm 0.9$	$23^{a} \pm 1.2$	$22^{a} \pm 1$

pH; electric conductivity (EC, µS/cm); water content (WC, %); water-soluble phenols (WSPs, µg TAE g^{-1} d.s); total organic carbon (TOC, %); total nitrogen (TN, %); carbon nitrogen ratio (C/N); soil organic matter (SOM, %); fluorescein hydrolase activity (FDA, µg fluorescein g^{-1} ds); dehydrogenase activity (DHA, µg INTF g^{-1} ds h^{-1}); microbial biomass carbon (MB µg C g^{-1} f.s); and cation exchange capacity (CEC, (cmol (+) Kg^{-1}).

Conversely, bacterial dominance was highest in soils treated with composts or with sulfur–bentonite formulations containing both composts. Actinomycetes were more abundant in compost-amended soils than in soil amended with vermicomposts or sulfur-based fertilizers.

The biomass of fungi, bacteria, and actinomycetes (expressed as µg C g^{-1} dry soil) mirrored their relative proportions within the total MBC (Figure 3). Colony-forming units (CFU g^{-1} dry soil) of fungi, bacteria, and actinomycetes also varied with treatment. Fungal colonies were most numerous in soils treated with SBOWC2 and SBOWV2, followed by SBOPV1 = C1 > C2 > V1 = V2. Bacterial colonies were most abundant in C1, C2, and V2, followed by V1 = SBOPC1 > SBOWC2 > SBOWV2 > SBOPV1. Actinomycete colonies were most numerous in C1 and C2, and least abundant in SBOPV1 and SBOPV2.

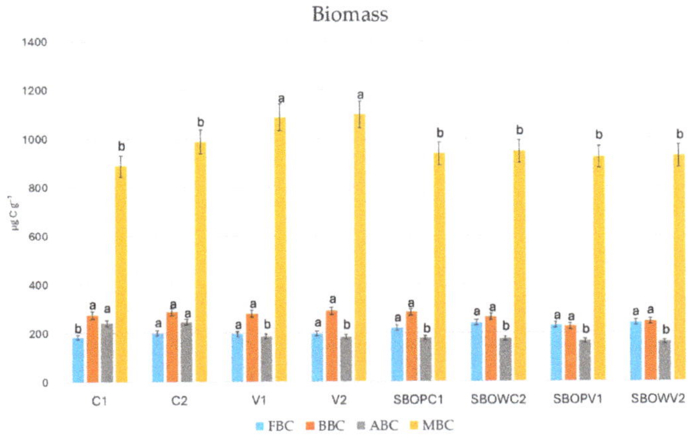

Figure 3. Fungi Biomass C (FBC, µg C g^{-1}); Bacteria Biomass C (BBC, µg C g^{-1}); and Actinomycete Biomass C (ABC, µg C g^{-1}). Data are the means of three replicates ± standard deviation (Different letters indicate significant differences. Tukey's test, $p \leq 0.05$).

https://doi.org/10.3390/environments13020093

The Principal Component Analysis (PCA) performed on the chemical and microbiological soil parameters explained a total variance of 76.1%, with PC1 and PC2 accounting for 47.1% and 29.0%, respectively (Figure 4). The distribution of treatments along PC1 clearly separated the organic fertilizers (SC1, SC2, SV1, and SV2) positioned on the right side of the biplot from the sulfur–bentonite-enriched formulations (SBOPC1, SBOWC2, SBOPV1, and SBOWV2), which clustered on the left side.

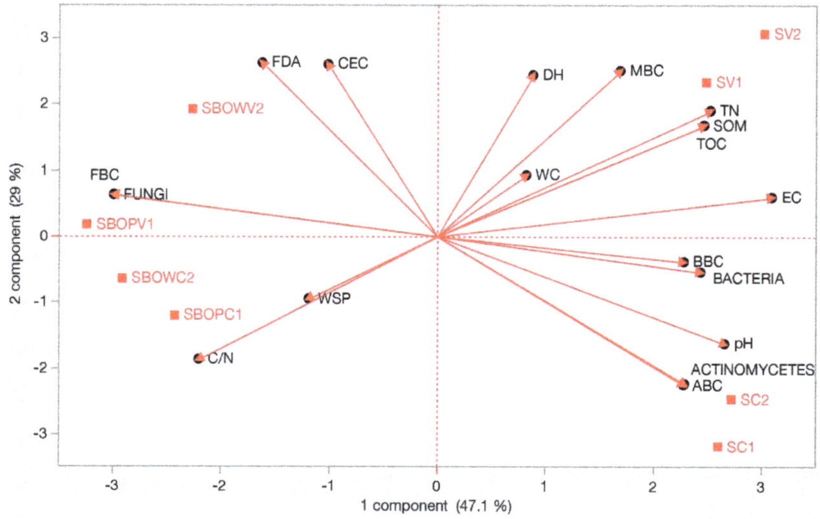

Figure 4. PCA of chemical and biological properties of soil treated with: SC1 (90% olive pomace + 10% straw); SC2 (90% orange waste and 10% of straw); SV1 vermicompost (olive pomace: sawdust: straw (45:45:10) earthworm 20%); SV2 vermicompost (orange waste: sawdust: straw (45:45:10) earthworm 20%); and sulfur–bentonite plus composted olive pomace (SBOPC1) and orange waste (SBOWC2).

Samples SV1 and SV2 were strongly associated with TOC, TN, SOM, MBC, FDA, DH, and CEC, indicating that soils treated with vermicompost exhibited higher organic matter content, total nitrogen, microbial biomass, and enzymatic activities, as well as greater cation exchange capacity. These parameters characterize a condition of enhanced biological and chemical fertility, confirming the beneficial effects of vermicompost on soil quality.

In contrast, treatments SBOPC1, SBOWC2, and SBOPV1 were located in the lower left quadrant and correlated with C/N and WSPs, suggesting a higher content of soluble phenolic compounds and a larger C/N ratio, indicative of less decomposed organic matter and slower microbial activity. Treatments SC1 and SC2 showed positive correlations with pH, EC, BBC, and actin, reflecting a more balanced microbial composition dominated by bacteria and actinomycetes and a slight alkalinizing effect typical of mature compost.

4. Discussion

4.1. Characteristics of Composts, Vermicomposts, and Sulfur–Bentonite-Enriched Amendments

The significant differences in pH among the various fertilizer formulations (ranging from 6.5–6.8 in C1 and SBOPC1 to 7.4–7.6 in V1, SBOPV1, and V2, SBOWV2) reflect the influence of amendment type and feedstock on soil reaction. The higher pH observed in the vermicompost-based formulations is consistent with previous findings indicating that vermicomposting typically produces more mineralized organic matter, enhances base cation release, and results in higher pH compared with traditional composting [23]. The

https://doi.org/10.3390/environments13020093

progressive increase in electrical conductivity (EC) from compost and sulfur–compost treatments to vermicompost and sulfur–vermicompost treatments also supports evidence that vermicomposts accumulate greater amounts of soluble salts and nutrients due to intensified biological transformation processes [24]. Earthworm activity alters the substrate composition, leading to higher concentrations of certain nutrients and organic matter compared with conventional composts [25]. Vermicompost differs from traditional compost in its composition and its influence on soil properties and plant performance. Unlike conventional compost, vermicompost is typically richer in readily available nutrients, plant growth-promoting substances (such as auxins and cytokinins), and beneficial microorganisms that can directly stimulate plant growth, enhance yields, and improve fruit quality and stress tolerance [25,26]. The presence of earthworms in the vermicomposting process also leads to improved soil aggregation and aeration, creating a more favorable environment for root development. In contrast, traditional compost primarily functions as a soil conditioner, adding organic matter that enhances soil structure and fertility over time [27]. Its primary role is to serve as a long-term nutrient source and enhance the overall quality of the growing medium. Both compost and vermicompost typically exhibit higher moisture content and greater concentrations of total carbon and nitrogen than sulfur-based formulations, reflecting their superior capacity for organic enrichment [28–30]. This observation is consistent with broader reviews indicating that organic amendments contribute significant amounts of organic matter and nitrogen, leading to improved soil fertility [31]. The observation that composts contained high potassium and calcium [32], whereas vermicomposts were richer in sodium and phosphate, suggests distinct nutrient release patterns: vermicomposting may favor P-solubilization and Na retention depending on feedstock and process conditions [33], and also the earthworm digestive process concentrates nutrients and breaks down complex organic matter into more plant-available forms for the gut microbiome rich in bacteria. The markedly higher sulfate concentrations in sulfur-based fertilizers are expected, given the added sulfur–bentonite matrix. Interestingly, the lower content of water-soluble phenolic compounds in the orange-waste-derived compost, its vermicompost, and in sulfur–vermicompost orange waste formulations may be beneficial. Phenolic compounds are known to inhibit microbial activity or seedling growth [34]. Lower phenolic loads could thus support improved microbial and soil responses. Taken together, these compositional distinctions provide a sound basis for interpreting the divergent soil responses observed in subsequent sections.

4.2. Soil Chemical and Biological Responses to Amendments

After fertilizer application, the soils treated with compost (SC1, SC2) exhibited the highest pH (8.2 and 8.3). Such an increase in pH may reflect the buffering capacity and release of base cations (e.g., Ca^{2+}, Mg^{2+}) from the compost material, in line with literature showing that mature composts can raise soil pH in neutral to alkaline soils [35]. The fact that EC values remained in the 220–335 $\mu S\ cm^{-1}$ range confirms that none of the treatments induced salinity stress, despite the higher soluble salt indices of some vermicomposts [36]. The highest total organic carbon (TOC) in SC2 and in SV1/SV2 implies that both compost and vermicompost amendments effectively increase soil organic matter. Vermicomposts reaching high total nitrogen (TN) in the SV treatments align with evidence that vermicompost enhances nitrogen availability more than raw compost [31]. The highest cation exchange capacity (CEC) in soils treated with vermicompost-containing fertilizers supports the view that increased organic matter and microbial by-products contribute additional exchange sites and improve soil ionic retention.

 https://doi.org/10.3390/environments13020093

Microbial Biomass, Enzyme Activity, and Implications for Soil Fertility and Sustainability

The highest microbial biomass carbon (MBC) and enhanced enzyme activities (fluorescein diacetate (FDA) hydrolysis and dehydrogenase (DH)) in soils treated with vermicompost coming from orange wastes underscore the strong stimulation of microbial functioning by these amendments. Orange waste tends to have higher pectin and soluble carbohydrate content and generally a more favorable short-term C:N for microbial growth after precomposting. This can explain the different effect it has on soil with respect to vermicompost from olive pomace [37]. Olive pomace often has higher levels of refractory lipids, certain polyphenols, and compounds that increase the recalcitrant fraction. Substrates that are easier to depolymerize promote faster microbial turnover and a broader niche space for different microbes. This corresponds with research showing that vermicompost with high soluble carbohydrate content promotes microbial biomass, metabolic activity, and enzyme function in soils [31]. The differential distribution of microbial groups is also of particular interest. The greatest fungal abundance (relative to MBC) in soils treated with sulfur–bentonite plus either compost or vermicompost suggests that the bentonite–sulfur matrix may favor fungal colonization—perhaps by enhancing micro-aggregates, creating moisture/structural niches, or modulating pH. Hammerschmiedt et al. (2024) [3] showed that fungal abundance was significantly affected by elemental sulfur. While some bacteria thrive and the total microbial biomass can increase, the oxidation of sulfur can also create a more favorable environment for fungal growth, particularly in terms of biomass. By contrast, bacterial dominance in compost-oriented treatments (and sulfur–bentonite + compost) is consistent with higher pH, greater labile carbon, and base cations favoring bacterial growth [35]. The elevated actinomycete populations in compost-amended soils reflect the role of actinomycetes in decomposing recalcitrant organic matter and their affinity for more mature compost substrates [38]. These patterns are corroborated by high-throughput studies showing that compost amendments increase Actinobacteria and Chloroflexi abundance [35]. Colony-forming unit (CFU) data further reinforce these biomass observations: the highest number of fungal colonies in SBOWC2 and SBOWV2; the highest bacterial colonies in C1, C2, and V2; and the highest actinomycete colonies in C1 and C2, with the lowest in SBOPV1/2. The alignment of CFU and biomass trends strengthens the biological inference. Overall, the vermicompost treatments appear superior in stimulating microbial biomass and activity, which is known to support rapid nutrient cycling, improved root–soil interactions, and enhanced fertility [31].

Although crop growth, yield, or quality parameters were not directly measured in the present study, the observed increases in total nitrogen, microbial biomass carbon, enzymatic activities, and cation exchange capacity are widely recognized as functional indicators of improved soil fertility and nutrient-use efficiency. These soil attributes are strongly associated with enhanced root development, nutrient availability, and crop productivity in agricultural systems. Therefore, the soil responses observed here suggest a clear agronomic potential of the tested circular fertilizers, particularly vermicompost-based formulations, which should be validated in future yield-oriented trials.

From the integrated perspective, vermicompost-based treatments (SV1, SV2, and vermicompost-containing sulfur–bentonite amendments) appear particularly effective in promoting microbial activation, nitrogen enrichment, and improved CEC. These traits support enhanced short- to medium-term soil fertility, nutrient availability, and likely crop productivity. The compost treatments (SC1, SC2) are especially effective in building TOC and raising soil pH—which may promote long-term soil structure, aggregation, and resilience, although microbial activity may lag relative to vermicompost. The sulfur–bentonite-enriched formulations add structural/chemical complexity and may serve special purposes (e.g., in soils needing sulfur or bentonite's retention properties), but their

 https://doi.org/10.3390/environments13020093

incremental benefits must be weighed against input cost and upstream footprint. In a production system context (especially Mediterranean-type soils where orange waste compost and sulfur–bentonite amendments are feasible), a strategic approach may be to use vermicompost amendments when rapid fertility enhancement and microbial activation are required, and compost amendments when the focus shifts to long-term soil carbon build-up and structural resilience. Moreover, the sulfur–bentonite options may offer added value where sulfur deficiency or moisture/retention issues prevail. From a green economy perspective, these results carry several implications. Vermicompost, by enhancing TN, microbial biomass, enzyme activity, and soil ionic retention, offers potential to reduce reliance on synthetic fertilizers, thus lowering upstream energy and greenhouse gas emissions associated with fertilizer manufacture and application [33]. Furthermore, increased microbial efficiency implies improved nutrient use efficiency and potentially reduced losses (leaching/volatilization).

The Principal Component Analysis (PCA) integrated soil chemical and biological parameters, explaining 76.1% of the total variance (PC1: 47.1%; PC2: 29.0%). The first principal component (PC1) was mainly driven by variables related to soil fertility and biological activity, including total organic carbon (TOC), total nitrogen (TN), soil organic matter (SOM), microbial biomass carbon (MBC), enzymatic activities (FDA and DH), and cation exchange capacity (CEC). This axis, therefore, represents a gradient of biologically mediated nutrient availability and microbial activation (Figure 4).

Vermicompost treatments (SV1 and SV2) clustered positively along PC1, indicating enhanced short-term fertility, accelerated nutrient cycling, and increased microbial functioning. In contrast, compost-based treatments (SC1 and SC2) were positioned toward higher pH, electrical conductivity (EC), and bacterial and actinomycete biomass, suggesting greater chemical buffering capacity and more stable, slower-decomposing organic matter pools.

The second principal component (PC2) was associated with variables describing organic matter quality and mineralization dynamics, including the C/N ratio and water-soluble phenols (WSPs). Sulfur–bentonite formulations (SBOP and SBOW) were negatively associated with PC2, reflecting higher C/N ratios and WSP content, indicative of slower mineralization rates and more recalcitrant organic substrates.

Overall, the PCA highlights two main functional pathways: vermicompost-based amendments promote rapid biological activation and short-term soil fertility enhancement, whereas compost and sulfur–bentonite formulations are associated with slower nutrient turnover and potentially greater long-term soil stability.

Finally, while microbial biomass, enzyme activity, and CFU-based plate counts provide robust functional indicators of soil microbial responses, they do not capture the diversity of unculturable microorganisms. High-throughput sequencing or metagenomic approaches (e.g., 16S rRNA and ITS analyses) could therefore elucidate key functional microbial groups (e.g., nitrifiers, P-solubilizing bacteria, and mycorrhizal fungi) and link microbial community structure to nutrient cycling more directly [39]. Finally, economic cost–benefit and scalability analyses for green economy implementation (e.g., local vermicompost production, use of waste streams, and regulatory incentives) should be integrated. The use of orange-waste-derived composts or vermicomposts aligns with circular economy principles (waste valorization, nutrient recycling, and local bio-fertilizer production), which bolster green economy outcomes. However, the embodied energy and emissions involved in processing (vermicomposting, sulfur–bentonite production) must be weighed: the net system benefit is a function of both amendment performance and upstream footprint. Therefore, selecting vermicompost-based amendments appears to offer the optimum trade-off between performance, circularity, and sustainability in many situations—but site-specific economic and environmental assessments remain advisable. The use of orange-waste-

https://doi.org/10.3390/environments13020093

derived composts or vermicomposts aligns with circular economy principles through waste valorization, nutrient recycling, and local bio-fertilizer production. The environmental relevance of these practices has been quantitatively documented in previous studies, which reported that one ton of untreated wet orange waste may emit approximately 0.130 kg CH_4, 30.9 kg CO_2, and 0.069 kg N_2O, while one ton of wet olive pomace can generate up to 1162.3 kg CO_2, 122 kg CH_4, and 0.12 kg N_2O when improperly disposed of [40]. Landfilling further exacerbates these impacts, with emissions ranging between 2603 and 2708 t CO_2e per dry ton of waste and CH_4 emissions of about 54 kg dry t^{-1} [41], and has been identified as a major contributor to global warming and photochemical ozone formation in life cycle assessment studies [42,43].

Additional studies have highlighted that compost production may require 1500–2000 MJ t^{-1} and involve average costs of approximately 130 € t^{-1} [44], whereas alternative recycling pathways based on low-input organic–mineral formulations may reduce both energy demand and production costs. Moreover, the partial replacement of mineral fertilizers with organic–mineral amendments has been associated with a reduction in greenhouse gas emissions of about 20% and a concurrent increase in soil organic matter by up to 50% [45].

Within this literature-based framework, the increases in soil organic carbon, total nitrogen, microbial biomass, and enzymatic activity observed in the present study provide functional evidence that supports the environmental and circular economy benefits reported by previous authors, although a direct quantification of emission savings and economic performance was beyond the scope of this work.

5. Conclusions

In conclusion, the results clearly demonstrate that the type of amendment plays a crucial role in determining soil quality outcomes. And vermicompost-based formulations stand out for their ability to enhance microbial activity and soil fertility, increasing microbial biomass, biological activity, and nutrient availability. In contrast, compost-based formulations provide significant benefits for organic matter accumulation and long-term soil resilience. Vermicompost is particularly effective for improving short-term fertility, whereas compost and sulfur–bentonite contribute to long-term soil structure and carbon stability. Although based on a two-growing-season study, the study shows clear short- to medium-term improvements in soil functionality after the application of circular fertilizers. Multi-year field trials are required to evaluate the long-term persistence of these effects, particularly regarding soil carbon stability and microbial succession.

Within the broader context of sustainable production systems and green economy objectives, vermicompost offers the most favorable balance between soil health improvement, nutrient cycling, and circular economy principles. Orange residue appears to be a more suitable feedstock for amendment production than olive pomace, although the latter also contributes positively to soil fertility. Overall, a balanced strategy integrating both compost and vermicompost—potentially supplemented with sulfur–bentonite where appropriate—may deliver the most resilient and sustainable outcomes.

Author Contributions: Conceptualization, A.M. (Adele Muscolo) and A.M. (Angela Maffia); methodology, F.A.M.; software, A.M. (Angela Maffia); validation, A.M. (Adele Muscolo) and C.M.; formal analysis, F.A.M.; investigation, S.B.; data curation, F.A.M.; writing—original draft preparation, A.M. (Angela Maffia); writing—review and editing, A.M. (Angela Maffia); visualization, A.M. (Adele Muscolo); project administration, A.M. (Adele Muscolo); funding acquisition, A.M. (Adele Muscolo). All authors have read and agreed to the published version of the manuscript.

 https://doi.org/10.3390/environments13020093

Funding: This work was supported by the Agritech National Research Center and received funding from the European Union Next-Generation EU (PIANO NAZIONALE DI RIPRESA E RESILIENZA (PNRR)—MISSIONE 4 COMPONENTE 2, INVESTIMENTO 1.4—D.D. 1032 17 June 2022, CN00000022).

Data Availability Statement: The original contributions presented in this study are included in thearticle. Further inquiries can be directed to the corresponding author.

Conflicts of Interest: The authors declare no conflicts of interest.

References

1. Zhang, X.; Davidson, E.A.; Mauzerall, D.L.; Searchinger, T.D.; Dumas, P.; Shen, Y. Managing nitrogen for sustainable development. *Nature* **2015**, *528*, 51–59. [CrossRef]
2. Chen, Y.; Sun, Z.; Zhou, Y.; Yang, W.; Ma, Y. The future of sustainable farming: An evolutionary game framework for the promotion of agricultural green production technologies. *J. Clean. Prod.* **2024**, *460*, 142606. [CrossRef]
3. Hammerschmiedt, T.; Holatko, J.; Bytesnikova, Z.; Skarpa, P.; Richtera, L.; Kintl, A.; Pekarkova, J.; Kucerik, J.; Jaskulska, I.; Radziemska, M.; et al. The impact of single and combined amendment of elemental sulfur and graphene oxide on soil microbiome and nutrient transformation activities. *Heliyon* **2024**, *10*, e38439. [CrossRef]
4. Voss, M.; Valle, C.; Gaudino, E.C.; Tabasso, S.; Forte, C.; Cravotto, G. Unlocking the potential of agrifood waste for sustainable innovation in agriculture. *Recycling* **2024**, *9*, 25. [CrossRef]
5. Badagliacca, G.; Testa, G.; La Malfa, S.G.; Cafaro, V.; Lo Presti, E.; Monti, M. Organic Fertilizers and Bio-Waste for sustainable soil management to support crops and control greenhouse gas emissions in Mediterranean agroecosystems: A review. *Horticulturae* **2024**, *10*, 427. [CrossRef]
6. Malik, K.M.; Khan, K.S.; Billah, M.; Akhtar, M.S.; Rukh, S.; Alam, S.; Munir, A.; Aulakh, A.M.; Rahim, M.; Qaisrani, M.M.; et al. Organic Amendments and Elemental Sulfur Stimulate Microbial Biomass and Sulfur Oxidation in Alkaline Subtropical Soils. *Agronomy* **2021**, *11*, 2514. [CrossRef]
7. Iqbal, A.; Ali, I.; Yuan, P.; Khan, R.; Liang, H.; Wei, S.; Jiang, L. Combined Application of Manure and Chemical Fertilizers Alters Soil Environmental Variables and Improves Soil Fungal Community Composition and Rice Grain Yield. *Front. Microbiol.* **2022**, *13*, 856355. [CrossRef]
8. Wang, J.; Xiao, S.; Hayat, K.; Liao, X.; Chen, J.; Zhang, L.; Xie, Y. Investigating the Effects of Elevation on Microbial Communities and Soil Properties at Fanjing Mountain, China. *Forests* **2024**, *15*, 1980. [CrossRef]
9. Liang, C.; Das, K.C.; McClendon, R.W. The influence of temperature and moisture contents regimes on the aerobic microbial activity of a biosolids composting blend. *Bioresour. Technol.* **2002**, *86*, 131–137. [CrossRef] [PubMed]
10. Maffia, A.; Marra, F.; Celano, G.; Oliva, M.; Mallamaci, C.; Hussain, M.I.; Muscolo, A. Exploring the Potential and Obstacles of Agro-Industrial Waste-Based Fertilizers. *Land* **2024**, *13*, 1166. [CrossRef]
11. Maffia, A.; Marra, F.; Canino, F.; Battaglia, S.; Mallamaci, C.; Oliva, M.; Muscolo, A. Humic Substances from Waste-Based Fertilizers for Improved Soil Fertility. *Agronomy* **2024**, *14*, 2657. [CrossRef]
12. Muscolo, A.; Papalia, T.; Settineri, G.; Mallamaci, C.; Panuccio, M.R. Sulfur bentonite-organic-based fertilizers as tool for improving bio-compounds with antioxidant activities in red onion. *J. Sci. Food Agric.* **2019**, *100*, 785–793. [CrossRef]
13. Muscolo, A.; Romeo, F.; Marra, F.; Mallamaci, C. Recycling agricultural, municipal and industrial pollutant wastes into fertilizers for a sustainable healthy food production. *J. Environ. Manag.* **2021**, *300*, 113771. [CrossRef] [PubMed]
14. European Compost Network (ECN). *ECN-QAS Quality Manual: Quality Assurance Scheme for Compost and Digestate*, 3rd ed.; European Compost Network e.V.: Köln, Germany, 2018; Volume 82, ISBN 978-3-9820826-0-2. Available online: https://www.compostnetwork.info/wordpress/wp-content/uploads/180711_ECN-QAS-Manual_3rd-edition_keyed-1.pdf (accessed on 12 December 2025).
15. AOAC. *Official Methods of Analysis*, 18th ed.; Association of Official Analytical Chemists: Arlington, VA, USA, 2005.
16. Walkley, A.; Black, I.A. An examination of the Degtjareff method for determining soil organic matter and a proposed modification of the chromic acid titration method. *Soil Sci.* **1934**, *37*, 29–38. [CrossRef]
17. Kjeldahl, J. Neue Methode zur Bestimmung des Stickstoff in organishen Kopern. *Anal. Chem.* **1883**, *22*, 354–358. [CrossRef]
18. Box, J.D. Investigation of the Folin-Ciocalteau phenol reagent for the determination of polyphenolic substances in natural waters. *Water Res.* **1983**, *17*, 511–525. [CrossRef]
19. Mehlich, A. Rapid Determination of Cation and Anion Exchange Properties and pHe of Soils. *J. Assoc. Off. Agric. Chem.* **1953**, *36*, 445–457. [CrossRef]
20. Adam, G.; Duncan, H. Development of a sensitive and rapid method for the measurement of total microbial activity using fluorescein diacetate (FDA) in a range of soils. *Soil Biol. Biochem.* **2001**, *33*, 943–951. [CrossRef]

 https://doi.org/10.3390/environments13020093

21. Von Mersi, W.; Schinner, F. An improved and accurate method for determining the dehydrogenase activity of soils with iodonitrotetrazolium chloride. *Biol. Fertil. Soils* **1991**, *11*, 216–220. [CrossRef]
22. Vance, E.D.; Brookes, P.C.; Jenkinson, D.S. An extraction method for measuring soil microbial biomass C. *Soil Biol. Biochem.* **1987**, *19*, 703–707. [CrossRef]
23. Majlessi, M.; Eslami, A.; Saleh, H.N.; Mirshafieean, S.; Babaii, S. Vermicomposting of food waste: Assessing the stability and maturity. *Iran. J. Environ. Health Sci. Eng.* **2012**, *9*, 25. [CrossRef]
24. Manzoor, A.; Naveed, M.S.; Ali, R.M.A.; Naseer, M.A.; Ul-Hussan, M.; Saqib, M.; Hussain, S.; Farooq, M. Vermicompost: A potential organic fertilizer for sustainable vegetable cultivation. *Sci. Hortic.* **2024**, *336*, 113443. [CrossRef]
25. Arancon, N.Q.; Edwards, C.A.; Bierman, P.; Welch, C.; Metzger, J.D. Influences of vermicomposts on field strawberries: 1. Effects on growth and yields. *Bioresour. Technol.* **2004**, *93*, 145–153. [CrossRef]
26. Lazcano, C.; Revilla, P.; Malvar, R.A.; Domínguez, J. Yield and fruit quality of four sweet corn hybrids (*Zea mays*) under conventional and integrated fertilization with vermicompost. *J. Sci. Food Agric.* **2011**, *91*, 1244–1253. [CrossRef] [PubMed]
27. Bernal, M.P.; Sommer, S.G.; Chadwick, D.; Qing, C.; Guoxue, L.; Michel, F.C. Current Approaches and Future Trends in compost Quality Criteria for agronomic, environmental, and Human Health benefits. In *Advances in Agronomy*; Elsevier: Amsterdam, The Netherlands, 2017; pp. 143–233. [CrossRef]
28. Edwards, C.A.; Arancon, N.Q.; Sherman, R.L. *Vermiculture Technology*; CRC Press: Boca Raton, FL, USA, 2010. [CrossRef]
29. Edwards, C.A.; Arancon, N.Q. *Biology and Ecology of Earthworms*; Springer: Berlin/Heidelberg, Germany, 2022. [CrossRef]
30. Vyas, P.; Sharma, S.; Gupta, J. Vermicomposting with microbial amendment: Implications for bioremediation of industrial and agricultural waste. *BioTechnologia* **2022**, *103*, 203–215. [CrossRef]
31. Rehman, S.U.; De Castro, F.; Aprile, A.; Benedetti, M.; Fanizzi, F.P. Vermicompost: Enhancing Plant Growth and Combating Abiotic and Biotic Stress. *Agronomy* **2023**, *13*, 1134. Correction in *Agronomy* **2024**, *14*, 1256. https://doi.org/10.3390/agronomy14061256. [CrossRef]
32. Geremu, T.; Hailu, H.; Diriba, A. Evaluation of Nutrient Content of Vermicompost Made from Different Substrates at Mechara Agricultural Research Center on Station, West Hararghe Zone, Oromia, Ethiopia. *Ecol. Evol. Biol.* **2020**, *5*, 125. [CrossRef]
33. Oyege, I.; Bhaskar, M.S.B. Effects of vermicompost on soil and plant health and promoting sustainable agriculture. *Soil Syst.* **2023**, *7*, 101. [CrossRef]
34. Gudeta, K.; Kumar, V.; Bhagat, A.; Julka, J.M.; Bhat, S.A.; Ameen, F.; Qadri, H.; Singh, S.; Amarowicz, R. Ecological adaptation of earthworms for coping with plant polyphenols, heavy metals, and microplastics in the soil: A review. *Heliyon* **2023**, *9*, e14572. [CrossRef]
35. Xu, D.; Yu, X.; Chen, J.; Li, X.; Chen, J.; Li, J. Effects of compost as a soil amendment on bacterial community diversity in saline–alkali soil. *Front. Microbiol.* **2023**, *14*, 1253415. [CrossRef]
36. Ismayilov, A.I.; Mamedov, A.I.; Fujimaki, H.; Tsunekawa, A.; Levy, G.J. Soil Salinity Type Effects on the Relationship between the Electrical Conductivity and Salt Content for 1:5 Soil-to-Water Extract. *Sustainability* **2021**, *13*, 3395. [CrossRef]
37. Borazan, A.A.; Değirmenci, L.; Değirmenci, Ö.C. Investigation of the effects of waste olive pomace on vermicompost. *Bilge Int. J. Sci. Technol. Res.* **2024**, *8*, 104–114. [CrossRef]
38. Sonia, M.-T.; Naceur, J.; Abdennaceur, H. Studies on the ecology of actinomycetes in an agricultural soil amended with organic residues: I. identification of the dominant groups of Actinomycetales. *World J. Microbiol. Biotechnol.* **2011**, *27*, 2239–2249. [CrossRef]
39. Wu, X.; Peng, J.; Liu, P.; Bei, Q.; Rensing, C.; Li, Y.; Yuan, H.; Liesack, W.; Zhang, F.; Cui, Z. Metagenomic insights into nitrogen and phosphorus cycling at the soil aggregate scale driven by organic material amendments. *Sci. Total Environ.* **2021**, *785*, 147329. [CrossRef] [PubMed]
40. Manfredi, S.; Tonini, D.; Christensen, T.H.; Scharff, H. Landfilling of waste: Accounting of greenhouse gases and global warming contributions. *Waste Manag. Res. J. A Sustain. Circ. Econ.* **2009**, *27*, 825–836. [CrossRef] [PubMed]
41. Lee, C.T.; Hashim, H.; Ho, C.S.; Van Fan, Y.; Klemeš, J.J. Sustaining the low-carbon emission development in Asia and beyond: Sustainable energy, water, transportation and low-carbon emission technology. *J. Clean. Prod.* **2016**, *146*, 1–13. [CrossRef]
42. Damgaard, A.; Manfredi, S.; Merrild, H.; Stensøe, S.; Christensen, T.H. LCA and economic evaluation of landfill leachate and gas technologies. *Waste Manag.* **2011**, *31*, 1532–1541. [CrossRef] [PubMed]
43. Manfredi, S.; Tonini, D.; Christensen, T.H. Environmental assessment of different management options for individual waste fractions by means of life-cycle assessment modelling. *Resour. Conserv. Recycl.* **2011**, *55*, 995–1004. [CrossRef]
44. Pergola, M.; Persiani, A.; Pastore, V.; Palese, A.M.; D'Adamo, C.; De Falco, E.; Celano, G. Sustainability Assessment of the Green Compost Production Chain from Agricultural Waste: A Case Study in Southern Italy. *Agronomy* **2020**, *10*, 230. [CrossRef]
45. Liu, H.; Li, J.; Li, X.; Zheng, Y.; Feng, S.; Jiang, G. Mitigating greenhouse gas emissions through replacement of chemical fertilizer with organic manure in a temperate farmland. *Sci. Bull.* **2015**, *60*, 598–606. [CrossRef]

 https://doi.org/10.3390/environments13020093

environments

MDPI

Article

Physicochemical Characterization of Vineyard Stump-Derived Hydrochars and Pyrochars and Preliminary Grapevine Tolerance Screening

José Manuel Díaz-Rasero [1], Teresa Sosa [2], Beatriz Ledesma [1] and Silvia Román [1,*]

[1] Department of Applied Physics, School of Industrial Engineering, University of Extremadura, 06006 Badajoz, Spain; josemanudr@unex.es (J.M.D.-R.); beatrizlc@unex.es (B.L.)
[2] Department of Plant Biology, Ecology and Earth Sciences, Faculty of Science, University of Extremadura, 06006 Badajoz, Spain; tesosa@unex.es
* Correspondence: sroman@unex.es

Abstract

This study explores a circular economy strategy for vineyard residue management through the conversion of pruning biomass into carbonaceous materials by hydrothermal carbonization (HTC) and pyrolysis (PYR), with and without iron (Fe) addition. A preliminary pot-based vegetation experiment was conducted as a screening assay to assess initial plant tolerance and exclude evident phytotoxic effects. Chlorophyll index values in grapevine leaves remained within physiological ranges across treatments and sampling dates, although no consistent treatment-related trends could be established. Overall, the results provide a physicochemical characterization of the carbonaceous materials derived from vineyard residues and demonstrate their initial compatibility with grapevine cultivation under controlled conditions. This work lays the groundwork at the material level for future, more comprehensive studies that integrate long-term soil, plant, and field assessments.

Keywords: hydrochars; pyrochars; vineyard waste; soil amendment

1. Introduction

The cultivation of vineyards for wine production is a major economic activity worldwide, and the recent expansion and intensification of viticultural practices are raising increasing concerns about their impact on biodiversity. Systematic reviews and meta-analyses indicate that organic viticulture and aboveground vegetation management (e.g., cover crops) enhance biodiversity and ecosystem services such as soil fertility, pest control, and carbon sequestration [1]. Maintaining inter-row vegetation is particularly beneficial, with no significant trade-off between grape yield and biodiversity [2]. Nevertheless, two fundamental challenges persist in vineyard systems: the impacts of climate change, including increasing drought frequency, and the progressive loss of soil nutrients and productivity under intensive management. Soil erosion caused by intense rainfall in Mediterranean vineyards leads to nutrient loss (as nitrogen, phosphorus, potassium, and organic carbon), degradation of soil structure, and reduced organic matter, all of which directly affect grape productivity and quality [3]. Studies in southern Italy confirm a negative correlation between soil erosion and vine productivity [4]. These effects are further exacerbated by intensive practices and excessive agrochemical use, which contribute to long-term soil degradation. In parallel, water scarcity associated with climate change represents an additional constraint on vineyard sustainability. In this context, alternatives are needed to sustain vineyard productivity and grape quality while addressing water and nutrient limitations. One potential strategy involves the valorization of vineyard residues to produce carbon-based materials with properties that may be relevant for agricultural soils. Within the framework of the circular economy, hydrothermal carbonization (HTC) offers a promising approach due to its low cost and ability to process high-moisture wastes. HTC generates both solid hydrochar (HC) and process water (PW), each with versatile applications [5]. Hydrochars are increasingly investigated in soil-related contexts, although their effects remain variable and strongly context-dependent, while pyrochars (PIRs) have a longer history of study, particularly in relation to carbon sequestration [6]. Unlike pyrolysis (PYR), HTC involves lower direct CO_2 emissions, typically reported in the range of 5–10% of the initial biomass carbon, which has increased interest in HTC within short-term decarbonization strategies.

The literature presents mixed results regarding the effectiveness of HCs in soils. Reported effects vary widely depending on the feedstock, hydrothermal carbonization conditions [7], crop type, and soil properties. For example, reductions in soil mineral nitrogen of 20–50% and increases in microbial biomass of up to 30% have been observed following HC application [8]. In some cases, HCs increase soil pH by 0.2–1.0 units and enhance biomass production in specific crops, whereas in others they may reduce nitrogen uptake [9]. Certain amino-functionalized HCs have been shown to decrease available heavy metals by 15–40% and improve plant growth under contaminated conditions [10]. Improvements in water retention (5–15%), reductions in bulk density, and increases in microbial abundance suggest that HCs may influence soil processes related to organic carbon stabilization [11,12]. However, these effects are not universal and should not be assumed without site-specific validation. Positive effects on plant growth and water-use efficiency under heat stress have also been documented in selected systems [13]. Recent studies have examined vineyard residues processed by HTC or PYR and their impact on biodiversity and ecosystem services. Grape pomace can be efficiently converted into hydrochar, with yield and quality depending on HTC conditions [14], although liquid fractions show lower biogas yields due to recalcitrant nitrogen compounds [15]. To date, no studies have specifically investigated the hydrothermal carbonization of vineyard stumps (VSs), as most existing research has focused on grape pulp [16], pruning residues for energy applications [17], or techno-economic assessments of HTC plants using vine shoots [18]. Pyrochars provide long-term benefits to vineyard soils by improving their physical, chemical, and biological fertility. They can also reduce soil erosion by enhancing aggregate stability, increasing infiltration through improved porosity, and modifying surface roughness, which slows overland flow. Field studies up to 10 years report increases in soil organic carbon, higher pH, reduced bulk density, and improved water retention, mitigating drought stress [19,20]. Other studies highlight enhanced cation exchange and microbial community shifts, with increased beneficial fungi and reduced pathogens [21,22]. Productivity outcomes are variable, with some reporting yield improvements but limited effects on wine quality. In this variability, in the case of vineyards, biochar amendment at rates up to 2.55 L m^{-2} resulted in a non-significant improvement in grape yield per plant. The high variability in response was largely due to the adverse climatic conditions typical of the Csa region's Mediterranean climate, which masked the observed positive trends. Such variability is common in biochar studies, as the agronomic response depends heavily on the application rate, soil characteristics, and the physicochemical properties of the biochar used. Conversely, under different climatic conditions and soil types, such as those found at two Oregon sites (Willamette Valley and Rogue Valley), biochar applied at rates up to 35 t ha^{-1} produced distinct results, highlighting the strong influence of climate–soil–biochar interactions on vineyard yield [23–25]. PIRs from vine shoots, cuttings, or canes also improve

 https://doi.org/10.3390/environments13020083

water retention and nutrient availability [26], though performance depends on pyrolysis temperature and raw material [27,28].

To date, no previous study has evaluated vineyard vines as a raw material for hydrothermal carbonization, nor has it assessed the properties of the resulting carbonaceous materials for potential soil applications. Against this background, the present study focuses on the production and physicochemical characterization of hydrochars and pyrochars derived from vineyard stumps (VSs) obtained from an agricultural cooperative in central-southern Spain. As a novel aspect, vineyard stumps were subjected to hydrothermal carbonization, a feedstock–process combination that has not been previously reported for viticultural systems. All prepared carbon materials were subsequently evaluated in a preliminary pot-based experiment designed as a tolerance screening, aimed at identifying potential phytotoxic or strongly negative effects on grapevine plants. FTIR analysis was employed to characterize the surface functional features of the carbon materials, providing indirect insight into properties that may be relevant for soil interactions, while chlorophyll index measurements were used as a non-destructive indicator of plant physiological status. In addition to comparing HCs and PIRs derived from vineyard stumps, this study also explores iron (Fe) incorporation via ferric chloride ($FeCl_3$) as a surface modification strategy. The rationale for Fe-enrichment is based on its ability to introduce Fe-O bonds and modify surface chemistry, which could broaden the functional properties of the materials and potentially lead to improvements in soil fertility. Such modification may broaden the range of potential applications of vineyard-derived carbonaceous materials, which warrant further investigation in future soil-based and field-scale studies. Table 1 below summarizes the main previous results related to HTC and pyrolysis related to grapevine.

Table 1. Summary of key previous results from HTC and pyrolysis in grapevines.

Topic/Study	Main Findings	Ref.
Effect of organic viticulture and vegetation cover	Increase biodiversity and ecosystem services (soil fertility, pest control, carbon sequestration).	[1,2]
HTC: products and applications	HTC produces solid hydrochar (HC) and process water (PW); viable for wet residues and circular economy uses.	[5]
Contrasting roles of HC and PIR	HC improves CEC and nutrient retention; PIR is more stable and favors carbon sequestration.	[6]
Variable HC outcomes in soils	Effects depend on feedstock and HTC conditions; HC can act as herbicide or fertilizer; results are heterogeneous	[7–10]
HC and soil properties	HC improves water retention, reduces bulk density, increases microbial biomass and carbon sequestration.	[11,12]
Plant responses under heat stress	HC showed positive effects on growth and water use efficiency under heat.	[13]
Conversion of vineyard residues by HTC	Grape pomace convertible to HC; yield and quality depend on HTC conditions.	[14,15]
Gap in vineyard stumps (VSs)	No specific studies applying VSs as HTC amendments identify as a knowledge gap.	[16–18]
Long-term benefits of PIR	Increases in soil organic C, pH, water retention and reduced bulk density in study up to 10 years.	[19,20]
Biological and productivity effects of PIR	Changes in CEC and microbial communities; effects on yield and wine quality are variable.	[21–25]
PIR and water/nutrient availability	PIR improves water retention and nutrient availability; performance depends on pyrolysis temperature and feedstock	[26–28]

https://doi.org/10.3390/environments13020083

2. Materials and Methods

2.1. Materials

Vineyard stumps (VSs), sourced from the Yuntero Cooperative, located in Manzanares, Castilla-La Mancha region (Spain), was used as feedstock in this study. This raw material was supplied in big pieces and needed mechanical conditioning: VS pieces were initially cut with a chainsaw then the sawdust generated during the cutting process was grinded. The remaining stump material was shredded using a GARLAND Chipper 790 Qg-V19. The shredded material was then thoroughly mixed with the sawdust to obtain a uniform appearance, producing a homogeneous size product. After shredding and mixing, the material was sieved to obtain a final particle size distribution between 1 and 4 mm (sieve, CISA 200/50, Norma ISO-3310.1) [29], which was selected to ensure homogeneous reaction kinetics during HTC and to reduce mass-transfer limitations.

2.2. Experimental Reactions

2.2.1. HTC Reactions

HTC reactions were carried out using a Parr 4848 reactor (Parr Instrument Company, Moline, IL, USA) with a total capacity of 1.2 L. The reaction mixture consisted of 720 g of water and 210 g of biomass (wet basis, without prior drying), corresponding to a solid-to-liquid ratio of 0.22. The components were thoroughly mixed and introduced into the reactor.

Heating to the target temperature was conducted over 90 min, including the heating ramp, and once the maximum temperature of 220 °C was reached, it was maintained for a residence time of 18.5 h. After completion of the reaction, the reactor was allowed to cool naturally to room temperature before opening. The resulting slurry was filtered, and the solid fraction was dried in an oven at 105 °C for 12 h. The obtained hydrochar (HC) was collected and stored in airtight containers. Multiple HTC runs were conducted to obtain sufficient material for subsequent analyses and preliminary plant assays. HTC reactions were also performed to enrich the HC obtained in Fe. To this end, the HTC reactions were carried out under the same temperature and time conditions but replacing the mentioned amount of water with a 0.25 M $FeCl_3 \cdot 6H_2O$ solution, as described in [30]. This solution was mixed directly with the biomass in the reactor before HTC process.

2.2.2. Pyrolysis Reactions

Pyrolysis was carried out in a stainless-steel tubular reactor at 600 °C for 1 h under a continuous N_2 flow of 100 mL/min to ensure an inert atmosphere. The starting material was 150 g of VS. After completion of the reaction, the reactor was allowed to cool in air while maintaining the N_2 flow to prevent combustion. The resulting PIR was then collected and stored for later use.

This reaction also produced Fe-enriched PIRs, following the procedure described in [31]. For this process, 10 g of dried and sieved biomass were added to 150 mL of a 0.25 M $FeCl_3 \cdot 6H_2O$ solution in a glass flask. The flask was capped and placed in a bath sonicator (40 kHz, 150 W) for 1 h 15 min at room temperature to enhance the interaction between the biomass and Fe^{3+} ions. Subsequently, the suspension was kept under magnetic stirring at 25 °C for 24 h, protected from light with aluminum foil to avoid Fe^{3+} photoreduction. After incubation, the solid was separated by drip filtration, washed with deionized water to remove unbound Fe, and dried in an oven at 105 °C for 48 h.

2.3. Characterization of Hydrochars, Pyrochars and Raw Material

The elemental composition (C, H, N, and S) of the HCs, PIRs and raw material were determined using a CHNS analyzer (LECO CHNS-932, LECO Corporation, St. Joseph,

 https://doi.org/10.3390/environments13020083

MI, USA). The O content was estimated as the difference between the sum of the above and the ash content at 100%. The fuel value of these materials was characterized by the higher heating value (HHV) measured in a calorimetric bomb (PARR), following the norm CEN/TS 14918. Lower heating value (LHV) was calculated by Equation (1):

$$LHV = HHV - 2.442 \cdot 9H, \quad (1)$$

where HHV is expressed in MJ kg^{-1} and H correspond to the hydrogen content (% dry basis) of the sample and 2.442 (MJ kg^{-1}) is the latent heat of vaporization of water associated with hydrogen combustion. The lower heating value (LHV) is expressed on a dry basis. Ion chromatography (IC) was applied to raw vineyard stump material to quantify the chloride content before thermochemical treatment, with the aim of evaluating the suitability of the raw material for subsequent conversion processes. The methodology consists of extracting the sample in an aqueous solution, filtering it, and then injecting it into an ion-exchange column where the anions present, including chlorides, are separated based on their affinity for the chromatographic resin. A conductivity detector measures chloride concentration, allowing its quantification with high sensitivity in the ppm (mg/kg) range. This technique is precise and effective for environmental studies and material characterization.

The proximate analysis (moisture, volatile matter, ash, and fixed carbon) was performed on the raw samples, using an STA 449 F3 Jupiter (Netzsch, Berlin, Germany) thermobalance in an oxidizing atmosphere (20% oxygen and 80% argon with a flow rate of 100 mL/min) using a ramp from 20 °C/min up to 700 °C.

For hydrocahrs, pyrochars, HC_Fe and PIR_Fe samples, FTIR and SEM analysis were made too. FTIR spectra were recorded with a Perkin Elmer model Paragon 1000PC spectrophotometer with a resolution of 4 cm^{-1} and 100 scans, equipped with a diamond crystal ATR accessory in attenuated total reflectance (ATR) mode.

The surface morphology of the samples was analyzed by scanning electron micrography (SEM, Hitachi, S-3600 N, Krefeld, Germany) observation. The SEM samples were prepared by depositing about 50 mg of sample on an aluminum stud covered with conductive adhesive carbon tapes, and then coating with Rd–Pd for 1 min to prevent charging during observations. Imaging was performed in the high vacuum mode at an accelerating voltage of 20 kV, using secondary electrons.

The ash content of all samples was measured directly after dry oxidation at 550 °C according to the DIN EN 14775:2010-04 standard.

2.4. *Analyses for Plant Evolution in Field Tests*

2.4.1. Chlorophyll Index

Chlorophyll index in the leaves of each plant was measured to gather information about the nutrient status and stress levels of the plants. Using a chlorophyll meter (MC-100, Apogee Instruments, Logan, UT, USA), measurements were taken three times on the following dates: 26-June 30-July and 10-Septempter. For each plant, three fully expanded leaves of similar size and developmental stage were selected from the middle part of the shoot, as recommended for consistency in physiological studies [32]. Chlorophyll index was recorded directly on each leaf, and the average of the three measurements was calculated to obtain a representative value per plant.

Subsequently, the average chlorophyll index was calculated for all plants with each type of carbons. All data are expressed in Relative Units (r.u.). No conversion to absolute chlorophyll concentration was performed.

2.4.2. Efficacy Trials on the Soil Application of HCs and PIRs

The plant trial was designed as a preliminary tolerance and non-phytotoxicity screening prior to field-scale experiments, rather than as an evaluation of agronomic performance.

 https://doi.org/10.3390/environments13020083

Pot-based systems were selected to allow controlled amendment concentrations and direct observation of early plant responses. However, these systems have known limitations, such as restricted root development, reduced soil aeration, limited water retention, and increased heat stress, especially under high summer temperatures. These limitations can increase plant mortality regardless of treatment effects and are recognized as an inherent disadvantage in early-stage analytical experiments.

This trial involved planting 96 vines from the Yuntero Cooperative, with the addition of prepared carbonaceous materials at 1% and 3% (v/v). The experiment was designed to evaluate HCs produced at 220 °C, pyrochars produced at 600 °C, and their iron Fe-enriched variants.

Figure 1 illustrates the spatial arrangement of the 96 pots according to treatment type and concentration. The grid-based layout was applied to ensure systematic distribution of treatments, with three replicates per treatment, although the design was not intended for inferential statistical comparison among treatments.

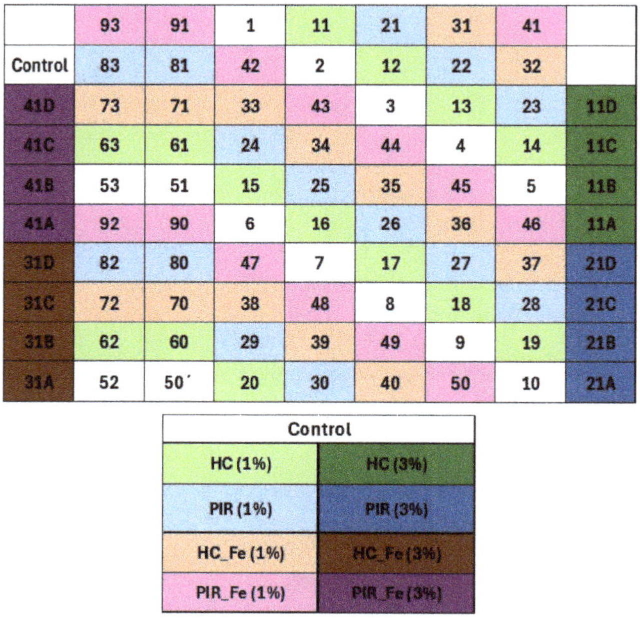

Figure 1. Spatial distribution of the 96 experimental pots according to treatment type and concentration. The layout follows a grid-based design inspired by Roman *centuriation*. Color coding indicates the type and percentage of carbon applied: HC (1% and 3%), PIR (1% and 3%), PIR_Fe (1% and 3%), HC_Fe (1% and 3%) and control. Labels correspond to pot identifiers used throughout the experiment.

Each pot was filled with soil collected from the cooperative's land. The experiment was conducted at the School of Agricultural Engineering in Badajoz, approximately 300 km from the sampling site. Both locations share comparable Mediterranean climatic conditions (Csa climate according to Köppen–Geiger classification).

Planting was initiated on 29 May 2024, in accordance with the established experimental design. A total of 52 sprouted plants were placed into pots and watered to field capacity, without the addition of PIRs or HC at this initial stage. The pots were then placed in the greenhouse, while the remaining plants were stored in a growth chamber for planting in the following days.

https://doi.org/10.3390/environments13020083

On 30 May 2024, a preliminary inspection confirmed that all pots contained sprouted plants. This early sprout was later determined not to interfere with the experimental outcomes.

On 12 June 2024, PIRs and HC were incorporated at a concentration of 1% into the 52 pots, followed by irrigation with 300 mL of water per pot. The pots were returned to the greenhouse and appropriately labeled. On 17 June 2024, an additional application of 3% HC was made to the pots labeled 11A, 11B, 11C, and 11D.

A second labeling phase took place on 20 June 2024, during which the pots were relocated outdoors. The remaining 44 pots received their respective carbon treatments and were irrigated with 300 mL of water. All the pots were watered weekly during the summer season.

Figure 2 shows the experimental pots after relocation outdoors, with visible labels and treatment assignments. The dimensions of the pots are: diameter 16.5 cm, height 41 cm, height of soil content: 34.5 cm. All the pots have drainage holes.

Figure 2. Example of experimental pots arranged outdoors following treatment application and labeling.

Despite manual watering throughout the summer, with applications of approximately 300 mL per pot, two to three times per week, depending on observed field capacity, plant survival was compromised by extreme environmental conditions. During the summer period, maximum temperatures reached up to 45 °C, along with episodes of very high solar radiation (UV index $\approx$ 10) and heat waves lasting up to a week. These factors were intensified by the high thermal absorption of the black pots used. Furthermore, in January 2025, minimum temperatures as low as −3 °C were recorded, which may have further affected plant development. These conditions, exacerbated by the high thermal absorption of black pots, are recognized as significant abiotic stress factors and are discussed as limitations of the experimental system, reinforcing the screening nature of the trial. Due to the exploratory nature of the vegetation experiment and the limited number of replicates per treatment, no inferential statistical analyses were applied.

3. Results and Discussion

3.1. Raw Materials

Table 2 presents the physicochemical properties of the raw material (VS) used in the experiment (all determined in dry basis). These parameters are relevant for contextualizing behavior during carbonization and its potential suitability for subsequent soil-related applications.

 https://doi.org/10.3390/environments13020083

Table 2. Characterization of vine stumps (VSs).

Proximate Analysis	%
Volatile matter	81.27
Fixed carbon	17.25
Ash	1.48
Elemental analysis	%
C	45.10
H	6.05
N	0.31
S	0.03
O [1]	47.03
HHV	MJ/kg
0% moisture	19.32
Moisture received	17.57
LHV	MJ/kg
0% moisture	17.99
Moisture received	16.37

[1] Calculated by difference.

The VS has a moderate moisture content (9.06%), which reduces the need for thermal pretreatment prior to carbonization. Its high volatile organic content (81.27%) can influence the porous structure of different PIRs or HCs, as during pyrolysis and HTC, molecules susceptible to elimination (mainly hemicellulose and cellulose) can open cavities in the material's structure. This porosity is related to its water adsorption and retention capacity during subsequent application. The fixed carbon content (17.25%), a typical value for lignocellulosic biomasses such as wood or fruit pits, suggests greater stability of the carbonized material, which is desirable for its persistence in the soil. Furthermore, its low ash content (1.48%) indicates a lower presence of mineral impurities, which can improve the quality of biochar for agricultural soils. For all samples in the following Table 2, the average values are shown in the table, with the deviation being less than 5% in all cases, as three repetitions were performed for the samples.

In terms of elemental composition, VS has a carbon content of 45.10%, characteristic of lignocellulosic biomasses, which can favor the chemical stability of biochar in the soil. Its hydrogen (6.05%) and oxygen (47.03%) content can influence the material's reactivity in carbonization processes. Furthermore, its low nitrogen content (0.31%) minimizes the formation of volatile nitrogen compounds during the thermochemical process, reducing pollutant emissions.

Overall, as shown in Table 2 the proximate and elemental analyses are typical of lignocellulosic biomasses, very similar to other woods. Thus, the carbon content and the volatile content indicate a high proportion of volatile carbon and high calorific value. The low ash content also suggests that this biomass generates little solid waste when burned, which is beneficial from both an environmental and equipment maintenance perspective. These similarities support the suitability of VS as a representative woody residue for comparative carbonization studies. Woods from pruned oak, almond, and walnut trees, previously studied by the research group, have shown similar results.

The calorific value of biomass reflects its energy potential; however, in the present work, the primary objective was not energy recovery but transformation into hydrochars and pyrochars for material characterization and preliminary plant compatibility assessment. From this perspective, the VS feedstock provides favorable conditions for obtaining carbon materials with relatively high carbon content and low ash fractions. Regarding chloride

https://doi.org/10.3390/environments13020083

content, the VS showed a higher concentration (249.90 mg/kg) compared to other residual material from the grape, such as the skins (from the crushing of the grape pulp). This material has a value of around 130 mg/kg. This difference may be due to the composition and structure of each material, since the VS, being more lignocellulosic than the grape, may have accumulated more mineral salts during the growth of the vine. Furthermore, the prolonged contact with the soil could have favored chloride retention compared to the skins, which come from the more superficial part of the fruit. The determination of chloride content is important because chlorine promotes the formation of corrosive compounds and deposits during thermochemical conversion, which can compromise equipment durability and process stability. It is also linked to regulated emissions, so quantifying chloride helps assess the suitability of the material as a biofuel beyond its energy potential.

3.2. Analysis of PIRs and HCs

The PIRs and HCs have been characterized according to their elemental analysis, HHV, and solid yield (Table 3). These results provide a comparative description of the materials obtained under different processing conditions and with or without iron enrichment. Regarding the HCs, a decrease in HHV and an increase in solid yield are notable for the iron-enriched HC compared to the iron-free HC. This may be because $FeCl_3$ creates a highly acidic environment that promotes the hydrolysis and solubilization of the biomass's organic components, reducing the carbonaceous fraction. However, this also causes some of the iron to precipitate and be retained in the hydrochar as oxides and hydroxides [33]. This iron incorporation increases the final mass of the solid, thus increasing the yield calculated as final mass divided by initial mass. Nevertheless, since the enriched hydrochar contains more ash and less organic carbon, its energy density decreases [34], the calorific value is reduced because the mineral fraction does not contribute energy to combustion and dilutes the available carbon content, (Table 3). For all samples in the following Table 3, the average values are shown in the table, with the deviation being less than 5% in all cases, as three repetitions were performed for each HC/PIR sample.

Table 3. Characterization of HCs and PIRs.

Sample	Solid Yield (%)	C (%)	N (%)	H (%)	S (%)	O [1] (%)	Ash (%)	HHV (MJ/kg)
PIR	27.13	68.64	0.67	3.32	0.10	18.32	8.95	30.11
HC	58.81	64.00	0.77	6.25	0.21	25.26	3.51	30.62
PIR_Fe	30.02	65.87	0.63	3.47	0.12	16.27	13.64	25.92
HC_Fe	62.90	54.30	0.50	5.27	0.18	36.21	3.54	21.06

[1] calculated by difference.

Something similar occurs with PIR samples. During pyrolysis, high temperatures in the absence of oxygen cause the thermal decomposition of the organic components of the biomass. If $FeCl_3$ is present in the mixture, the iron acts as a catalyst, promoting the breaking of bonds in cellulose, hemicellulose, and lignin, and accelerating the formation of volatile products and gases. This means that some of the carbon that would normally remain in the biochar is transferred to the gaseous or liquid phase, reducing the organic fraction of the solid. However, at the same time, iron is incorporated into the biochar in the form of oxides or inorganic compounds, increasing the final mass of the solid and making the apparent yield higher. The result is a biochar with more ashes and less available carbon, which explains the decrease in calorific value [35].

Another aspect to analyze is the apparent increase in nitrogen content in the HC and PIR samples compared to the initial biomass. This increase does not imply the formation

 https://doi.org/10.3390/environments13020083

of new nitrogen during carbonization but should be interpreted as a concentration effect resulting from the loss of volatile mass during the HTC or pyrolysis processes. The nitrogen present in the biomass, mainly in the form of proteins, can be transformed during HTC by hydrolysis at relatively low temperatures (≈120 °C) [36], generating amino acids that can subsequently undergo decarboxylation, deamination, or participate in Maillard reactions, producing amines, ammonia, organic acids, or N-cyclic compounds. However, since this study did not provide data on the protein content of the original biomass, it is not possible to accurately determine which pathways dominate in N retention in the final solid. Even so, regardless of the initial protein content, the resulting biochar typically retains a fraction of the original nitrogen, possibly incorporated through Maillard reactions, as described in previous studies [37].

Regarding the surface morphology (Figure 3) of the different samples, the greatest differences are observed between HC and HC_Fe. The presence of iron during hydrothermal carbonization has a clear influence on the surface morphology of the resulting hydrochars. Without iron, the hydrochar typically exhibits a compact and relatively homogeneous structure, characterized by smooth or only mildly textured surfaces and limited visible porosity. The carbon matrix tends to form dense agglomerates with few open cavities, reflecting the milder progression of dehydration, depolymerization, and structural rearrangement when no catalytic species are present. As a result, the material displays a more closed morphology with a lower degree of surface development. In contrast, the incorporation of iron leads to pronounced modifications in the microstructure. Iron acts as a catalytic agent, promoting more extensive bond cleavage, dehydration, and condensation reactions during the hydrothermal process. Consequently, the hydrochars produced in the presence of iron show markedly rougher and more heterogeneous surfaces, with the formation of pores, fissures, and irregular cavities distributed throughout the matrix. This enhanced textural development indicates a deeper structural transformation of the precursor material. The presence of secondary hydrochar can be seen in both HC samples, with the shape of spheres on the surface of both samples.

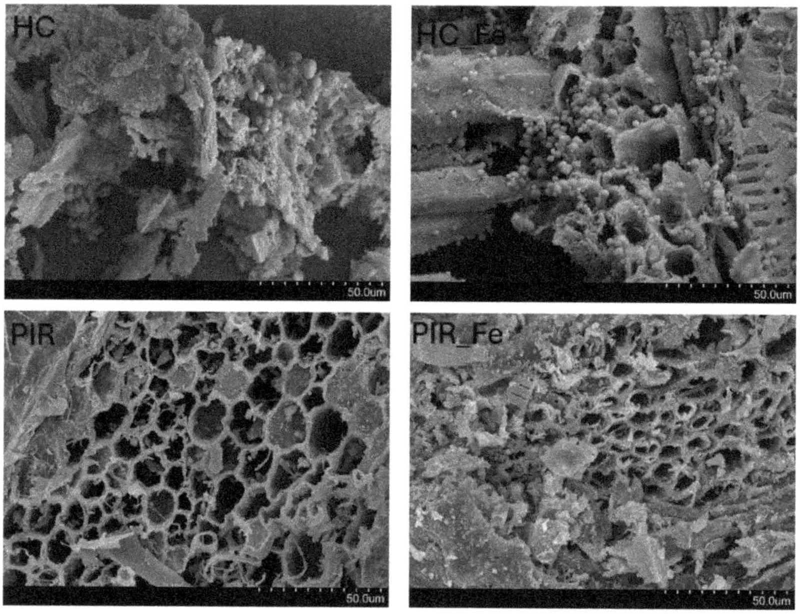

Figure 3. SEM of HC, HC_Fe, PIR and PIR_Fe samples.

 https://doi.org/10.3390/environments13020083

In pyrolysis, the thermal severity of the process already produces a highly devolatilized and porous carbon matrix, which means that the material naturally develops a rough, fractured surface regardless of whether iron is present. Because the structural breakdown, aromatization, and pore formation are largely driven by high temperature, the additional catalytic effect of iron has a comparatively smaller impact on the final morphology. As a result, both PIR and PIR_Fe tend to exhibit similar surface textures in SEM, with only subtle differences such as slightly enhanced porosity or localized iron-rich domains in the Fe-treated samples. The dominant influence of temperature during pyrolysis therefore reduces the visual contrast between samples produced with and without iron (Figure 3).

Thus, the differences between HCs and PIRs can be attributed to several factors. Thus, HTC involves a series of temperature and pressure driven reactions, including hydrolysis, dehydration, decarboxylation, and condensation, which progressively transform wet biomass into a carbon rich solid. As a result of these reaction pathways, HCs typically exhibit higher oxygen content, lower aromaticity, and greater surface functionality than PIRs produced through dry thermal decomposition. These compositional differences suggest potentially distinct behaviors in soil environments. From an environmental perspective, HTC offers advantages such as the ability to process wet feedstocks without prior drying and the potential for lower emissions, although HCs generally show lower long term carbon persistence compared to PIRs.

3.3. *Analysis of Chlorophyll Index*

The effect of the different treatments on the chlorophyll index of each plant was evaluated over three measurement dates: 26 June 2024, 3 July 2024, and 10 September 2024. The average chlorophyll index recorded for each treatment and date is summarized in Table 4. Overall, values ranged between 22 and 31 (r.u.) which are within the physiological range typically reported for grapevine [38]. Across all pots, an increase in chlorophyll index was observed from June to July, followed by a decline in September. This seasonal pattern has been described in grapevine and other woody species and reflects leaf maturation followed by the onset of senescence and summer stress conditions [39]. This trend was observed across treatments, including controls, indicating a dominant influence of seasonal and environmental factors. Some Fe-enriched treatments (e.g., 3%PIR_Fe) showed comparatively higher chlorophyll values at certain time points, while others (e.g., 3% HC_Fe) exhibited lower values at the final sampling. However, no consistent or systematic pattern associated with a specific treatment or concentration was identified. Given the preliminary nature of the trial and the absence of inferential statistical analysis, these observations should be interpreted as descriptive trends rather than treatment effects.

Table 4. Chlorophyll index (r.u.) measured in vine leaves at different sampling dates under the different treatments.

	Control	1%HC	1%PIR	1%HC_Fe	1%PIR_Fe	3%HC	3%PIR	3%HC_Fe	3%PIR_Fe
26 June 2024	27.2	28.6	28.1	28.3	29.1	29.3	-	-	-
3 July 2024	30.0	30.6	29.7	30.5	29.0	25.0	24.0	29.6	31.4
10 September 2024	24.0	24.3	22.5	25.7	23.3	-	31.4	17.3	26.1
Average	27.1	27.9	26.8	28.2	27.1	27.2	27.7	23.4	28.7

Average values presented in Table 4 correspond to deviations below 4%, based on repeated measurements. The chlorophyll index data was used solely as a non-destructive indicator of plant physiological status to screen for evident phytotoxic responses.

 https://doi.org/10.3390/environments13020083

3.4. HTC Reaction Pressure

During the HTC reaction, differences in the pressure generated during the process were observed. Thus, in HTC reactions using iron-enriched biomass, significantly lower pressures were obtained than in reactions with the same biomass and under the same conditions, but without iron enrichment (23.8–35.3 bar respectively). This pressure reduction upon addition of $FeCl_3$ during HTC is primarily due to three interrelated effects: lower net generation of gaseous moles due to the diversion of matter to the solid phase (more hydrochar formation), capture/transformation of volatile species (especially sulfur compounds), and changes in the chemistry of the medium (pH and redox) that alter decomposition pathways.

Chemically, Fe^{3+} acts as an acid/oxidizing catalyst that promotes dehydration, condensation, and aromatization reactions in the lignocellulosic matrix, increasing carbon retention in the hydrochar and reducing the volatile fraction susceptible to gas formation during the HTC [33].

Additionally, Fe can precipitate or form complexes with sulphur species and soluble fractions of the biomass, causing coagulation and solid-phase sequestration that decreases the contribution of compounds such as H_2S to the total pressure. $FeCl_3$ hydrolysis also acidifies the medium, shifting equilibria toward less volatile products and promoting greater stabilization in the form of solids or non-gaseous liquids [40].

To quantitatively attribute the drop of the pressure, it is useful to measure the gas composition (GC for CO_2, H_2, CH_4, H_2S, CO), the yield and elemental characterization of the hydrochar, sulphur and Fe–S phase analysis in the solids (XRD/SEM EDX), and pH/Fe dissolved in the liquor. Tests with different FeCl doses and time-based replicates allow for the dose-effect dependence to be traced and mass balances to be established that explain the extent to which solid sequestration or volatile reduction is responsible for the observed pressure decrease [41]. However, this would exceed the objectives of this work, so further studies would be necessary in this area and with other biomasses.

3.5. FTIR Analysis

The FTIR spectra of the original samples and their evolution over time were analyzed in open-air conditions (19 July 2024, 29 November 2024 and 27 February 2025) to identify signs of degradation and changes in surface functional groups.

The study established key differences in the chemical composition and stability of the materials: HC_original; PIR_original (original pyrochar); HC_Fe_original (Fe-doped HC); PIR_Fe_original (Fe-doped PIR).

The spectra obtained were compared with the same samples analyzed after months of exposure under ambient conditions. The nomenclature use for each sample is: HC/PIR _day–month–year.

First, the differences between HC and PIR were analyzed. The HC samples (Figure 4) present a higher content of surface oxygenated groups compared to the PIR, reflected in more intense bands in the following regions:

3200–3400 cm^{-1} (O-H): Greater intensity in HC, indicating the presence of hydroxyl groups. 1700 cm^{-1} (C=O): Notable in HC, attributed to carbonyls of carboxylic acids and esters. 1600–1500 cm^{-1} (aromatic C=C/COO^-): More pronounced in PIR, suggesting a more graphitic and less functionalized structure [42]. 1200–1000 cm^{-1} (C-O): More prominent in HC, reflecting the presence of alcohols and ethers.

The PIR samples (Figure 5), in contrast, show lower intensity in these bands and a more pronounced signal at 1580 cm^{-1}, indicative of more condensed aromatic structures. This confirms that pyrolysis largely eliminates the oxygenated groups present in hydrothermal

 https://doi.org/10.3390/environments13020083

carbonization. Therefore, HC has greater surface functionalization and chemical reactivity, while PIR is more graphitic and stable.

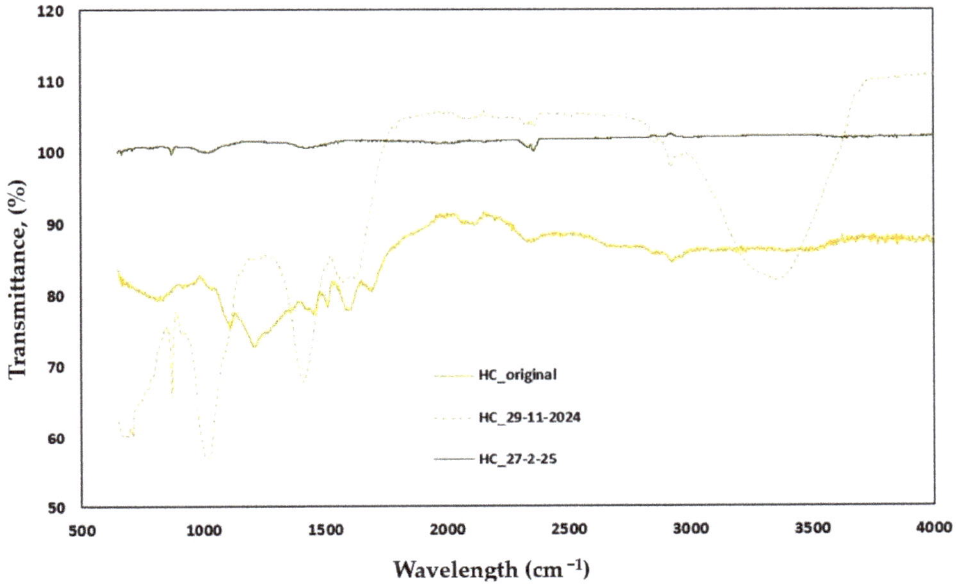

Figure 4. FTIR of HC during experimental time in soil.

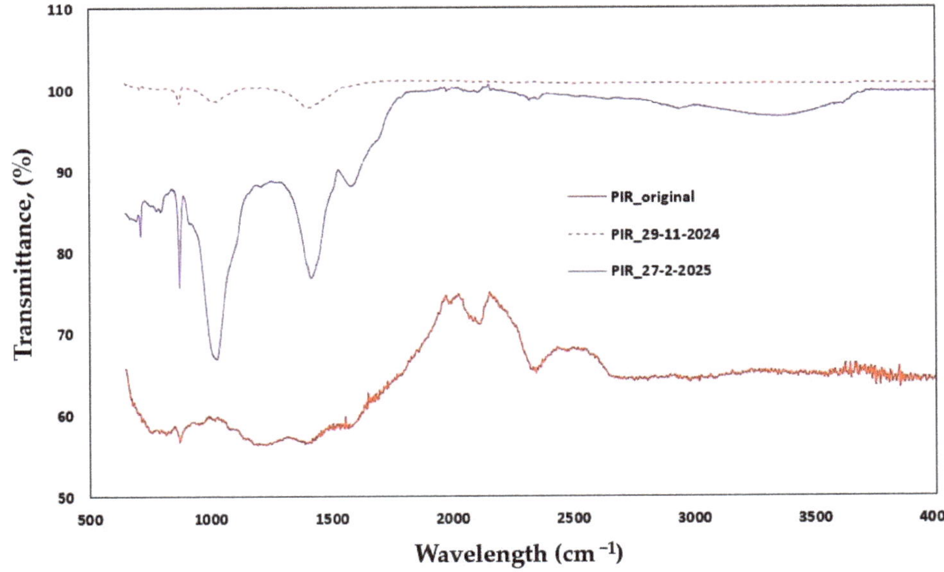

Figure 5. FTIR of PIR during experimental time in soil.

Secondly, the effect of iron (Fe) doping is analyzed (Figures 6 and 7). The Fe-doped samples show an additional band at 540–580 cm^{-1}, attributed to the formation of Fe-O bonds, confirming the presence of iron oxides on the surface. In addition, the following effects are observed:

 https://doi.org/10.3390/environments13020083

1. The O-H and C=O bands do not decrease, indicating that the oxygenated functional groups remain after doping.
2. There is a slight decrease in the aliphatic C-H bands (2920 and 2850 cm^{-1}), possibly due to interactions between Fe and the carbon structure [43].
3. The aromatic C-H bands are less intense (~870, 810, and 750 cm^{-1}), suggesting a change in the structural organization of the carbon.

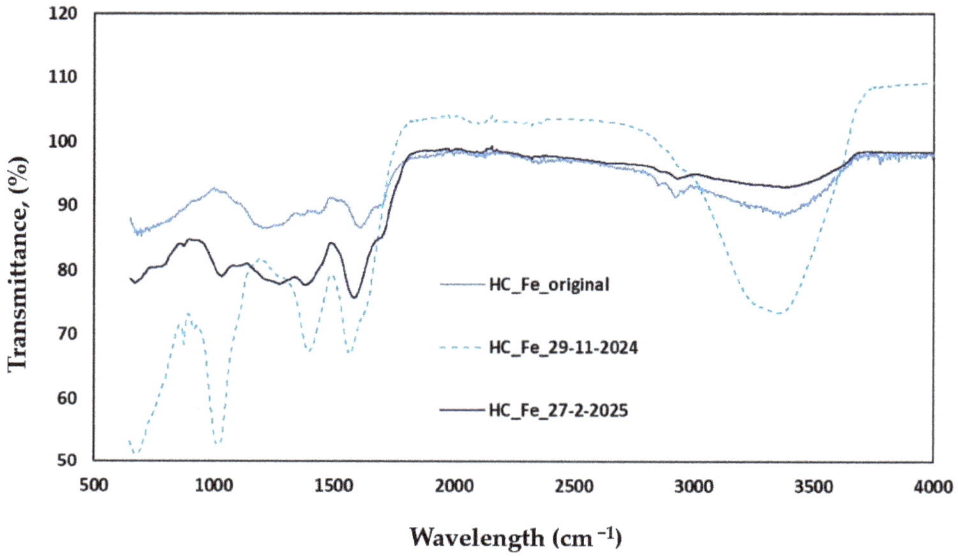

Figure 6. FTIR of HC_Fe during experimental time in soil.

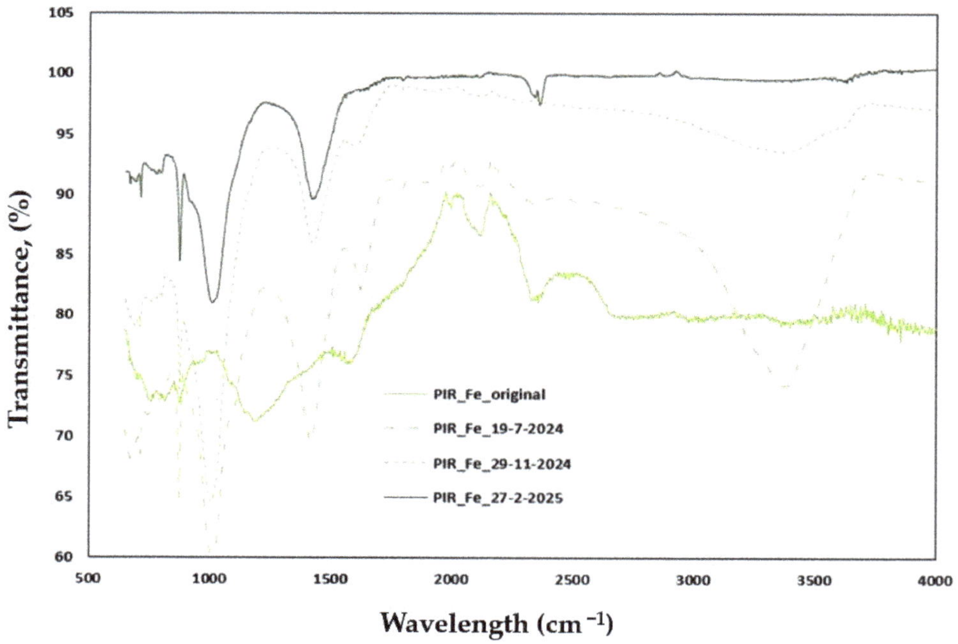

Figure 7. FTIR of PIR_Fe during experimental time in soil.

https://doi.org/10.3390/environments13020083

This confirms that Fe doping does not significantly alter surface functionalization but introduces magnetic properties due to the presence of iron oxides.

Table 5 summarizes the different peaks and functional groups found. Comparing the above with other studies, which also use vine residues versus other biomasses for soil amendment, yields different results. First, a study such as [26], focuses on the evaluation of a biochar with low aromatic and high aliphatic content, highlighting its lower structural stability but greater presence of oxygenated functional groups. Furthermore, this work directly compares two types of carbon obtained by different techniques: HTC and pyrolysis and analyzes the effect of iron doping on their structural properties.

Table 5. Key peaks and functional groups observed.

Functional Group	Peak (cm^{-1})	Observation
O-H (hydroxyls)	3200–3400	Higher intensity in HC; indicates presence of hydroxyl groups.
C=O (carbonyls: carboxylic acids and esters)	1700	Prominent in HC; attributed to carbonyl groups.
Aromatic C=C/COO^-	1600–1500	More pronounced in PIR; suggests more condensed aromatic/graphitic structure.
Aromatic specific signal	1580	Stronger in PIR; indicative of condensed aromatic structures.
C-O (alcohols and ethers)	1000–1200	More prominent in HC; reflects alcohol and ether functionalities.
Aliphatic C-H	2920; 2850	Slight decrease after Fe doping; possible interactions between Fe and carbon structure.
Aromatic C-H	~870; 810; 750	Reduced intensity after Fe doping; suggests changes in structural organization.

A key similarity between the two studies lies in the importance attributed to oxygenated functional groups (such as O-H, C=O, and C-O) [44], which are identified using Fourier transform infrared spectroscopy (FTIR). In both cases, the intensity of the bands corresponding to these groups is directly related to the surface functionalization of carbon. This could suggest that the presence of these groups improves the interaction of biochar with the soil, for which further research is needed; while the second study demonstrates that HC samples, obtained using HTC, exhibit greater intensity in these bands, indicating a surface richer in reactive groups compared to PIR samples.

The differences observed between the materials can be mainly attributed to experimental techniques. HTC tends to retain more oxygenated groups on the carbon surface, resulting in greater chemical reactivity. In contrast, pyrolysis, operating at higher temperatures and under drier conditions, favors the formation of condensed aromatic and graphitic structures, eliminating a large portion of the oxygenated groups. This technical difference explains why the PIR samples present a more intense signal in the aromatic bond region (1580 cm^{-1}) and lower intensity in the oxygenated functional group bands.

Another relevant aspect is the raw material used. The chemical composition of vineyard biomass, rich in lignin and phenolic compounds, can influence the formation of aromatic structures during pyrolysis, accentuating carbon graphitization. In contrast, if HTC is applied to the same raw material, the original functional groups are more likely to be preserved, which is reflected in the higher intensity of the O-H, C=O and C-O bands.

Finally, iron doping introduces an additional dimension as a novelty in this work. Although it does not significantly alter the presence of oxygenated groups, it does modify the structural organization of carbon, as indicated by the decrease in aliphatic and aromatic

https://doi.org/10.3390/environments13020083

bands. Furthermore, the appearance of bands attributable to Fe-O bonds confirms the incorporation of metal oxides, which could have implications for catalytic or magnetic applications, beyond the agronomic scope addressed in the first study [26]. Overall, comparing this study with previous ones, they agree that the surface functionalization of carbon is a determining factor in its properties and applications. The differences observed in the surface chemistry of grapevine-derived biochars are explained by the production techniques (HTC vs. PYR), the nature of the raw material (such as grapevine), and possible post-treatment chemical modifications (such as Fe doping). These variations confer distinct properties to the materials that could be relevant for different potential uses, ranging from exploratory soil studies to more advanced technological applications, the suitability of which will need to be evaluated through further specific research

4. Conclusions

This study provides a comparative physicochemical characterization of hydrochars (HCs) and pyrochars (PIRs) derived from vineyard stumps, including iron-enriched variants, and a preliminary assessment of plant tolerance based on chlorophyll index measurements. The results confirm that hydrothermal carbonization and pyrolysis generate carbonaceous materials with clearly distinct chemical structures. HCs retained a higher abundance of oxygen-containing functional groups, while PIRs exhibited a more condensed and aromatic carbon structure, consistent with their respective formation mechanisms. Iron incorporation was successfully achieved in both materials, as evidenced by FTIR analysis, introducing Fe–O bonds without eliminating surface oxygenated functionalities.

The vegetation experiment conducted in pots, designed as a preliminary screening to rule out obvious phytotoxic effects, showed that chlorophyll index values remained within the physiological ranges typically reported for grapevines across all treatments and sampling dates, and that the observed temporal variability was mainly due to seasonal and environmental factors. Although consistent trends related to the treatments cannot be established, these preliminary results highlight the potential of vineyard residues to be transformed into sustainable soil amendments that contribute to climate resilience, circular economy strategies, and improved soil quality.

In summary, the findings support the suitability of vineyard stumps as a feedstock for producing carbonaceous materials with differentiated properties through HTC and pyrolysis. The work establishes a solid material-level foundation and demonstrates initial plant compatibility under controlled conditions, which is a prerequisite for subsequent agronomic evaluation. Future research should involve replicated field trials with appropriate statistical designs, direct measurements of soil properties (e.g., pH, CEC, nutrient availability, moisture), and integrated plant physiological assessments to elucidate amendment–soil–plant interactions and medium- to long-term material stability. Additionally, iron-doped materials warrant further investigation for applications beyond agronomy, including nutrient management and environmental remediation.

Author Contributions: Conceptualization, S.R. and B.L.; methodology, J.M.D.-R., S.R. and T.S.; formal analysis, J.M.D.-R. and S.R.; investigation, J.M.D.-R., S.R., B.L. and T.S.; writing—original draft preparation, J.M.D.-R.; writing—review and editing, J.M.D.-R., B.L. and T.S.; visualization, J.M.D.-R., S.R. and B.L.; supervision, S.R.; project administration, S.R. and B.L.; funding acquisition, J.M.D.-R. and S.R. All authors have read and agreed to the published version of the manuscript.

Funding: This research was funded by JUNTA DE EXTREMADURA PhD Fellowship, grant number PD23085-5939110060220230002690, co-financed through the ESF+ fund. AGENCIA ESPAÑOLA DE INVESTIGACIÓN contributed with the financial help through project PID2024-157952OB-I00, funded under the ORGÁNICA program 18. GA.54.FA.

https://doi.org/10.3390/environments13020083

Data Availability Statement: The original contributions presented in the study are included in the article, further inquiries can be directed to the corresponding author.

Acknowledgments: The authors are thankful to SAIUEX (Servicios de Apoyo a la Investigación de la Universidad de Extremadura) have also contributed to this study by their support on analytical tests. The authors have reviewed and edited the output and take full responsibility for the content of this publication.

Conflicts of Interest: The authors declare no conflicts of interest.

Abbreviations

The following abbreviations are used in this manuscript:

VSs	Vineyard Stumps
PIR	Pyrochar
HC	Hydrochar
HTC	Hydrothermal carbonization
PYR	Pyrolysis
PIR_Fe	Pyrochars used in soil
HC_Fe	Hydrochar used in soil
HHV	Higher heating value
LHV	Lower heating value

References

1. Winter, S.; Bauer, T.; Strauss, P.; Kratschmer, S.; Paredes, D.; Popescu, D.; Landa, B.; Guzmán, G.; Gómez, J.A.; Guernion, M.; et al. Effects of vegetation management intensity on biodiversity and ecosystem services in vineyards: A meta-analysis. *J. Appl. Ecol.* **2018**, *55*, 2484–2495. [CrossRef]
2. Paiola, A.; Assandri, G.; Brambilla, M.; Zottini, M.; Pedrini, P.; Nascimbene, J. Exploring the potential of vineyards for biodiversity conservation and delivery of biodiversity-mediated ecosystem services: A global-scale systematic review. *Sci. Total Environ.* **2020**, *706*, 135839. [CrossRef]
3. Lazzari, M.; Piccarreta, M. Soil Erosion vs. Vineyard Productivity: The Case of the Aglianico del Vulture DOC and DOCG Areas (Southern Italy). *Sustainability* **2023**, *15*, 15700. [CrossRef]
4. Ferreira, C.S.S.; Seifollahi-Aghmiuni, S.; Destouni, G.; Ghajarnia, N.; Kalantari, Z. Soil degradation in the European Mediterranean region: Processes, status and consequences. *Sci. Total Environ.* **2022**, *805*, 150106. [CrossRef]
5. Cano, B.L.; Sánchez, M.A.; Rasero, J.M.D.; Suero, S.R.; Delgado, S.N. *Introduction to Hydrocarbonization: Principles and Applications*, 1st ed.; Springer: Cham, Switzerland, 2024; pp. 1–14.
6. Chen, W.H.; Biswas, P.P.; Zhang, C.; Kwon, E.E.; Chang, J.S. Achieving carbon credits through biomass torrefaction and hydrothermal carbonization: A review. *Renew. Sustain. Energy Rev.* **2025**, *208*, 115056. [CrossRef]
7. Wang, X.; Duo, J.; Jin, Z.; Yang, F.; Lai, T.; Collins, E. Effects of Hydrothermal Carbonization Conditions on the Characteristics of Hydrochar and Its Application as a Soil Amendment: A Review. *Agronomy* **2025**, *15*, 327. [CrossRef]
8. Bargmann, I.; Martens, R.; Rillig, M.C.; Kruse, A.; Kücke, M. Hydrochar amendment promotes microbial immobilization of mineral nitrogen. *J. Plant Nutr. Soil Sci.* **2014**, *177*, 59–67. [CrossRef]
9. Bargmann, I.; Rillig, M.C.; Kruse, A.; Greef, J.M.; Kücke, M. Effects of hydrochar application on the dynamics of soluble nitrogen in soils and on plant availability. *J. Plant Nutr. Soil Sci.* **2014**, *177*, 48–58. [CrossRef]
10. Xia, Y.; Luo, H.; Li, D.; Chen, Z.; Yang, S.; Liu, Z.; Yang, T.; Gai, C. Efficient immobilization of toxic heavy metals in multi-contaminated agricultural soils by amino-functionalized hydrochar: Performance, plant responses and immobilization mechanisms. *Environ. Pollut.* **2020**, *261*, 114217. [CrossRef] [PubMed]
11. Islam, M.A.; Limon, M.S.H.; Romić, M.; Islam, M.A. Hydrochar-based soil amendments for agriculture: A review of recent progress. *Arab. J. Geosci.* **2021**, *14*, 102. [CrossRef]
12. Sun, K.; Han, L.; Yang, Y.; Xia, X.; Yang, Z.; Wu, F.; Li, F.; Feng, Y.; Xing, B. Application of hydrochar altered soil microbial community composition and the molecular structure of native soil organic carbon in a paddy soil. *Environ. Sci. Technol.* **2020**, *54*, 2715–2725. [CrossRef]
13. Battipaglia, G.; Niccoli, F.; Kabala, J.P.; Marzaioli, R.; Di Santo, T.; Strumia, S.; Castaldi, S.; Rutigliano, F.A. Hydrochar application improves growth and intrinsic water use efficiency of *Populus alba*, especially during hot season. *Forests* **2023**, *14*, 658. [CrossRef]
14. Basso, D.; Weiss-Hortala, E.; Patuzzi, F.; Baratieri, M.; Fiori, L. In deep analysis on the behavior of grape marc constituents during hydrothermal carbonization. *Energies* **2018**, *11*, 1379. [CrossRef]

 https://doi.org/10.3390/environments13020083

15. Pagés-Díaz, J.; Alvarado, A.O.C.; Montalvo, S.; Diaz-Robles, L.; Curio, C.H. Anaerobic bio-methane potential of the liquors from hydrothermal carbonization of different lignocellulose biomasses. *Renew. Energy* **2020**, *157*, 182–189. [CrossRef]
16. Zhang, B.; Heidari, M.; Regmi, B.; Salaudeen, S.; Arku, P.; Thimmannagari, M.; Dutta, A. Hydrothermal carbonization of fruit wastes: A promising technique for generating hydrochar. *Energies* **2018**, *11*, 2022. [CrossRef]
17. Hoffmann, V.; Jung, D.; Zimmermann, J.; Rodriguez Correa, C.; Elleuch, A.; Halouani, K.; Kruse, A. Conductive carbon materials from the hydrothermal carbonization of vineyard residues for the application in electrochemical double-layer capacitors (EDLCs) and direct carbon fuel cells (DCFCs). *Materials* **2019**, *12*, 1703. [CrossRef] [PubMed]
18. Lorero, I.; Vizcaíno, A.J.; Alguacil, F.J.; López, F.A. Activated carbon from winemaking waste: Thermoeconomic analysis for large-scale production. *Energies* **2020**, *13*, 6462. [CrossRef]
19. Baronti, S.; Magno, R.; Maienza, A.; Montagnoli, A.; Ungaro, F.; Vaccari, F. Long term effect of biochar on soil plant water relation and fine roots: Results after 10 years of vineyard experiment. *Sci. Total Environ.* **2022**, *838*, 158225. [CrossRef]
20. Giagnoni, L.; Maienza, A.; Baronti, S.; Vaccari, F.; Genesio, L.; Taiti, C.; Martellini, T.; Scodellini, R.; Cincinelli, A.; Costa, C.; et al. Long-term soil biological fertility, volatile organic compounds and chemical properties in a vineyard soil after biochar amendment. *Geoderma* **2019**, *337*, 122–131. [CrossRef]
21. Idbella, M.; Baronti, S.; Giagnoni, L.; Renella, G.; Becagli, M.; Cardelli, R.; Maienza, A.; Vaccari, F.P.; Bonanomi, G. Long-term effects of biochar on soil chemistry, biochemistry, and microbiota: Results from a 10-year field vineyard experiment. *Appl. Soil Ecol.* **2024**, *195*, 105217. [CrossRef]
22. Maienza, A.; Baronti, S.; Cincinelli, A.; Martellini, T.; Grisolia, A.; Miglietta, F.; Renella, G.; Stazi, S.; Vaccari, F.; Genesio, L. Biochar improves the fertility of a Mediterranean vineyard without toxic impact on the microbial community. *Agron. Sustain. Dev.* **2017**, *37*, 1–10. [CrossRef]
23. García-Jaramillo, M.; Meyer, K.; Phillips, C.; Acosta-Martínez, V.; Osborne, J.; Levin, A.; Trippe, K. Biochar addition to vineyard soils: Effects on soil functions, grape yield and wine quality. *Biochar* **2021**, *3*, 565–577. [CrossRef]
24. Aguirre, J.; González-Egido, S.; González-Lucas, M.; González-Pernas, F. Medium-term effects and economic analysis of biochar application in three Mediterranean crops. *Energies* **2023**, *16*, 4131. [CrossRef]
25. Schmidt, H.; Kammann, C.; Niggli, C.; Evangelou, M.; Mackie, K.; Abiven, S. Biochar and biochar-compost as soil amendments to a vineyard soil: Influences on plant growth, nutrient uptake, plant health and grape quality. *Agric. Ecosyst. Environ.* **2014**, *191*, 117–123. [CrossRef]
26. Uras, Ü.; Carrier, M.; Hardie, A.G.; Knoetze, J.H. Physico-chemical characterization of biochars from vacuum pyrolysis of South African agricultural wastes for application as soil amendments. *J. Anal. Appl. Pyrolysis* **2012**, *98*, 207–213. [CrossRef]
27. Marshall, J.; Muhlack, R.; Morton, B.J.; Dunnigan, L.; Chittleborough, D.; Kwong, C.W. Pyrolysis temperature effects on biochar–water interactions and application for improved water holding capacity in vineyard soils. *Soil Syst.* **2019**, *3*, 27. [CrossRef]
28. Videgain, M.; Marco, P.; Martí, C.; García-Ramos, F.J.; Manyà, J.J.; Jaizme-Vega, M.C. Evaluación de los efectos de la aplicación de biochar como enmienda orgánica en suelos agrícolas. In Proceedings of the Congreso Ibérico Agroingeniería 2019, Zaragoza, Spain, 3–6 September 2019; pp. 1–14. [CrossRef]
29. *ISO 3310-1*; Test Sieves—Technical Requirements and Testing—Part 1: Test Sieves of Metal Wire Cloth. ISO: Geneva, Switzerland, 2016.
30. Olivares-Marin, M.; Román, S.; Ledesma, B.; Álvarez, A. Optimizing Al and Fe Load during HTC of Water Hyacinth: Improvement of Induced HC Physicochemical Properties. *Catalysts* **2023**, *13*, 506. [CrossRef]
31. Gai, C.; Zhang, F.; Lang, Q.; Liu, T.; Peng, N.; Liu, Z. Facile One-Pot Synthesis of Iron Nanoparticles Immobilized into the Porous Hydrochar for Catalytic Decomposition of Phenol. *Appl. Catal. B Environ.* **2017**, *204*, 566–576. [CrossRef]
32. Uddling, J.; Gelang-Alfredsson, J.; Piikki, K.; Pleijel, H. Evaluating the relationship between leaf chlorophyll concentration and SPAD-502 chlorophyll meter readings. *Photosynth. Res.* **2007**, *91*, 37–46. [CrossRef]
33. Li, Q.; Zhang, Y.; Song, Y.; Yang, H.; Yang, L.; Bai, L.; Wei, D.; Wang, W.; Liang, Y.; Chen, H. Apple residues derived porous carbon nanosheets synthesized with $FeCl_3$ assisted hydrothermal carbonization for supercapacitors with high rate performance. *Carbon Lett.* **2023**, *33*, 549–560. [CrossRef]
34. MacDermid-Watts, K.; Adewakun, E.; Norouzi, O.; Deb, A.T.; Pradhan, R.; Dutta, A. Effects of $FeCl_3$ Catalytic Hydrothermal Carbonization on Chemical Activation of Corn Wet Distillers' Fiber. *ACS Omega* **2021**, *6*, 14875–14886. [CrossRef]
35. Liu, B.; Xing, Z.; Xue, Y.; Zhang, J.; Zhai, J. Effect of Pyrolysis Temperature on the Carbon Sequestration Capacity of Spent Mushroom Substrate Biochar in the Presence of Mineral Iron. *Molecules* **2024**, *29*, 5712. [CrossRef]
36. Ekpo, U.; Ross, A.B.; Camargo-Valero, M.A.; Fletcher, L.A. Influence of pH on hydrothermal treatment of swine manure: Impact on extraction of nitrogen and phosphorus in process water. *Bioresour. Technol.* **2016**, *214*, 637–644. [CrossRef]
37. Kruse, A.; Krupka, A.; Schwarzkopf, V.; Gamard, C.; Henningsen, T. Influence of Proteins on the Hydrothermal Gasification and Liquefaction of Biomass. 1. Comparison of Different Feedstocks. *Ind. Eng. Chem. Res.* **2005**, *44*, 3013–3020. [CrossRef]
38. Cerovic, Z.G.; Masdoumier, G.; Ghozlen, N.B.; Latouche, G. A new optical leaf-clip meter for simultaneous non-destructive assessment of leaf chlorophyll and epidermal flavonoids. *Physiol. Plant.* **2012**, *146*, 251–260. [CrossRef]

 https://doi.org/10.3390/environments13020083

39. Canton, M.; Borghezan, M.; Silva, T.C.; Welter, J.F.; Villar, L.; Rosa, D.; Silva, A.L.; Pescador, R. Chlorophyll evaluation on leaves of 'Sauvignon Blanc' during vegetative growth in São Joaquim, Santa Catarina, Brazil. *Acta Hortic.* **2017**, *1188*, 15–20. [CrossRef]
40. Petrović, J.; Ercegović, M.; Simić, M.; Koprivica, M.; Dimitrijević, J.; Jovanović, A.; Janković Pantić, J. Hydrothermal Carbonization of Waste Biomass: A Review of Hydrochar Preparation and Environmental Application. *Processes* **2024**, *12*, 207. [CrossRef]
41. Lin, Q.; Yang, X.; Ma, W.; Zahoor, A.; Jin, F.; Chen, X.; Tai, L.; de Caprariis, B.; De Filippis, P.; Damizia, M. Hydrothermal carbonization of fish sludge: Effect of FeCl3 on the hydrochar properties and phosphorus and heavy metal transformation. *Environ. Technol. Innov.* **2025**, *40*, 104355. [CrossRef]
42. Zhang, X.; Wang, Y.; Li, J.; Chen, H. FTIR Analysis of Biomass and Its Application in Soil Studies. *IOP Conf. Ser. Earth Environ. Sci.* **2019**, *239*, 012035. [CrossRef]
43. Zhang, X.; Li, Y.; Wang, L.; Chen, H. Structural Characterization of Biomass Components Using FTIR Spectroscopy. *Int. J. Mol. Sci.* **2023**, *24*, 8936. [CrossRef]
44. Kocsis, T.; Buzás, N.; Farkas, I.; Kocsis, Z. Investigation of Biomass-Derived Materials Using FTIR and Thermal Analysis. *J. Agric. Environ. Sci.* **2025**, *8*, 870. [CrossRef]

 https://doi.org/10.3390/environments13020083

environments

MDPI

Article

Vermicompost and Leachate from Sewage Sludge and Vineyard Pruning Residues: An Exploratory Assessment of Cucumber Germination and Early Seedling Growth

Elisabete Nascimento-Gonçalves [1,2], Tiago Azevedo [1], Henda Lopes [1], Catarina Medeiros [1], Virgílio Falco [1,2,3], Ana Cláudia Coelho [4], João R. Sousa [1,2], Ana M. Coimbra [1,2], Marta Roboredo [1,2], Paula A. Oliveira [1,2] and Maria C. Morais [1,2,*]

[1] Centre for the Research and Technology of Agro-Environmental and Biological Sciences (CITAB), University of Trás-os-Montes and Alto Douro (UTAD), 5000-801 Vila Real, Portugal; egoncalves@utad.pt (E.N.-G.); tiagoaazevedo99@gmail.com (T.A.); hendalopes@utad.pt (H.L.); cmedeiros@utad.pt (C.M.); vfalco@utad.pt (V.F.); jricardo@utad.pt (J.R.S.); acoimbra@utad.pt (A.M.C.); roboredo@utad.pt (M.R.); pamo@utad.pt (P.A.O.)

[2] Institute for Innovation, Capacity Building and Sustainability of Agri-Food Production (Inov4Agro), University of Trás-os-Montes and Alto Douro (UTAD), 5000-801 Vila Real, Portugal

[3] LAQV-REQUIMTE—Laboratory for Green Chemistry (LAQV) of the Network of Chemistry and Technology (REQUIMTE), University of Porto, 4050-313 Porto, Portugal

[4] Animal and Veterinary Research Centre (CECAV), University of Trás-os-Montes and Alto Douro (UTAD), 5000-801 Vila Real, Portugal; accoelho@utad.pt

* Correspondence: cmorais@utad.pt

Abstract

The sustainable management of organic residues remains a major challenge in agriculture. Vermicomposting offers an environmentally friendly strategy to convert organic waste into nutrient-rich, biologically stable biofertilizers. This exploratory study evaluated the effects of vermicompost and its leachate, produced from sewage sludge and vineyard pruning residues, on cucumber (*Cucumis sativus* L.) germination and 25-day early seedling growth. Treatments included a control (peat and perlite, CNT), two vermicompost doses, 20 g kg^{-1} and 40 g kg^{-1} (VC_D1 and VC_D2, respectively) and a 5% (v/v) vermicompost leachate (VC_L) applied as the sole irrigation source. Foliar nutrient contents and physicochemical properties of the substrate and leachate were determined. Germination was not significantly affected ($p > 0.05$), but VC_D1 promoted slightly faster and more uniform seed emergence. Growth responses were dose dependent, with VC_D1 significantly enhancing shoot biomass (approximately 15% than the CNT and VC_D2) and providing a balanced foliar nutrient profile, whereas VC_D2 significantly reduced growth, promoted excessive foliar K and P, and lower Ca, Fe, and Mn contents. VC_L enhanced foliar N accumulation but did not significantly ($p > 0.05$) increase biomass. Both vermicompost and its leachate were pathogen-free, with metal concentrations below regulatory limits. Overall, these findings suggest that, under the tested conditions, vermicomposting these residues can generate potentially safe amendments for cucumber seedling growth, though dose optimization is essential. This exploratory approach supports residue valorization and contributes to circular economy principles and sustainable agriculture goals.

Keywords: circular economy; *Cucumis sativus*; organic amendments; residue valorization; sustainable agriculture; vermicompost leachate

1. Introduction

Wine production generates large amounts of organic residues during vineyard management, as well as during winemaking operations. Vineyard residues include pruning wood, leaves and shoots, while from the winemaking processes, grape pomace, stems, lees, and wastewater are the main residues produced [1,2]. These are typically rich in organic matter, mineral nutrients, and polyphenols compounds that can be recovered and transformed into value-added products [3–5]. However, improper management of these residues poses significant environmental risks, such as air pollution from greenhouse gas emissions during the open-field burning of pruning waste and the contamination of soil and water by phenolic-rich effluents [6,7].

Sewage sludge is a major residue resulting from urban and industrial wastewater treatment processes and is produced in large quantities due to the increasing implementation of wastewater treatment plants worldwide [8,9]. These sludges are rich in organic matter, macronutrients and micronutrients, making them potentially valuable to be used as soil amendments or organic fertilizers [10,11]. However, their direct application to soils may be limited by the presence of contaminants such as metals, persistent organic pollutants, microplastics, personal care products, and pathogenic microorganisms [12,13]. Therefore, sustainable strategies for sewage sludge valorization, which include composting, anaerobic digestion, vermicomposting, or advanced stabilization processes, are being developed to ensure environmental and sanitary safety, widely promoting this residue integration into circular economy frameworks [14–17].

The increasing demand for sustainable agriculture has intensified the development of effective methods to recycle organic residues, while minimizing their environmental footprint [18]. Among biological recycling processes, vermicomposting has emerged as a promising, low-cost, and environmentally friendly approach [19]. Through the synergistic action of earthworms and microorganisms, organic matter is transformed into a stable, humus-like product known as vermicompost [20]. This product is characterized by its high nutrient content, homogeneity, porosity, water-holding capacity, stability, and low carbon-to-nitrogen (C/N) ratio [21]. Earthworm activity enhances both microbial proliferation and enzymatic activity by fragmenting, ingesting and digesting fresh organic material, while simultaneously reducing the bioavailability of toxic metals such as cadmium, zinc, copper, and lead, thereby limiting their bioaccumulation [19]. In addition to the solid product, the liquid fraction obtained from vermicompost, referred to as leachate [22], contains soluble nutrients, beneficial microorganisms, and plant growth regulators that can stimulate plant development and nutrient uptake [23–25].

Although some studies have explored the valorization of viticulture residues and sewage sludge separately through composting and vermicomposting [26–32], the combination of vine pruning residues with sewage sludge for composting or vermicomposting remains scarcely explored. The mixture of these materials offers complementary benefits in terms of nutritional balance, physical structure, and the presence of bioactive compounds [2,12]. Therefore, exploring such synergistic effects could provide an efficient and sustainable pathway for the joint management of agro-industrial and urban organic residues. Beyond local and regional contexts, the management and valorization of these residues represent a global challenge with increasing relevance for sustainable agriculture and waste management.

Cucumber (*Cucumis sativus* L.) is one of the most widely cultivated vegetables worldwide [33]. This vegetable is valued for its higher water content, dietary fiber, B-complex vitamins, and vitamin C, making it suitable for low-calorie diets and functional food applications [34]. Moreover, its bioactive compounds, particularly antioxidants, enhance its nutritional value and potential applications in the cosmetic and pharmaceutical in-

 https://doi.org/10.3390/environments13010059

dustries [34]. Cucumber is established as a bioindicator species due to its rapid and uniform germination, combined with high sensitivity to phytotoxic compounds even at low concentrations [34].

In this context, this study aimed to evaluate the effects of vermicompost and its leachate, derived from sewage sludge and vineyard pruning residues, on cucumber germination and early seedling growth. Conducted as an exploratory assessment, the study focused on their agronomic and environmental potential as sustainable organic amendments, within a conceptual circular economy framework centered on nutrient recycling and waste valorization.

2. Materials and Methods

2.1. Vermicompost Production

Vine prunings were collected from local vineyards in Vila Real, Portugal, and shredded into 1–2 cm pieces. Sewage sludge was obtained from the Urban Wastewater Treatment Plant in Sanfins do Douro, Portugal. The two residues were mixed in a 1:1 (w/w) ratio, corresponding to 1.5 kg of each residue, and placed in 5 L plastic container. *Eisenia fetida* earthworms were introduced in sufficient numbers to ensure effective decomposition, with a total biomass of 75 g, and were provided with a net (1 cm) limited separated comfort zone. The vermicomposting process was conducted as a single experimental batch maintained for nine months under controlled temperature and moisture conditions in darkness. Reproducibility across independent batches was not evaluated. Leachate was collected through a perforated bottom connected to a drainage outlet, allowing the liquid fraction to percolate into a separate collection vessel placed below each container. Collection was performed every two weeks. At the end of the vermicomposting process, the vermicompost material was pooled and thoroughly homogenized, and three subsamples were collected for further analysis. Similarly, the leachate was collected as a pooled composite sample, obtained by homogenizing the total volume produced at the end of the process. All samples were sent to the Soil and Plant Analysis Laboratory of UTAD for chemical analyses (Table 1) and the analytical procedures followed standard methods as described in [35].

The vermicompost exhibited a C/N ratio of 21, slightly alkaline pH, 7.6, and high organic matter content, 849 g kg^{-1} DW, all indicative of a well-matured and stable product. In terms of nutrients, Ca, N and K were the most abundant macroelements, while Fe and Zn showed the highest concentrations among the micronutrients. Metal contents in the vermicompost were below the legal thresholds (Portaria n° 185/21 July 2022). The leachate presented a similar pH, 7.7, but slightly higher electrical conductivity than the solid vermicompost. Its nutrient concentrations were lower, with K and Ca as the most abundant macronutrients and Fe as the most abundant micronutrient. Metal contents in the leachate were also below regulatory limits (Portaria n° 185/21 July 2022). The total polyphenol content determined using the Folin–Ciocalteu reagent method [36], corresponded to 2.91 ± 0.22 mg gallic acid equivalents (GAE) g^{-1} DW for vermicompost and 67.27 ± 0.42 mg GAE L^{-1} for leachate. The phytotoxicity of vermicompost and leachate was evaluated through germination and radicle elongation tests using garden cress (*Lepidium sativum* L.) seeds, and phytotoxicity responses were evaluated as the germination index (GI) [37]. GI values confirmed the absence of phytotoxicity for both fractions, with the vermicompost exhibiting a very high GI ($198.29 \pm 25.96\%$) and the leachate also showing a value well above the 80% threshold generally considered safe for plant growth ($139.33 \pm 60.44\%$). Basal respiration, indicative of microbial activity, was estimated from C-CO_2 emissions, according to Anderson [38] and was 3.33 ± 1.57 mg CO_2 g^{-1} OM d^{-1} in the vermicompost, reflecting a moderate level of biological activity consistent with a

https://doi.org/10.3390/environments13010059

stabilized organic amendment (Table 1). The vermicompost and leachate were analyzed for total and fecal coliforms (*Escherichia coli*) and *Salmonella* spp. [39] and both microorganisms were found absent in the vermicompost and leachate.

Table 1. Chemical properties, total polyphenols, germination index (GI) of garden cress, and basal respiration rates of vermicompost and leachate derived from sewage sludge and vineyard pruning residues. Data are presented as mean ± SEM (n = 3).

	Vermicompost	Leachate
Chemical properties		
C/N ratio	20.56 ± 0.28	-
EC (dS m^{-1}) #	0.52 ± 0.01	0.65
OM (g kg^{-1} DW)	849 ± 11	-
pH (H_2O) #	7.6 ± 0.1	7.70
$N\text{-}NH_4^+/N\text{-}NO_3^-$ ratio	3.13 ± 1.24	-
Macronutrients	g kg^{-1} DW	g L^{-1}
N	23.93 ± 0.56	0.07
P	5.53± 0.35	0.18
K	21.61 ± 1.43	1.79
Ca	26.39 ± 2.76	1.07
Mg	4.60 ± 0.25	0.04
S	3.52 ± 0.21	0.02
Micronutrients	mg kg^{-1} DW	mg L^{-1}
B	26.94 ± 1.69	0.59
Fe	2633.09 ± 162.19	1.79
Mn	236.04 ± 15.06	0.38
Metals	mg kg^{-1} DW	mg L^{-1}
Cd	0.281 ± 0.004	0.007
Cr	9.218 ± 0.345	0.342
Cu	110.020 ± 6.094	0.243
Pb	7.740 ± 0.359	0.183
Hg	0.074 ± 0.010	-
Ni	6.490 ± 0.218	0.158
Zn	329.339 ± 19.817	0.587
Total polyphenols	mg GAE g^{-1} DW	mg GAE L^{-1}
	2.91 ± 0.22	67.27 ± 0.40
GI (%)	198.29 ± 25.96	139.33 ± 60.44
Basal respiration (mg CO_2 g^{-1} OM d^{-1})	3.33 ± 1.57	-

C/N: carbon/nitrogen ratio; DW: dry weight; EC: electrical conductivity; GAE: gallic acid equivalents; OM: organic matter; # (1:5).

2.2. Experimental Design

The experiment was conducted outdoors at the University of Trás-os-Montes and Alto Douro, Vila Real, Portugal (41°30′2″ N, 7°38′55″ W; 770 m altitude) during summer of 2024 under natural environmental conditions. The region has a Mediterranean climate, with average minimum and maximum temperatures of 15.2 °C and 25.0 °C, respectively, during the experimental period (27 July–20 August), and no rainfall was recorded during this period. Cucumber (*Cucumis sativus* L. cv. 'Market', Eurosementes, Riachos, Portugal) seeds were sown in polystyrene trays with 30 × 30 × 65 mm cells at a depth of approximately 1–2 cm below the substrate surface, and four treatments were established. All treatments used the same base substrate, composed of a commercial peat-based mix (BIO Siro, Leal &

 https://doi.org/10.3390/environments13010059

Soares, Mira, Portugal) with 20% perlite to enhance aeration. The chemical properties of this base substrate are presented in Table 2. No additional fertilizers were applied during the experimental period. The control (CNT) consisted of only the commercial peat-based substrate amended with 20% perlite. Two vermicompost treatments included incorporation of vermicompost (VC) into the substrate at rates equivalent to 20 g kg^{-1} (VC_D1) and 40 g kg^{-1} (VC_D2). These application rates were selected to comply with Portuguese legislation (Portaria n.° 185/21 July 2022), considering metal concentrations and substrate pH, and were intended to represent low and moderate amendments levels for assessing dose-dependent effects. A fourth treatment (VC_L) consisted of CNT substrate with seedling irrigated exclusively with vermicompost leachate diluted to 5% (*v*/*v*), following recommended dilution rates for commercial vermicompost leachates. Given the lack of established fertilization guidelines specifically for vermicompost leachate in cucumber, this concentration was selected as a conservative, safety-oriented rate to test its agronomic potential without toxicity issues. For the solid treatments (VC_D1 and VC_D2), vermicompost was thoroughly mixed with the substrate prior to sowing. For VC_L, the diluted leachate replaced tap water throughout the experiment. Plants were irrigated daily with tap water in CNT, VC_D1, and VC_D2, or with the leachate diluted to 5% (*v*/*v*) in VC_L, to maintain adequate moisture level. Each treatment consisted of three replicates with 12 plants per replicate. Seedlings were grown for 25 days. At the end of the experiment, the bulk substrate from all experimental units within each treatment was thoroughly mixed to obtain a homogenous material, from which one composite sample per treatment was collected for chemical analysis.

Table 2. Chemical properties of the base substrate used in all treatments.

Chemical properties	
C/N ratio	55.82
EC (dS m^{-1}) #	0.52
OM (g kg^{-1} DW)	535.80
pH (H_2O) #	5.78
$N\text{-}NH_4^+/N\text{-}NO_3^-$ ratio	0.17
Macronutrients (g element kg^{-1} DW)	
N	5.568
P	0.950
K	2.859
Ca	11.263
Mg	1.131
S	1.109
Micronutrients (mg element kg^{-1} DW)	
B	7.154
Fe	2879.596
Mn	116.671
Metals (mg kg^{-1} DW)	
Cd	0.087
Cr	7.616
Cu	9.734
Pb	4.911
Hg	0.01
Ni	3.596
Zn	29.409

C/N: carbon/nitrogen ratio; DW: dry weight; EC: electrical conductivity; OM: organic matter; # (1:5).

https://doi.org/10.3390/environments13010059

2.3. Germination, Growth and Physiological Measurements

Germinated seeds were counted daily from day 2 to day 18 after sowing. Seeds were considered germinated when seedlings emerged above the substrate surface. Several germination-related parameters, including the germination percentage (G, %), mean germination time (MGT, day), coefficient of velocity of germination (CVG, %) and synchrony of germination (Z, without units) were calculated [40,41]. Morphological and physiological parameters were recorded at 10, 15 and 25 days after sowing. The shoot length (cm) was measured from the substrate surface to the tip of the longest leaf using a graduated ruler. The number of true leaves were counted manually for each seedling. The chlorophyll content index of the leaves was measured by a Minolta SPAD-502 chlorophyll meter (Minolta corporation, Ltd., Osaka, Japan). At 25 days, cucumber seedlings were carefully removed from the trays, and the following growth parameters were determined: stem diameter (mm) was measured at the base of the stem using a digital caliper (Fine Science Tools #30087-00, Heidelberg, Germany); root length (cm) was measured from the collar region to the tip of the longest root; and biomass (g) was determined using a scale (Sartorius Entris II, BCE, Lemgo, Germany), after oven dried at 60 °C to a constant weight. Additionally, for each treatment, shoot composite samples (n = 3/treatment) were analyzed for macronutrient and micronutrient content using standard methods at Soil and Plant analysis Laboratory of UTAD [35].

2.4. Statistical Analysis

Statistical analyses were performed using GraphPad Prism Version 10.3.1 (GraphPad Software, San Diego, CA, USA). Growth results over time were subjected to a one-way repeated measures ANOVA, with day of observation as a within-factor and treatments as between-group factor. Germination and morphological parameters were analyzed by one-way analysis of variance (ANOVA). Mean differences were separated using Tukey's HSD test at a 5% level of significance. All data sets satisfied the ANOVA assumptions of homogeneity of variance (Levene's test) and normality of errors (Shapiro–Wilk test). Data are presented as mean ± standard error of the mean (SEM).

3. Results

3.1. Chemical Properties of Treatment Mixtures at the End of the Experiment

At the end of the experiment, the chemical properties of the substrate mixtures provide an overview of the substrate properties and nutrient availability under the tested conditions (Table 3). Vermicompost addition reduced the C/N ratio and increased electrical conductivity (EC). Organic matter (OM) content was slightly higher in vermicompost mixtures and pH values remained close to neutral. Macronutrients content was consistently higher in vermicompost treatments, particularly in VC_D2. Micronutrients also increased with the incorporation of vermicompost and leachate, especially Cu and Zn in VC_D1. Metal concentrations remained low and below accepted safety thresholds in all treatments.

Table 3. Chemical characteristics of the treatment's mixtures (CNT—control, VC_D1—vermicompost at 20 g Kg^{-1}, VC_D2—vermicompost at 40 g Kg^{-1}, VC_L—vermicompost leachate) at the end of the experiment.

	CNT	VC_D1	VC_D2	VC_L
Chemical properties				
C/N ratio	43.32	35.81	29.58	40.27
EC (dS m^{-1}) #	0.13	0.18	0.28	0.08
OM (g kg^{-1} DW)	475.05	492.81	497.03	448.16

https://doi.org/10.3390/environments13010059

Table 3. *Cont.*

	CNT	VC_D1	VC_D2	VC_L
pH (H_2O) #	6.51	6.56	6.62	6.66
$N\text{-}NH_4^+/N\text{-}NO_3^-$ ratio	3.56	4.92	1.83	5.41
Macronutrients (g kg^{-1} DW)				
N	6.36	7.98	9.75	6.45
P	0.67	0.99	1.22	0.62
K	1.61	2.81	3.44	1.50
Ca	12.05	12.45	13.92	10.92
Mg	1.20	1.51	1.89	1.04
S	0.77	1.05	1.27	0.74
Micronutrients (mg kg^{-1} DW)				
B	6.79	8.99	11.45	5.98
Fe	3625.61	3833.66	4356.79	2912.74
Mn	134.73	146.35	161.90	119.73
Metals (mg kg^{-1} DW)				
Cd	<0.01	<0.01	<0.01	<0.01
Cr	0.35	0.72	0.32	0.33
Cu	11.14	25.67	31.70	10.24
Pb	0.07	0.08	0.08	0.06
Hg	0.01	0.01	0.01	0.01
Ni	0.12	0.22	0.11	0.10
Zn	37.65	104.49	110.96	33.19

C/N: carbon/nitrogen ratio; DW: dry weight; EC: electrical conductivity; OM: organic matter; # (1:5).

3.2. Effect of Vermicompost and Leachate on Cucumber Germination

Overall, all treatments supported successfully cucumber germination (Figure 1). In the early stages (days 2–5), VC_D1 reached 83–94% germination, slightly higher than the CNT (78–94%) and VC_D2 (78–92%). VC_L exhibited slower initial germination, reaching 72–81% in the first days. At eight days after sowing, VC_D1 reached 100% germination, whereas VC_D2 achieved full germination (100%) only at 16 days, and the CNT at 14 days.

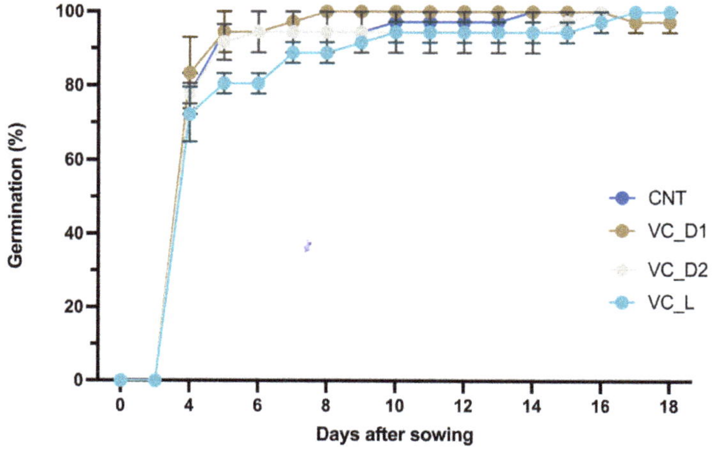

Figure 1. Cumulative germination percentage of cucumber seeds under different treatments (CNT—control, VC_D1—vermicompost at 20 g Kg^{-1}, VC_D2—vermicompost at 40 g Kg^{-1}, VC_L—vermicompost leachate) during the first 18 days after sowing. Data are presented as mean ± SEM (n = 3).

https://doi.org/10.3390/environments13010059

At the end of the germination period, calculated germination parameters such as germination percentage (G), mean germination time (MGT), coefficient of velocity of germination (CVG), and synchrony index (Z) (Figure 2) did not show significant differences among treatments ($p > 0.05$). Although no statistically significant differences were detected ($p > 0.05$), some trends in germination parameters were apparent. VC_D1 tended to promote slightly faster germination (MGT = 4.31 ± 0.23 days) and higher uniformity (CVG = 23.35 ± 1.19%; Z = 0.71 ± 0.16) compared to VC_D2 and VC_L.

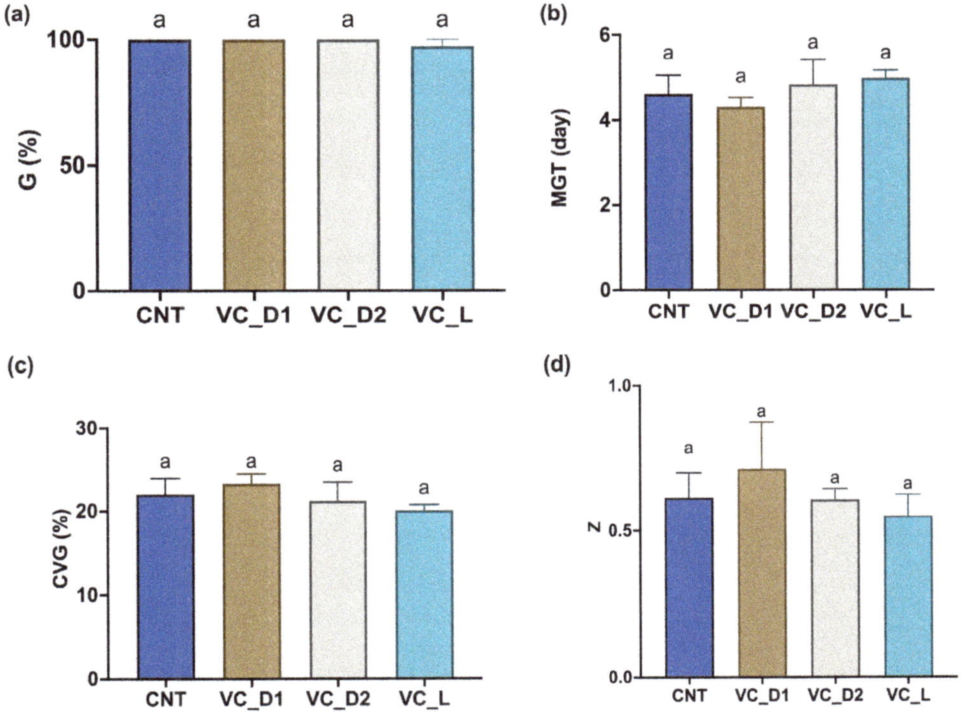

Figure 2. (**a**) Germination percentage (G, %), (**b**) mean germination time (MGT, day), (**c**) coefficient of velocity of germination (CVG, %) and (**d**) synchrony of germination (Z) determined 18 days after sowing in all treatments (CNT—control, VC_D1—vermicompost at 20 g Kg^{-1}, VC_D2—vermicompost at 40 g Kg^{-1}, VC_L—vermicompost leachate). Data are presented as mean ± SEM ($n = 3$). Bars with the same letters indicate non-significant differences ($p > 0.05$).

3.3. Effect of Vermicompost and Leachate on Early Cucumber Morphological and Biochemical Parameters

Shoot length increased gradually across all treatments between 10 and 25 days after sowing (Figure 3a). VC_D2 consistently showed the lowest shoot length values over time, with significant differences at 10 and 25 days when compared with CNT, VC_D1 and VC_L ($p > 0.05$). Leaf number (Figure 3b) followed a uniform pattern across treatments from three leaves at 10–15 days to approximately four leaves at 25 days, with no significant differences being observed among treatments ($p > 0.05$). SPAD chlorophyll values (Figure 3c) decreased progressively over time in all treatments and, at the end of the experiment, VC_D2 showed significantly lower SPAD values than CNT and VC_L ($p < 0.05$).

 https://doi.org/10.3390/environments13010059

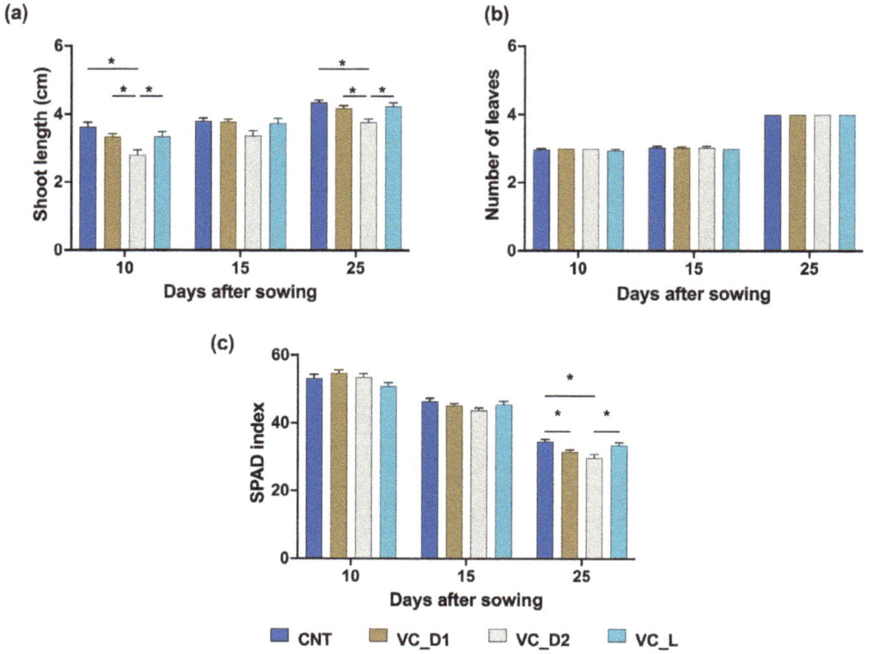

Figure 3. Shoot height (**a**), number of leaves (**b**) and SPAD chlorophyll values index (**c**) in all treatments (CNT—control, VC_D1—vermicompost at 20 g Kg^{-1}, VC_D2—vermicompost at 40 g Kg^{-1}, VC_L—vermicompost leachate). Data are presented as mean ± SEM (n = 3). Statistical analysis was performed using a repeated measures two-way ANOVA * represents significant differences between the treatments ($p < 0.05$).

At the end of the experiment, the morphological aspect of cucumber seedlings are visible in Figure 4.

Figure 4. Representative cucumber seedlings from each treatment (CNT—control, VC_D1—vermicompost at 20 g Kg^{-1}, VC_D2—vermicompost at 40 g Kg^{-1}, VC_L—vermicompost leachate) at 25 days after sowing.

 https://doi.org/10.3390/environments13010059

Stem diameter (Figure 5a) was significantly reduced under VC_D2 treatment compared to VC_D1 and VC_L ($p < 0.05$). VC_D1 significantly increased biomass (Figure 5c) compared with the CNT and VC_D2 ($p < 0.05$), while VC_L showed an intermediate value that did not differ significantly from VC_D1 or CNT ($p > 0.05$). Although root height (Figure 5b) and biomass (Figure 5c) did not differ statistically from other treatments, VC_D1 consistently showed a trend towards higher values ($p > 0.05$).

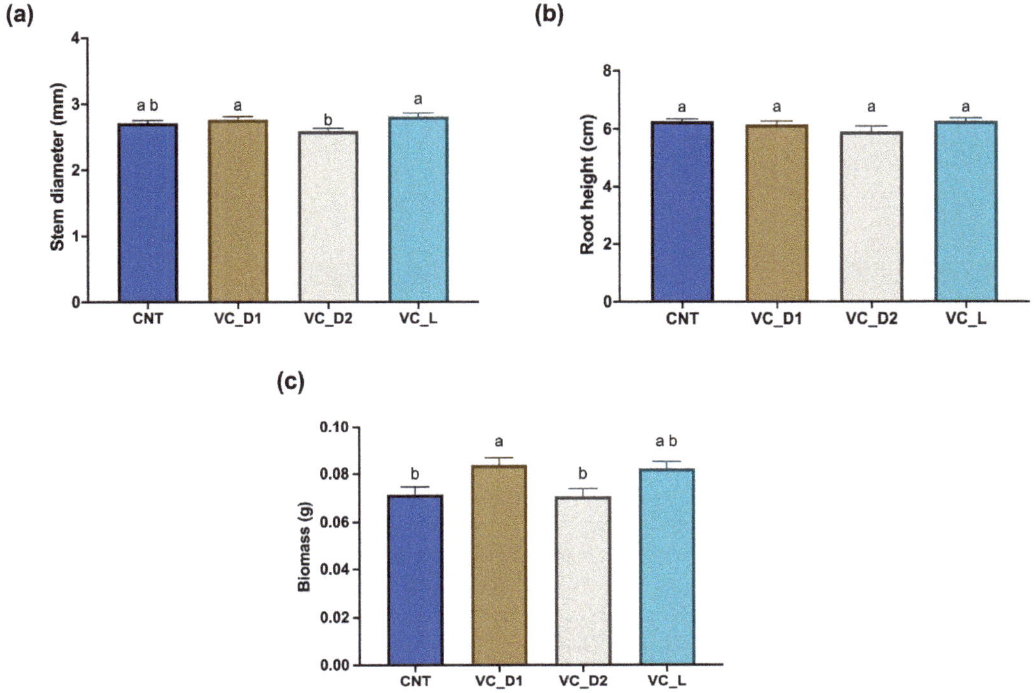

Figure 5. Stem diameter (**a**), root height (**b**) and biomass (**c**) measured at the end of experiment in all treatments (CNT—control, VC_D1—vermicompost at 20 g Kg^{-1}, VC_D2—vermicompost at 40 g Kg^{-1}, VC_L—vermicompost leachate). Data are presented as mean ± SEM (n = 3). Statistical analysis was performed using a one-way ANOVA. For each parameter different lowercase letters indicate significant differences among the treatments ($p < 0.05$).

3.4. Effect of Vermicompost and Leachate on Cucumber Foliar Nutrient Content

The macronutrient and micronutrient contents of cucumber leaves are summarized in Table 4. Plants from VC_L treatment exhibited the highest nitrogen concentration (27.4 g kg^{-1} DW), whereas VC_D1 and VC_D2 showed the lower foliar nitrogen values. Phosphorus and potassium contents were higher in all VC treatments, particularly in VC_D2, compared with the control (CNT). In contrast, calcium content was lower across all treatments compared with the CNT. Concerning micronutrients, zinc and copper concentrations were slightly higher in VC treatments, especially in VC_D2. While iron and manganese concentrations were higher in the CNT plants than in plants grown under VC or VC_L treatments.

https://doi.org/10.3390/environments13010059

Table 4. Macro and micronutrients contents in cucumber leaves at harvest (n = 3 per treatment) in all treatments (CNT—control, VC_D1—vermicompost at 20 g Kg^{-1}, VC_D2—vermicompost at 40 g Kg^{-1}, VC_L—vermicompost leachate).

	CNT	VC_D1	VC_D2	VC_L
Macronutrients (g element kg^{-1} DW)				
N	24.114	21.828	21.898	27.392
P	5.139	5.844	6.481	6.423
K	45.513	49.810	64.741	55.297
Ca	20.344	17.957	14.964	16.646
Mg	5.896	6.285	6.209	5.510
S	10.51	8.82	8.60	10.48
Micronutrients (mg element kg^{-1} DW)				
B	41.567	42.363	40.074	38.035
Fe	139.440	130.026	118.292	121.898
Mn	63.433	54.874	54.232	62.123
Cu	10.652	11.348	12.283	11.902
Zn	70.803	74.687	79.227	75.308

DW: dry weight.

4. Discussion

This study provides preliminary insight into the effects of vermicompost and its leachate, both obtained from the vermicomposting of sewage sludge and vineyard pruning residues, on cucumber (*Cucumis sativus* L.) germination and early growth. The discussion integrates chemical, germination, and early growth responses to elucidate plant performance under the tested experimental conditions. Both the vermicompost and leachate presented metal concentrations below legal thresholds and showed no evidence of phytotoxicity (germination index > 80%), indicating effective stabilization of the original organic materials. In this context, the results are consistent with circular economy principles by contributing to nutrient recycling and reducing the environmental burden of organic waste management.

Chemical characterization of the final substrate mixtures confirmed that vermicompost addition to peat-based commercial substrate consistently enhanced macro- and micronutrients contents. The decrease in the C/N ratio in vermicompost treatments indicates higher availability of mineral nitrogen, that is likely supported by the superior shoot biomass observed in VC_D1. However, the elevated electrical conductivity detected in VC_D2 indicates possible salinity stress, which may have impaired water uptake and plant growth. These results highlight the role of dose optimization [42,43]. Similar observations have been reported in cucumber varieties [44] marigold [45], and peas [46], where moderate vermicompost doses enhanced seedling growth, whereas excessive applications reduced seedling performance due to increased salinity or nutrient imbalance.

Although germination parameters did not differ significantly among treatments, some trends were observed, and seedling growth seem to exhibit dose-dependent responses. Overall, neither vermicompost nor its leachate negatively affected cucumber seed germination under the present experimental conditions. All treatments supported high and uniform germination, reaching values close to 100% by the tenth day after sowing. These findings are consistent with the preliminary phytotoxicity tests (germination index > 80%), which had already suggested the absence of inhibitory compounds in both vermicompost and leachate. Nevertheless, VC_D1 promoted a slightly faster and more synchronized germination, whereas the leachate treatment displayed a slower onset. This suggests that moderate

https://doi.org/10.3390/environments13010059

doses of the tested vermicompost may enhance seedling emergence, likely due to a balanced supply of essential nutrients and the presence of bioactive humic substances [24,47–49].

Regarding cucumber early growth, VC_D1 significantly increased biomass compared to the CNT, indicating that this vermicompost dose provided a favorable nutrient balance for early plant development. Conversely, VC_D2 reduced shoot length, stem diameter, and SPAD values. Previous studies have similarly reported that excessive doses of VC may negatively affect seedling performance [44,50]. For instance, Ceritoglu et al. [50] observed that while 10% and 20% concentrations of vermicompost promoted seedling growth in chickpea, the higher concentrations led to inhibited germination and reduced seedling vigor. Another study found that cucumber seedlings grown in substrates with more than 30% vermicompost exhibited increased leaf area and mineral content; however, higher vermicompost concentrations also resulted in higher nitrate levels in the fruits, indicating potential quality concerns [44].

In contrast, seeds irrigated with leachate exhibited a slower initial germination, which may be attributed to residual salts or phenolic compounds that can transiently delay emergence [51]. As highlighted by other studies, the effectiveness of vermicompost-leachate strongly depends on its concentration and application frequency [51–54]. For instance, a study that evaluated the effects of various leachate concentrations on lettuce seed germination, vegetative development, photosynthetic pigments, and phytotoxicity concluded that lower concentrations correlated positively with germination and growth parameters but higher concentrations of leachate exhibited negative correlations, indicating greater phytotoxicity [51]. Similarly, another study found that while low concentrations of liquid vermicompost promoted bean seedling growth, higher doses inhibited development and caused chlorosis [54]. It should be noted that the observed leachate effects are concentration-dependent and are based solely on the 5% (v/v) dilution tested in this study.

Leaf nutrient analysis provided further insight into the mechanisms underlying the observed growth patterns. Vermicompost treatments, especially VC_D2, strongly increased potassium and phosphorus concentrations (64.7 and 6.5 g kg^{-1} DW, respectively), while reducing calcium, iron, and manganese relative to the CNT. Such nutrient imbalances may explain the limited growth performance in VC_D2, as excessive K can antagonize Ca and Mg uptake, while reduced Fe and Mn availability may hinder chlorophyll biosynthesis and photosynthetic efficiency [55,56]. The VC_D1 treatment, on the other hand, maintained a more balanced nutrient profile, with moderate increases in P, K, Zn, and Cu, avoiding the extreme shifts observed in VC_D2. This balance likely contributed to its superior shoot biomass, suggesting that nutrient equilibrium, rather than total nutrient content, is key to promoting optimal growth.

Overall, the findings of this exploratory study indicate that vermicompost dose strongly determine cucumber performance by modulating nutrient dynamics, ionic balance, and bioactive compound effects. Excessive application rates may induce nutritional antagonisms rather than stimulating cucumber growth. As vermicompost not only supplies mineral nutrients but also contains humic substances and plant growth regulators that can modulate physiological processes [24,57,58]; the improved performance of VC_D1 observed under the tested experimental conditions likely resulted from the combined effect of nutrient availability and bioactive stimulation.

5. Conclusions

This exploratory study indicates that vermicompost and leachate derived from sewage sludge and vineyard pruning residues are chemically safe and potentially valuable fertilizers for cucumber seedling growth. The results are aligned with residue valorization and nutrient recycling, thereby contributing to the United Nations Sustainable Development

https://doi.org/10.3390/environments13010059

Goals, particularly SDG 12 (Responsible Consumption and Production), as well as SDG 2 (Zero Hunger) and SDG 6 (Clean Water and Sanitation).

All treatments supported high germination rates without evidence of phytotoxicity, indicating the maturity and environmental safety of the products. Nonetheless, plant growth responses depended on the application rate and amendment form. However, due to the limited number of application rates evaluated, the present study does not allow the establishment of robust dose–response relationships. Vermicompost addition at 20 g kg^{-1} enhanced seedling biomass and maintained a balanced foliar nutrient profile, whereas a higher rate, 40 g kg^{-1}, induced nutrient imbalances and reduced growth. Leachate application increased foliar nitrogen concentrations but did not significantly improve biomass accumulation at the tested dilution. These results underscore the importance of dose optimization, as excessive application rates may offset the potential benefits of vermicompost-based amendments.

From a sustainability standpoint, the co-vermicomposting of sewage sludge and vineyard pruning residues can be a promising strategy for residue valorization within circular economy frameworks, transforming waste materials into agronomically valuable resources. Nevertheless, given the use of a single experimental batch, further research is required to validate the agronomic relevance of these amendments beyond the tested experimental conditions. Future studies should encompass the complete crop cycle, including flowering, fruiting, and yield quality, as well as detailed characterization of leachate composition—particularly phenolic and humic fractions—to better elucidate their roles in plant development. In addition, intermediate vermicompost application rates and/or alternative application methods should be tested to reduce the nutrient imbalances observed at the highest rate. Field trials across different soil types and crop species will also be essential to confirm the scalability and practical applicability under real agricultural conditions.

Author Contributions: Conceptualization, T.A., E.N.-G., M.C.M. and P.A.O.; methodology, T.A., E.N.-G., C.M., H.L., A.C.C. and M.C.M.; formal analysis, T.A. and M.C.M.; writing—original draft preparation, T.A., E.N.-G. and M.C.M.; writing—review and editing, T.A., E.N.-G., A.C.C., V.F., J.R.S., A.M.C., M.R., P.A.O. and M.C.M.; supervision, J.R.S., A.M.C., M.R., P.A.O. and M.C.M.; funding acquisition, P.A.O. All authors have read and agreed to the published version of the manuscript.

Funding: The Vine&Wine PT project has received funding from the NextGeneration EU program, through Portugal's Recovery and Resilience Plan (project n. ° C644866286-00000011). This work is supported by National Funds by FCT–Portuguese Foundation for Science and Technology, under the projects UID/04033/2025: Center for the Research and Technology of Agro-Environmental and Biological Sciences (https://doi.org/10.54499/UID/04033/2025) and LA/P/0126/2020 (https://doi.org/10.54499/LA/P/0126/2020). T.A. and H.L. thank FCT for their PhD grants (2023. 01329.BD and PRT/BD/154380/2023, respectively), while E.N.-G. and C.M. acknowledge PRR and European Funds NextGeneration EU for their fellowship grants (BI/UTAD/41/2025 and BI/UTAD/17/2025, respectively).

Data Availability Statement: The data that support the findings of this study are available upon reasonable request.

Conflicts of Interest: The authors declare no conflicts of interest.

Abbreviations

The following abbreviations are used in this manuscript:

CVG	Coefficient of velocity of germination
DW	Dry weight
EC	Electrical conductivity
GAE	Gallic acid equivalents
GI	Germination index
MGT	Mean germination time
OM	Organic matter
SEM	Standard error of the mean
SPAD	Soil–plant analysis development chlorophyll index
VC	Vermicompost

References

1. Niculescu, V.-C.; Ionete, R.-E. An Overview on Management and Valorisation of Winery Wastes. *Appl. Sci.* **2023**, *13*, 5063. [CrossRef]
2. Nascimento-Gonçalves, E.; Azevedo, T.; Lopes, H.; Sousa, J.R.; Oliveira, P.A.; Roboredo, M.; Coimbra, A.M.; Morais, M.C. Vermicomposting as a Valorization Solution to the Winery Sector By-Products. *Agronomy* **2024**, *14*, 1111. [CrossRef]
3. Barbanera, M.; Cardarelli, A.; Carota, E.; Castellini, M.; Giannoni, T.; Ubertini, S. Valorization of Winery and Distillery By-Products by Hydrothermal Carbonization. *Sci. Rep.* **2021**, *11*, 23973. [CrossRef]
4. Rodrigues Machado, A.; Atatoprak, T.; Santos, J.; Alexandre, E.M.C.; Pintado, M.E.; Paiva, J.A.P.; Nunes, J. Potentialities of the Extraction Technologies and Use of Bioactive Compounds from Winery By-Products: A Review from a Circular Bioeconomy Perspective. *Appl. Sci.* **2023**, *13*, 7754. [CrossRef]
5. Rodrigues, R.P.; Gando-Ferreira, L.M.; Quina, M.J. Increasing Value of Winery Residues through Integrated Biorefinery Processes: A Review. *Molecules* **2022**, *27*, 4709. [CrossRef]
6. Jesus, M.S.; Romaní, A.; Genisheva, Z.; Teixeira, J.A.; Domingues, L. Integral Valorization of Vine Pruning Residue by Sequential Autohydrolysis Stages. *J. Clean. Prod.* **2017**, *168*, 74–86. [CrossRef]
7. Provenzano, M.; Pacchera, F.; Silvestri, C.; Ruggieri, A. From Vineyard to Value: A Circular Economy Approach to Viticulture Waste. *Resources* **2024**, *13*, 172. [CrossRef]
8. Mahmoudi, A.; Mousavi, S.A.; Darvishi, P. Performance and Recent Development in Sewage Sludge-to-Bioenergy Using Microbial Fuel Cells: A Comprehensive Review. *Int. J. Hydrogen Energy* **2024**, *50*, 1432–1455. [CrossRef]
9. Salva, J.; Sečkár, M.; Schwarz, M.; Samešová, D.; Mordáčová, M.; Poništ, J.; Veverková, D. Analysis of the Current State of Sewage Sludge Treatment from the Perspective of Current European Directives. *Environ. Sci. Eur.* **2025**, *37*, 59. [CrossRef]
10. Głąb, T.; Żabiński, A.; Sadowska, U.; Gondek, K.; Kopeć, M.; Mierzwa-Hersztek, M.; Tabor, S.; Stanek-Tarkowska, J. Fertilization Effects of Compost Produced from Maize, Sewage Sludge and Biochar on Soil Water Retention and Chemical Properties. *Soil Tillage Res.* **2020**, *197*, 104493. [CrossRef]
11. Siddiqui, M.I.; Rameez, H.; Farooqi, I.H.; Basheer, F. Recent Advancement in Commercial and Other Sustainable Techniques for Energy and Material Recovery from Sewage Sludge. *Water* **2023**, *15*, 948. [CrossRef]
12. Molaey, R.; Appels, L.; Yesil, H.; Tugtas, A.E.; Çalli, B. Sustainable Heavy Metal Removal from Sewage Sludge: A Review of Bioleaching and Other Emerging Technologies. *Sci. Total Environ.* **2024**, *955*, 177020. [CrossRef]
13. Grobelak, A.; Całus-Makowska, K.; Jasińska, A.; Klimasz, M.; Wypart-Pawul, A.; Augustajtys, D.; Baor, E.; Sławczyk, D.; Kowalska, A. Environmental Impacts and Contaminants Management in Sewage Sludge-to-Energy and Fertilizer Technologies: Current Trends and Future Directions. *Energies* **2024**, *17*, 4983. [CrossRef]
14. Hajam, Y.A.; Kumar, R.; Kumar, A. Environmental Waste Management Strategies and Vermi Transformation for Sustainable Development. *Environ. Chall.* **2023**, *13*, 100747. [CrossRef]
15. Manea, E.E.; Bumbac, C.; Dinu, L.R.; Bumbac, M.; Nicolescu, C.M. Composting as a Sustainable Solution for Organic Solid Waste Management: Current Practices and Potential Improvements. *Sustainability* **2024**, *16*, 6329. [CrossRef]
16. Giwa, A.S.; Maurice, N.J.; Luoyan, A.; Liu, X.; Yunlong, Y.; Hong, Z. Advances in Sewage Sludge Application and Treatment: Process Integration of Plasma Pyrolysis and Anaerobic Digestion with the Resource Recovery. *Heliyon* **2023**, *9*, e19765. [CrossRef]
17. Kacprzak, M.; Baran, J.; Fijalkowski, K. Emerging Sewage Sludge Treatment Technologies for Land Carbon Sequestration: A Comprehensive Review. *J. Mater. Cycles Waste Manag.* **2025**, *27*, 2866–2886. [CrossRef]
18. Dominguez, J.; Edwards, C.A. Vermicomposting Organic Wastes: A Review. In *Soil Zoology for Sustainable Development in the 21st Century*; Geocities: Cairo, Egypt, 2004; pp. 369–395.
19. Gupta, R.; Garg, V.K. Stabilization of Primary Sewage Sludge during Vermicomposting. *J. Hazard. Mater.* **2008**, *153*, 1023–1030. [CrossRef]

https://doi.org/10.3390/environments13010059

20. Singh, A.; Karmegam, N.; Singh, G.S.; Bhadauria, T.; Chang, S.W.; Awasthi, M.K.; Sudhakar, S.; Arunachalam, K.D.; Biruntha, M.; Ravindran, B. Earthworms and Vermicompost: An Eco-Friendly Approach for Repaying Nature's Debt. *Environ Geochem Health* **2020**, *42*, 1617–1642. [CrossRef]
21. Ahmed, N.; Al-Mutairi, K.A. Earthworms Effect on Microbial Population and Soil Fertility as Well as Their Interaction with Agriculture Practices. *Sustainability* **2022**, *14*, 7803. [CrossRef]
22. Quaik, S.; Ibrahim, M.H. A Review on Potential of Vermicomposting Derived Liquids in Agricultural Use. *Int. J. Sci. Res.* **2013**, *3*, 1–6.
23. Pant, A.P.; Radovich, T.J.; Hue, N.V.; Talcott, S.T.; Krenek, K.A. Vermicompost Extracts Influence Growth, Mineral Nutrients, Phytonutrients and Antioxidant Activity in Pak Choi (*Brassica rapa* Cv. Bonsai, Chinensis Group) Grown under Vermicompost and Chemical Fertiliser. *J. Sci. Food Agric.* **2009**, *89*, 2383–2392. [CrossRef]
24. Atiyeh, R.M.; Lee, S.; Edwards, C.A.; Arancon, N.Q.; Metzger, J.D. The Influence of Humic Acids Derived from Earthworm-Processed Organic Wastes on Plant Growth. *Bioresour. Technol.* **2002**, *84*, 7–14. [CrossRef]
25. Pathma, J.; Sakthivel, N. Microbial Diversity of Vermicompost Bacteria That Exhibit Useful Agricultural Traits and Waste Management Potential. *Springerplus* **2012**, *1*, 26. [CrossRef]
26. Gómez-Brandón, M.; Fornasier, F.; de Andrade, N.; Domínguez, J. Influence of Earthworms on the Microbial Properties and Extracellular Enzyme Activities during Vermicomposting of Raw and Distilled Grape Marc. *J. Environ. Manag.* **2022**, *319*, 115654. [CrossRef]
27. Domínguez, J.; Martínez-Cordeiro, H.; Álvarez-Casas, M.; Lores, M. Vermicomposting Grape Marc Yields High Quality Organic Biofertiliser and Bioactive Polyphenols. *Waste Manag. Res.* **2014**, *32*, 1235–1240. [CrossRef] [PubMed]
28. Santana, N.A.; Jacques, R.J.S.; Antoniolli, Z.I.; Martínez-Cordeiro, H.; Domínguez, J. Changes in the Chemical and Biological Characteristics of Grape Marc Vermicompost during a Two-Year Production Period. *Appl. Soil Ecol.* **2020**, *154*, 103587. [CrossRef]
29. Dume, B.; Hanc, A.; Svehla, P.; Michal, P.; Chane, A.D.; Nigussie, A. Vermicomposting Technology as a Process Able to Reduce the Content of Potentially Toxic Elements in Sewage Sludge. *Agronomy* **2022**, *12*, 2049. [CrossRef]
30. Dume, B.; Hanc, A.; Svehla, P.; Michal, P.; Chane, A.D.; Nigussie, A. Composting and Vermicomposting of Sewage Sludge at Various C/N Ratios: Technological Feasibility and End-Product Quality. *Ecotoxicol. Environ. Saf.* **2023**, *263*, 115255. [CrossRef]
31. Ludibeth, S.-M.; Marina, I.-E.; Vicenta, E.M. Vermicomposting of Sewage Sludge: Earthworm Population and Agronomic Advantages. *Compos. Sci. Util.* **2012**, *20*, 11–17. [CrossRef]
32. Azevedo, T.; Nascimento-Gonçalves, E.; Lopes, H.; Medeiros, C.; Falco, V.; Sousa, J.R.; Coimbra, A.M.; Roboredo, M.; Oliveira, P.A.; Morais, M.C. Valorization of Vineyard By-Products Through Vermicomposting: A Comparative Pilot-Scale Study with *Eisenia fetida* and *Eisenia andrei*. *Agronomy* **2025**, *15*, 1340. [CrossRef]
33. Samba, N.; Nunomura, O.; Lu, N.; Johkan, M.; Nakano, A.; Tsukagoshi, S. Cucumber (*Cucumis sativus* L.) Growth and Productivity under Solar Radiation-Based Quantitative Nutrient Management in Hydroponic System. *Agronomy* **2024**, *14*, 296. [CrossRef]
34. Mukherjee, P.K.; Nema, N.K.; Maity, N.; Sarkar, B.K. Phytochemical and Therapeutic Potential of Cucumber. *Fitoterapia* **2013**, *84*, 227–236. [CrossRef] [PubMed]
35. Sá, C.; Trindade, A.; Coutinho, J. *Manual de Colheita de Amostras e de Dados No Campo: Sistema de Monitorização Nacional Do Solo*; DGADR: Lisboa, Portugal, 2024.
36. Singleton, V.L.; Rossi, J.A. Colorimetry of Total Phenolics with Phosphomolybdic-Phosphotungstic Acid Reagents. *Am. J. Enol. Vitic.* **1965**, *16*, 144–158. [CrossRef]
37. Zucconi, F.; Forte, M.; Monaco, A.; Bertoldi, M. Biological Evaluation of Compost Maturity. *Biocycle* **1981**, *22*, 27–29.
38. Anderson, J.P.E. Soil Respiration. In *Methods of Soil Analysis*; John Wiley & Sons, Ltd.: Hoboken, NJ, USA, 1982; pp. 831–871.
39. Soares, A.S.; Miranda, C.; Teixeira, C.A.; Coutinho, J.; Trindade, H.; Coelho, A.C. Impact of Different Treatments on Escherichia Coli during Storage of Cattle Slurry. *J. Environ. Manag.* **2019**, *236*, 323–327. [CrossRef]
40. Ranal, M.A.; de Santana, D.G.; Ferreira, W.R.; Mendes-Rodrigues, C. Calculating Germination Measurements and Organizing Spreadsheets. *Braz. J. Bot.* **2009**, *32*, 849–855. [CrossRef]
41. Ranal, M.A.; de Santana, D.G. How and Why to Measure the Germination Process? *Braz. J. Bot.* **2006**, *29*, 1–11. [CrossRef]
42. Méndez-Cifuentes, A.; Valdez-Aguilar, L.A.; Cadena-Zapata, M.; Alvarado-Camarillo, D.; González-Fuentes, J.A. Nutrient Solution Electrical Conductivity Affects Yield and Growth of Sub-Irrigated Tomatoes. *Horticulturae* **2023**, *9*, 826. [CrossRef]
43. Ding, X.; Jiang, Y.; Zhao, H.; Guo, D.; He, L.; Liu, F.; Zhou, Q.; Nandwani, D.; Hui, D.; Yu, J. Electrical Conductivity of Nutrient Solution Influenced Photosynthesis, Quality, and Antioxidant Enzyme Activity of Pakchoi (*Brassica campestris* L. ssp. *Chinensis*) in a Hydroponic System. *PLoS ONE* **2018**, *13*, e0202090. [CrossRef] [PubMed]
44. Jankauskienė, J.; Laužikė, K.; Kavaliauskaitė, D. Effects of Vermicompost on Quality and Physiological Parameters of Cucumber (*Cucumis sativus* L.) Seedlings and Plant Productivity. *Horticulturae* **2022**, *8*, 1009. [CrossRef]
45. Shafique, I.; Andleeb, S.; Aftab, M.S.; Naeem, F.; Ali, S.; Yahya, S.; Ahmed, F.; Tabasum, T.; Sultan, T.; Shahid, B.; et al. Efficiency of Cow Dung Based Vermi-Compost on Seed Germination and Plant Growth Parameters of *Tagetes erectus* (Marigold). *Heliyon* **2021**, *7*, e05895. [CrossRef] [PubMed]

46. Akinnuoye-Adelabu, D.B.; Steenhuisen, S.; Bredenhand, E. Improving Pea Quality with Vermicompost Tea and Aqueous Biochar: Prospects for Sustainable Farming in Southern Africa. *South Afr. J. Bot.* **2019**, *123*, 278–285. [CrossRef]
47. Ye, T.; Ijenyo, M.; Abbey, L. A Preliminary Study on Emergence and Growth of Carrot Seedlings in Response to Varying Proportions of Vermicompost and Copper. *Hortic. Int. J.* **2022**, *6*, 62–65. [CrossRef]
48. Ravindran, B.; Lee, S.R.; Chang, S.W.; Nguyen, D.D.; Chung, W.J.; Balasubramanian, B.; Mupambwa, H.A.; Arasu, M.V.; Al-Dhabi, N.A.; Sekaran, G. Positive Effects of Compost and Vermicompost Produced from Tannery Waste-Animal Fleshing on the Growth and Yield of Commercial Crop-Tomato (*Lycopersicon esculentum* L.) Plant. *J. Environ. Manag.* **2019**, *234*, 154–158. [CrossRef]
49. Hussain, N.; Abbasi, T.; Abbasi, S.A. Generation of Highly Potent Organic Fertilizer from Pernicious Aquatic Weed Salvinia Molesta. *Environ. Sci. Pollut. Res.* **2018**, *25*, 4989–5002. [CrossRef]
50. Ceritoglu, M.; Erman, M.; Ceritoglu, F.; Bektas, H. The Response of Grain Legumes to Vermicompost at Germination and Seedling Stages. *Legume Res.* **2021**, *44*, 936–941. [CrossRef]
51. Chávez, A.G.; Hernández, L.R.; Ramírez, R.I.; Gardea, M.C.E.D.; Huerta, J.H. Efecto del lixiviado de vermicompost en la germinación de lechuga in vitro. *Nexo Rev. Cient.* **2024**, *37*, 74–88. [CrossRef]
52. Kosem, H.; Kocak, M.Z.; Kaysim, M.G.; Celikcan, F.; Kulak, M. Liquid Leachate Produced from Vermicompost Effects on Some Agronomic Attributes and Secondary Metabolites of Sweet Basil (*Ocimum basilicum* L.) Exposed to Severe Water Stress Conditions. *Horticulturae* **2022**, *8*, 1190. [CrossRef]
53. Bidabadi, S.S.; Afazel, M.; Poodeh, S.D. The Effect of Vermicompost Leachate on Morphological, Physiological and Biochemical Indices of *Stevia rebaudiana* Bertoni in a Soilless Culture System. *Int. J. Recycl Org Waste Agric.* **2016**, *5*, 251–262. [CrossRef]
54. Öztürkci, Y.; Akköprü, A. Effects of Solid and Liquid Vermicompost Application on Bean Growth and Common Bacterial Blight Disease in Different Growth Medium. *IJAWS* **2021**, *7*, 30–40. [CrossRef]
55. Fageria, N.K. *The Use of Nutrients in Crop Plants*; CRC Press: Boca Raton, FL, USA, 2016.
56. White, P.J.; Brown, P.H. Plant Nutrition for Sustainable Development and Global Health. *Ann. Bot.* **2010**, *105*, 1073–1080. [CrossRef] [PubMed]
57. Arancon, N.Q.; Edwards, C.A.; Lee, S.; Byrne, R. Effects of Humic Acids from Vermicomposts on Plant Growth. *Eur. J. Soil Biol.* **2006**, *42*, S65–S69. [CrossRef]
58. Oyege, I.; Balaji Bhaskar, M.S. Effects of Vermicompost on Soil and Plant Health and Promoting Sustainable Agriculture. *Soil Syst.* **2023**, *7*, 101. [CrossRef]

environments

MDPI

Article

Effect of Organic Soil Amendments and Vineyard Topographic Position on the Chemical Composition of Syrah, Trincadeira, Alicante Bouschet, and Antão Vaz Grapes (*Vitis vinifera* L.) in the Alentejo Wine Region

Matteo Pierini [1,†], Shrika G. Harjivan [1,†], Nicolò Sieli [1], Maria João Cabrita [2,3], Sérgio Prats [2,4], Sofia Catarino [1,5,6,7] and Jorge M. Ricardo-da-Silva [1,5,6,*]

[1] Instituto Superior de Agronomia, Universidade de Lisboa, Tapada da Ajuda, 1349-017 Lisboa, Portugal; sofiacatarino@isa.ulisboa.pt (S.C.)
[2] MED—Mediterranean Institute for Agriculture, Environment and Development & CHANGE—Global Change and Sustainability Institute, Universidade de Évora, Pólo da Mitra, Ap. 94, 7006-554 Évora, Portugal; mjbc@uevora.pt (M.J.C.)
[3] Departamento de Fitotecnia, Escola de Ciências e Tecnologia, Universidade de Évora, Pólo da Mitra, Ap. 94, 7006-554 Évora, Portugal
[4] Misión Biológica de Galicia—National Spanish Research Council (MBG-CSIC), 36143 Salcedo, Pontevedra, Spain
[5] LEAF—Linking Landscape Environment Agriculture and Food—Research Center, Instituto Superior de Agronomia, Universidade de Lisboa, Tapada da Ajuda, 1349-017 Lisboa, Portugal
[6] Associate Laboratory TERRA, Instituto Superior de Agronomia, Universidade de Lisboa, Tapada da Ajuda, 1349-017 Lisboa, Portugal
[7] CeFEMA—Centre of Physics and Engineering of Advanced Materials, Instituto Superior Técnico, Universidade de Lisboa, Av. Rovisco Pais, 1, 1049-001 Lisboa, Portugal
* Correspondence: jricardosil@isa.ulisboa.pt
† These authors contributed equally to this work.

Abstract

Climate change and unsustainable agricultural practices are triggering land degradation in semi-arid Mediterranean regions. Organic amendments, such as mulching materials, have shown promising potential to mitigate these impacts by improving soil chemical, physical, and biological properties, while enhancing grapevine growth and productivity. This study evaluated the effects of wheat straw mulch (M) and wheat straw combined with biochar (MB), together with vineyard topography (bottom vs. top), on grape chemical and phenolic composition in four *Vitis vinifera* L. cultivars (Syrah, Trincadeira, Alicante Bouschet, and Antão Vaz) grown in the Alentejo wine region. Grapes were sampled separately at top and bottom topographic positions, and classical and phenolic parameters were analyzed. The application of M and MB significantly modified must composition, mainly through changes in nitrogen and sugar levels across topographic positions. Only MB exhibited stronger effects, enhancing must quality, while MB and M reduced bottom–top variability. Similar patterns and positional effects were observed for phenolic and color parameters. Both organic treatments lowered total monomeric anthocyanin concentrations, although positional differences with wheat straw mulch were found. The results highlight that combining soil management with topography and variety response can optimize grape phenolic composition and promote sustainable viticulture through targeted, site-specific mulching strategies.

Keywords: organic amendment; biochar; mulch; grape composition; polyphenols

1. Introduction

The European Commission has reported that, under the present conditions, if gas emissions are not reduced, global drylands will expand by about 10% by the end of this century [1]. The steady rise in temperature and the increasing frequency of extreme weather events, combined with unsustainable agricultural practices (such as excessive tillage, the absence of cover crops, and deforestation), are reducing food availability by compromising soil fertility, soil biodiversity, the land's natural resilience, and water quality [2]. Moreover, desertification is a process that, once it occurs, cannot be reversed except over hundreds of years (500 years for 2.5 cm of soil to form) [3].

The management of agricultural practices has a direct impact on preserving fertile soils, combating desertification, and mitigating climate change, an essential task for ensuring our food supply [4]. In this context, "sustainable agriculture" has been developed to promote agricultural practices that produce collective benefits beyond the final product, for the environment and human health [5]. One practice is soil organic mulching, in which organic materials are applied and retained on the soil surface. This application leads to an increase in soil organic matter (SOM), soil coverage, and carbon sequestration depending on the organic material used. SOM is a vast mixture of carbon compounds that helps maintain soil fertility and participates in nutrient, water, and biological cycles [4]. Moreover, soil cover with organic materials can control weed growth, minimize water loss, improve soil infiltration, decrease soil compaction, buffer extreme temperatures, and increase vine health [6]. Instead, carbon sequestration occurs when carbon-rich industrial by-products are used as mulch or organic amendments, consequently increasing soil carbon pool in place of the atmospheric carbon pool, which is the main cause of climate change [7].

A representative example is biochar, a carbon-rich organic amendment (70–80% carbon) obtained from slow pyrolysis of organic waste in an oxygen-limited environment process that makes it a slow carbon-releasing material. By applying it to the soil, the soil C pool is increased. Large-scale application programs estimate that biochar could sequester up to 9.50 billion tons of CO_2 in the soil by 2100 [8]. Lippi et al., 2025 [8], provided an excellent summary of the effects of biochar on soil and its influence on grapevines. Their findings highlight that the impact of biochar on soil and plants varies significantly depending on its characteristics, soil type and condition, application rate and frequency, and other factors. Biochar has been reported to positively affect several soil properties: (i) it promotes the formation of stable soil aggregates and reduces soil erosion [9]; (ii) it improves soil water-holding capacity and increases water retention [10,11]; and (iii) it enhances the soil's cation exchange capacity and reduces soil phytotoxicity [8]. Overall, biochar contributes to carbon sequestration by increasing soil organic carbon, is capable of improving soil habitat and fertility, and can immobilize heavy metals, limiting their mobility and potential toxicity in the soil system [12].

Biochar has been shown to increase root growth and mitigate drought stress in plants, enhancing their resilience under water-limited conditions [13]. Despite these positive effects, no significant impact on grape quality and wine sensory attributes has been observed [14]. In certain cases, biochar application has also led to increases in vegetative growth and, occasionally, improvements in yield.

A comprehensive review by Diacono and Montemurro (2010) [15] examined various crops, not just vines, summarizing different scientific studies that evaluate the impact of organic amendments on plant yield. In some cases, yields increased; in others, they decreased or showed no differences between control and amended soil, highlighting the importance of the other factors involved. Other studies, using mulch in the vineyard, observed increases in yield and vine growth [16], as well as increases in leaf N and K content. These results are normally observed starting from the third year of application [6].

https://doi.org/10.3390/environments13010044

A study published by Pinamonti et al. (1998) [17] reported an increase in the content of K and tartaric acid; a decrease in P, Ca, and Mg; and unchanged content of N, Fe, and Mn in grapes obtained from vineyards with OA treatment. This was attributed to the different depths at which roots absorb nutrients, with deeper root absorption occurring in treatments without amendments [17]. The mineral composition of wine after OA application changes extremely depending on the type of amendment used; however, it does not seem to cause any issues [18]. Some authors have reported an increase in total acidity after soil mulching, but only in dry years was an increase in total soluble solids (TTS) obtained [19], while in other conditions, a decrease in TSS was observed and total acidity remained constant [20]. Also, in a study conducted in New Zealand, researchers observed an increase in grape K and yeast assimilable nitrogen (YAN) after mulching with compost made from vineyards and winery waste [21]. It is believed that favorable thermal environment induced by a mulch layer can stimulate K uptake by the vine roots [6], which can explain the increase in K content registered by different authors after mulch application. Moreover, research on phenol composition following mulching revealed different responses depending on the *terroir*. In particular, the initial soil conditions and the type of OAs used were the major cause of variability on the responses [11]. Other studies investigating the effects of mulching on berry phenolic composition have reported contrasting results: some observed reductions or no changes [20,22–24], while others noted increases [25]. Most existing studies have focused on the nutrient improvement due to OA application, a few of them on soil temperature and almost none on the different water availability due to mulching.

This study aims to evaluate the combined effects of two organic treatments, wheat straw alone (M) and wheat straw combined with commercial biochar (MB), and vineyard topography (bottom vs. top hillslope position) on the chemical composition of grapes from four *Vitis vinifera* L. cultivars (Syrah, Trincadeira, Alicante Bouschet, and Antão Vaz), which were monitored over one harvest in Portuguese vineyards located in the Alentejo wine region. The specific objectives were the following: (a) to evaluate the classical chemical analyses (pH, total acidity, TSS, and YAN), to determine color parameters (color intensity and tonality), and to quantify phenolic compounds (total phenols, non-flavonoids, flavonoids, monomeric/oligomeric/polymeric proanthocyanidins, and total anthocyanins) of grapes collected separately from the top (t) and the bottom (b) topographic positions for each cultivar, and (b) to quantify monoglucosylated anthocyanins and their derivatives present in the samples by high-performance liquid chromatography (HPLC).

By assessing how two different soil organic treatments (M and MB) and vineyard topographic position interact to influence grape chemical composition, this study aims to provide new insights into soil–landscape features, with promising implications for both wine quality enhancement and the long-term sustainability of viticulture in Mediterranean and other water-limited regions.

2. Materials and Methods

2.1. Vineyard Locations and Harvest

In this study, four different cultivars of *Vitis vinifera* L.—three red varieties (Syrah (224 m above sea level; masl), Trincadeira (219 masl), and Alicante Bouschet (215 masl)) and one white grape variety (Antão Vaz (222 masl))—were present in two different vineyards, Fundação Eugénio de Almeida and Herdade do Mouchão, both located in the Alentejo wine region (Portugal) and separated by around 40 kms.

The first study area, Fundação Eugénio de Almeida (FEA), is in the municipality of Évora, São Manços. It lies between latitudes 38 and 39° (GPS coordinates 38°31′03″ N 7°44′17″ W). The soils are classified as Leptosol sandy-loam with 66% rock content (World Reference Base for Soil Resources, WRB), 10 cm soil depth, topsoil pH of 6.7, WHCmax of

 https://doi.org/10.3390/environments13010044

47.7 mL (maximum water-holding capacity), and soil organic carbon (SOC) content of 5.5%. The climate is classified as Csa (temperate, with dry, hot summers). The average annual precipitation is 609 mm, mainly from October to April. The average annual temperature is 16 °C, with a minimum temperature above 5.8 °C (January) and a maximum temperature of 32.3 °C (August). Significant temperature variations are expected between day and night and between winter and summer. The vineyards were planted with Syrah (two sites, Syrah I and II) and Trincadeira (Trincadeira site), both grapevines grafted on 1103P, SO4, and 110R rootstocks in 2018, trained to an ascending vertical system with 2.5 m spacing (in-row). The vines were trained using a spur pruning system. Vine rows were oriented south–southwest for Syrah I (N200E), east–southeast for Syrah II (N100E), and northwest for Trincadeira (N300E). The steepness was 11° for Syrah I, 14° for Syrah II, and 15° for the Trincadeira site. Winter pruning was performed with mechanical pre-pruning, followed by manual finishing. Drip irrigation was performed during the summer. The harvest was manual, conducted on 5 September 2024 (Figure 1).

(A)

(B)

Figure 1. Location of the vineyard study areas: (**A**) Fundação Eugénio de Almeida (FEA) in the São Manços, accounting for the Syrah I (blue), Syrah II (green), and Trincadeira sites (purple), and (**B**) Herdade do Mouchão in the Casa Branca, including the Alicante Bouschet (yellow) and Antão Vaz (orange) sites. The view in the picture is oriented to the north.

The second study area, Herdade do Mouchão (MOU), is in the municipality of Sousel, Casa Branca. It lies between latitudes 38 and 39° (38°55′03″ N 7°47′20″ W). The soils are classified as Cambisol, a more developed soil compared to Leptosol (WRB), deeper, and characterized by a cambic horizon (Bw) with 58% of rock content, 10 cm soil depth, topsoil pH of 8.1, WHCmax of 48.5 mL, and soil organic carbon (SOC) content of 5.7%. The climate is classified as Csa (temperate, with dry, hot summers). The vineyard's proximity to a reservoir creates a meso-climate, which could influence their growing conditions (Figure 1). Alicante Bouschet and Antão Vaz grapevines were grafted on 99R rootstocks in 1995, trained to an ascending vertical system with 2.5 m spacing (in-row). The grapevines were trained using a bilateral cordon. Vine rows are faced east–southeast for the Alicante Bouschet site (N100E) and west–northwest for the Antão Vaz site (N2900E), and the steepness was 14° and 11° (for Alicante Bouschet and for Antão Vaz, respectively). The difference in altitude between the Antão Vaz top and bottom positions in the hillslope was 10 m. Winter pruning was performed with mechanical pre-pruning, followed by manual finishing. Drip irrigation was performed during the summer. The harvest was manual, conducted on 19 September 2024.

 https://doi.org/10.3390/environments13010044

A total of 5 sites were divided into three test plots with different organic amendments: no organic treatment (C), wheat straw mulch (M treatment), and wheat straw combined with biochar (MB treatment). Each plot was further divided into two sections corresponding to the bottom (b) and top (t) topographic positions within the hillslope, with a 10 m difference in altitude between them (Figure 2).

(A)
(B)

Figure 2. Example of the experimental design for Syrah (**A**) and Trincadeira (**B**) cultivars at FEA: bottom (blue part of the hill); top (brown part of the hill). M (mulched), C (control, untreated), and MB (mulched and biochar). The view of the picture is oriented to the north.

2.2. Treatments

This study used wheat straw from Agriloja (Évora, Portugal) and "Ecochar", a commercial biochar from Ibero Massa Florestal, produced from shredded wood of invasive Acacia species (*A. dealbata* Link and *A. melanoxylon* R. Br), that are of significant ecological concern in Portugal. The detailed properties of both organic amendments are listed in Table 1. The mulch treatment (M) consisted in the manual application of a layer of wheat straw mulch over the soil surface, without incorporation into the soil, at an application rate of 200 g m^{-2}. The mulch and biochar treatment (MB) consisted of manual application, without incorporation into the soil, of a layer of biochar over the soil surface (1000 g of biochar per m^2), which was subsequently covered by a layer of straw mulch (200 g m^{-2}), in order to prevent biochar mobilization [26]. The treatments were applied between 9–10 November 2023 for the FEA and MOU sites.

Table 1. Specific characteristics of the treatments applied in soils of FEA and MOU vineyards. Values between brackets are the standard deviation over the mean values (n = 3); polycyclic aromatic hydrocarbons (PAHs); water drop penetration time (WDPT) [10].

Material	Application Rate (g m^{-2})	Cover Distribution (%)	Layer Thickness (mm)	Organic Matter Content (%)	Water-Holding Capacity (mL)	Bulk Density (g/cm^{-3})	pH	Electrical Conductivity (mS cm^{-1})	PAH Content (mg kg^{-1})	Repellency WDPT (s)
Biochar	1000 (0)	100 (0)	5 (2)	93.8 (5)	50 (0)	0.6 (0.1)	9 (0.3)	0.5 (0.1)	14.9	3 (2)
Straw	200 (0)	100 (0)	40 (10)	92.5 (5)	50 (0)	0.3 (0.1)	5.5 (0.2)	0.3 (0)	<0.04	6 (4)

2.3. Sampling

Grape samples (200 berries per sample) were collected for each variety and treatment from six distinct vines. Sampling was stratified by hillslope position, with two grapes samples picked separately from the top (t) and another two from the bottom (b) of the vine row to account for the different heights and levels of sunlight exposure. The grape samples

https://doi.org/10.3390/environments13010044

were stored at −4 °C until further processing. Samples were organized and designated according to their treatments and hillslope locations: C b (control bottom position), C t (control top position), M b (mulch bottom position), M t (mulch top position), MB b (mulch–biochar bottom position), and MB t (mulch–biochar top position). A single composite collection was performed for each treatment and topographic position, and all analyses were conducted on those corresponding samples.

2.4. Grape Preparation for Analysis

According to the methodology proposed by Carbonneau and Champagnol (1993) [27], the grape extracts were obtained after maceration of seeds and skins in an ethanol and buffer solution at pH 3.20 for 24 h at 25 °C, followed by clarification of the suspension by centrifugation. The extracts were used to measure phenolic composition, while the classical analysis was performed on the must obtained from the pulp. The analyses were conducted in triplicate for each sample. For samples prepared for HPLC analysis, 2 mL of the clarified solution was filtered with a Whatman-Cytiva Europe (Velizy-Villacoublay, France) cellulose filter with a pore diameter of 0.45 μm and frozen at −4 °C until processing.

2.5. Classical Chemical Analysis

The basic chemical parameters were obtained directly from the corresponding must of each sample using an analytical method recommended by the OIV: weight berry, pH using pH meter (Orion Star A211, Thermocientific, Waltham, MA, USA) following reference protocol OIV-MA-AS313-15, titratable acidity (OIV-MA-AS313-01), and total soluble solid compounds (TSSs) measured using a digital refractometer (Atal PAL-1, Atago Co., Ltd., Tokyo, Japan). The instrument was calibrated with distilled water prior to analysis, and measurements were performed at room temperature. Results are expressed in °Brix. The determination of yeast assimilable nitrogen (YAN) was obtained according to an adapted methodology proposed by Gump et al. (2002) [28], which takes into account the quantity of the total nitrogen derived from free amino acids and ammonium forms. All the chemical analyses were performed in triplicate.

2.6. Global Phenolic Compounds and Chromatic Characteristics

In this study, the global phenolic compounds (total phenols, non-flavonoids, flavonoids, and anthocyanins) and chromatic characteristics (color and tonality) in the red grape varieties (Syrah I, Syrah II, Trincadeira, and Alicante Bouschet) were measured.

Total phenolic compounds were determined by measuring at 280 nm a diluted sample of the clarified extract prepared as previously described in Section 2.4, according to the protocol described by Ribéreau-Gayon (1970) [29]. Non-flavonoids were quantified by measuring the absorbance at 280 nm after the precipitation of flavonoids by reaction with formaldehyde over 72 h at specific conditions (at low pH, at room temperature, and in darkness) as reported by Kramling and Singleton (1969) [30]. Flavonoids were obtained by subtracting non-flavonoids from the total phenols. Total anthocyanins were evaluated using the protocol developed by Ribéreau-Gayon and Stonestreet (1965) [31] based on the SO_2 bleaching method. The color intensity (Abs 420 + 520 + 620 nm) and tonality (ratio of Abs 420 nm and 520 nm) were estimated according to OIV methods (OIV, 2014) [32].

2.7. Separation and Quantification of Individual Monomeric Anthocyanins by HPLC

To identify and quantify the individual monomeric anthocyanins present in the skin and seeds extract, a LC 300 HPLC system (PerkinElmer, Waltham, MA, USA) was used. The PerkinElmer LC 300 HPLC was equipped with a high-pressure quaternary pump with an adjustable flow rate of 0.001 to 10 mL/min, an autosampler, a reverse phase column, and an automatic injector coupled with photodiode array (PDA) detector covering

 https://doi.org/10.3390/environments13010044

a wavelength ranging from 190 to 700 nm. The separation occurred on a column C18 (150 × 4.6 mm), Gemini® NX-C18 110A with internal particles of 3 μm purchased from Phenomenex (Torrance, CA, USA). The oven temperature was maintained at 30 °C, the injection volume was 40 μL (the extracts and solvents were previously filtered through a 0.45 μm membrane), and the flow rate was 0.7 mL/min. The solvents were A (40% formic acid and 60% bi-distilled water), B (acetonitrile PA), and C (bi-distilled water). Methanol–water (50:50, *v:v*) was used to wash the column after the analyses. The initial conditions used were 25% A, 6% B, and 69% C for 15 min, followed by a 25% linear gradient of A, 25.5% B, and 49.5% C for 70 min, finishing with 20 min of 25% A 25.5% B and 49.5% C. The PDA was set at 520 nm, and the chromatographic runs were performed in triplicate. Data collection and analyses were carried out using SimplicityChrom software (version 1.0, PerkinElmer, Waltham, MA, USA). Identification followed the method described by Roggero et al. (1986) [33], and quantification was based on the corresponding calibration curves of six anthocyanin standards: delfinidin 3-O-glucoside, cyanin 3-O-glucoside, petunin 3-O-glucoside, pelargonidin 3-O-glucoside, peonidin 3-O-glucoside, and malvidin 3-O-glucoside curve. For each anthocyanin standard, a calibration curve was prepared using five standard solutions (25, 50, 75, 100, and 250 mg/L), which were obtained from successive dilutions of a stock solution of the anthocyanin (1.25 mg dissolved in 5 mL of 0.1% HCl in MeOH). The six anthocyanin standards were purchased from Extrasynthese. All the chemical analyses were performed in triplicate.

2.8. Separation and Quantification of Proanthocyanidins According to Their Degree of Polymerization

The separation of monomeric flavanols and oligomeric/polymeric proanthocyanidin fractions were carried out using a Sep-Pak® C18 solid-phaser extraction cartridge (Waters, Milford, MA, USA) based on their size according to the methodology proposed by Sun et al. (1998a) [34]. The content of each fraction was determined using the vanillin assay according to the method of Sun et al. (1998b) [35]. The quantification was carried out using standard curves prepared from monomeric flavanols and oligomeric- and polymeric-proanthocyanidin of grape-seed isolates as described by Sun et al. (2001) [36]. Extractions on the C18 cartridge and the readings after reactions with vanillin were performed in triplicate. All the chemical analyses were performed in triplicate.

2.9. Statistical Analysis

A two-way analysis of variance (ANOVA) was performed to evaluate the effects of the soil management treatment (factor 1: three levels—control, mulch, and mulch+ biochar treatments) and topographic position (factor 2: two levels—top and bottom positions), as well as their interaction. Where the ANOVA indicated significant effects, post hoc comparisons were conducted using Tukey's Honest Significant Difference (HSD) test at a 5% significance level. All analyses were performed using RStudio software 2025 (version 4.4.3).

3. Results

3.1. Classical Chemical Parameters

Significant differences in must composition parameters were obtained across grape varieties (Syrah, Trincadeira, Alicante Bouschet, and Antão Vaz), topographic positions (bottom—b or top—t), and organic treatments (control—C, wheat straw mulch—M treatment, or wheat mulch + biochar—MB treatment). The main classical parameters measured from musts were pH, total acidity, total soluble solids (TSSs) expressed in °Brix, and yeast assimilable nitrogen (YAN). The mean values and standard errors for each cultivar, includ-

 https://doi.org/10.3390/environments13010044

ing the interaction between organic treatments and vineyard topographic positions, are summarized in Table 2.

For **Syrah I**, significant differences for pH, total acidity, and TSS values were shown for both factors, while YAN concentrations were only affected by treatment. The pH values ranged from 4.01 (M b) to 4.40 (C t), with higher values under mulch top position (M t 4.24) and mulch-biochar bottom position (MB b 4.21). Total acidity varied between 3.78 and 4.58 g tartaric acid/L, and TSS ranged from 24.83 to 26.93 °Brix. TSS increased under M t when compared to the control top position (C t 24.83 vs. M t 26.20 °Brix). YAN ranged from 102 to 207 mg/L, with higher concentration shown for MB b.

For **Syrah II**, treatment factor had a significant effect on all parameters (Table 2). The pH values ranged from 4.17 to 4.24, with higher values under MB treatments at both positions. Total acidity varied from 2.98 to 3.70 g tartaric acid/L and TSS content ranged from 25.67 to 27.30 °Brix. YAN concentrations were between 61.25 and 195.42 mg/L, with the highest values observed in MB t.

The classical analysis of **Trincadeira**, shown in Table 2, indicated that pH, TSS, and YAN concentrations differed mainly between the two vineyard topographic positions, while treatment effects on these parameters were limited. The pH values ranged from 3.90 to 4.04, with higher pH values shown in the M b sample. Total acidity varied from 3.80 to 4.38 g tartaric acid/L, with highest value obtained for MB b treatments. YAN concentrations were between 70.00 and 102.08 mg/L, with increases observed under M t and MB b when compared to the control (C t = 70.00 vs. M t = 90.42 mg/L; C b = 96.25 vs. MB b = 102.08 mg/L).

For **Alicante Bouschet** grapes, the pH levels differed significantly by both position and treatment, ranging from 4.16 to 4.00, with higher values shown for the MB b sample. Total acidity values were similar across treatments and positions, varying from 3.28 to 3.38 g tartaric acid/L. TSS concentrations ranged from 25.87 to 29.40 °Brix, and the highest TSS concentrations were observed at the bottom positions (C b = 29.40; M b = 27.50° and MB b = 29.27 °Brix). YAN concentrations in the must differed significantly for both position and treatment factors, ranging from 191.04 to 246.46 mg/L, with the highest values shown for M t (246.46 mg/L) and MB b (211.46 mg/L).

In **Antão Vaz**, the pH values, ranging from 3.95 to 4.16, were significantly different by both position and treatment, with no position × treatment interaction. The M b sample accounted for the highest pH level when compared to MB-treated samples. Total acidity ranged from 3.03 to 3.24 g tartaric acid/L across all samples. Treated samples yielded slightly higher values of total acidity (M t = 3.24, MB b = 3.20 and MB t = 3.23 g tartaric acid/L), when compared to the corresponding control samples (C b = 3.15 and C t = 3.18 g tartaric acid/L) with the exception of M b (3.03 g tartaric acid/L). TSS concentrations varied between 20.85 and 23.18°Brix. YAN concentrations differed significantly by treatment and by position × treatment interaction, ranging from 126.90 to 159.70 mg/L, with the highest values shown in MB t in treated samples (MB t = 132.0 mg/L).

3.2. Chromatic Characteristics and General Phenolic Composition Assessment

Considering the results summarized in Table 3, the color parameters and total phenolic content for **Syrah I** samples were significantly affected by both treatment and vineyard topographic position factors, with the exception of monomeric flavan-3-ol content.

 https://doi.org/10.3390/environments13010044

Table 2. General classical chemical parameters of Syrah I, Syrah II, Trincadeira, Alicante Bouschet, and Antão Vaz grapes. Control (C), wheat straw mulch (M), and wheat straw mulch + biochar (MB). Bottom topographic position (b) and top topographic position (t); total soluble solid (TSS), yeast assimilable nitrogen (YAN).

Analytical Parameter		C b	C t	M b	M t	MB b	MB t	Position	Treatment	Position-Treatment
Syrah I	pH	$4.35^{b} \pm 0.02$	$4.40^{a} \pm 0.05$	$4.01^{f} \pm 0.00$	$4.24^{c} \pm 0.00$	$4.21^{d} \pm 0.00$	$4.15^{e} \pm 0.01$	***	***	***
	Total acidity	$4.58^{a} \pm 0.08$	$4.30^{b} \pm 0.04$	$4.10^{bc} \pm 0.12$	$3.83^{de} \pm 0.00$	$4.03^{cd} \pm 0.04$	$3.78^{e} \pm 0.16$	***	***	n.s.
	TSS (°Brix)	$26.93^{a} \pm 0.06$	$24.83^{e} \pm 0.06$	$24.90^{e} \pm 0.00$	$26.20^{b} \pm 0.00$	$25.83^{c} \pm 0.12$	$25.20^{d} \pm 0.00$	***	***	***
	YAN (mg/L)	$137.08^{c} \pm 0.06$	$188.13^{b} \pm 0.05$	$102.08^{d} \pm 0.06$	$123.96^{c} \pm 0.03$	$207.08^{a} \pm 0.08$	$135.63^{c} \pm 0.05$	n.s.	***	***
Syrah II	pH	$4.19^{b} \pm 0.01$	$4.19^{b} \pm 0.01$	$4.17^{c} \pm 0.01$	$4.18^{bc} \pm 0.01$	$4.24^{a} \pm 0.01$	$4.24^{a} \pm 0.01$	n.s.	***	n.s.
	Total acidity	$3.03^{c} \pm 0.04$	$3.05^{c} \pm 0.16$	$2.98^{c} \pm 0.04$	$3.00^{c} \pm 0.00$	$3.70^{a} \pm 0.09$	$3.40^{b} \pm 0.04$	*	***	**
	TSS (°Brix)	$26.97^{b} \pm 0.06$	$25.67^{d} \pm 0.06$	$26.60^{c} \pm 0.10$	$25.73^{d} \pm 0.06$	$26.5^{c} \pm 0.96$	$27.3^{a} \pm 0.06$	***	***	***
	YAN (mg/L)	$61.25^{d} \pm 0.00$	$61.25^{d} \pm 0.00$	$80.21^{c} \pm 2.53$	$86.04^{c} \pm 2.53$	$135.63^{b} \pm 0.00$	$195.42^{a} \pm 6.70$	***	***	***
Trincadeira	pH	$3.94^{c} \pm 0.01$	$4.04^{a} \pm 0.01$	$3.99^{b} \pm 0.00$	$3.95^{c} \pm 0.01$	$3.95^{c} \pm 0.01$	$3.90^{d} \pm 0.01$	***	n.s.	***
	Total acidity	$4.30^{a} \pm 0.04$	$3.80^{d} \pm 0.04$	$3.93^{c} \pm 0.04$	$4.13^{b} \pm 0.00$	$4.38^{a} \pm 0.04$	$3.83^{d} \pm 0.00$	***	**	***
	TSS (°Brix)	$20.90^{d} \pm 0.00$	$22.10^{b} \pm 0.00$	$20.73^{e} \pm 0.06$	$21.20^{c} \pm 0.00$	$22.40^{a} \pm 0.00$	$20.63^{f} \pm 0.06$	***	n.s.	***
	YAN (mg/L)	$96.25^{a} \pm 0.00$	$70.00^{b} \pm 0.00$	$87.51^{ab} \pm 0.00$	$90.42^{ab} \pm 0.03$	$102.08^{a} \pm 0.03$	$87.50^{ab} \pm 0.00$	*	n.s.	*
Alicante Bouschet	pH	$4.06^{b} \pm 0.00$	$4.00^{d} \pm 0.01$	$4.04^{c} \pm 0.02$	$4.06^{bc} \pm 0.01$	$4.16^{a} \pm 0.01$	$4.00^{d} \pm 0.01$	***	***	***
	Total acidity	$3.30^{a} \pm 0.08$	$3.33^{a} \pm 0.09$	$3.33^{a} \pm 0.04$	$3.38^{a} \pm 0.08$	$3.28^{a} \pm 0.11$	$3.35^{a} \pm 0.04$	n.s.	n.s.	n.s.
	TSS (°Brix)	$29.40^{a} \pm 0.00$	$26.40^{e} \pm 0.00$	$27.50^{c} \pm 0.00$	$26.90^{d} \pm 0.00$	$29.27^{b} \pm 0.06$	$25.87^{f} \pm 0.06$	***	***	***
	YAN (mg/L)	$191.04^{c} \pm 0.03$	$192.50^{c} \pm 0.00$	$195.42^{c} \pm 0.03$	$246.46^{a} \pm 0.03$	$211.46^{b} \pm 0.03$	$208.54^{b} \pm 0.03$	***	***	***
Antão Vaz	pH	$4.16^{a} \pm 0.06$	$4.06^{ab} \pm 0.08$	$4.15^{a} \pm 0.07$	$3.95^{b} \pm 0.05$	$4.03^{ab} \pm 0.14$	$3.95^{b} \pm 0.02$	***	**	n.s.
	Total acidity	$3.15^{a} \pm 0.09$	$3.18^{a} \pm 0.26$	$3.03^{a} \pm 0.33$	$3.24^{a} \pm 0.13$	$3.20^{a} \pm 0.36$	$3.23^{a} \pm 0.20$	n.s.	n.s.	n.s.
	TSS (°Brix)	$23.18^{a} \pm 0.97$	$22.40^{ab} \pm 1.31$	$21.53^{bc} \pm 0.63$	$20.95^{c} \pm 0.14$	$22.55^{ab} \pm 0.05$	$20.85^{c} \pm 0.57$	***	***	n.s.
	YAN (mg/L)	$159.72^{ab} \pm 0.10$	$139.31^{b} \pm 0.00$	$126.92^{ab} \pm 2.51$	$128.33^{a} \pm 2.53$	$129.82^{ab} \pm 0.00$	$132.01^{ab} \pm 6.70$	n.s.	*	*

Means followed by the same letter in the lines did not differ by Tukey's test at 5% ($p \leq 0.05$). Standard deviation of triplicate analysis; n.s. (not significant); * (significant differences at a 95% confidence level); ** (significant differences at a 99.9% confidence level); *** (significant differences at a 99.99% confidence level); total acidity expressed as g of tartaric acid per liter.

https://doi.org/10.3390/environments13010044

Table 3. General phenolic composition, color parameters, and condensed tannins in skin and seed extracts of Syrah I, Syrah II, Trincadeira, and Alicante Bouschet.

Analytical Parameter	Syrah I								
	C b	C t	M b	M t	MB b	MB t	Position	Treatment	Position–Treatment
Color and global phenolic compounds									
Total phenols (mg/L)	1438.58 ^{a} ± 7.88	984.03 ^{c} ± 8.41	992.63 ^{c} ± 0.90	1225.54 ^{b} ± 1.61	989.09 ^{c} ± 2.33	1265.54 ^{b} ± 14.86	**	***	***
Non-flavonoids (mg/L)	60.83 ^{bc} ± 1.85	55.65 ^{c} ± 1.17	63.46 ^{b} ± 2.72	62.64 ^{b} ± 3.76	72.40 ^{a} ± 1.54	59.49 ^{bc} ± 0.56	***	***	**
Flavonoids (mg/L)	1378.44 ^{a} ± 9.89	927.87 ^{c} ± 9.51	930.48 ^{c} ± 1.24	1162.03 ^{b} ± 6.49	917.55 ^{c} ± 2.89	1206.38 ^{b} ± 14.93	n.s.	**	***
Total anthocyanins (mg/L of malvidin)	610.72 ^{a} ± 8.4	463.23 ^{c} ± 5.55	545.03 ^{ab} ± 7.17	560.65 ^{b} ± 12.67	460.10 ^{c} ± 7.51	481.69 ^{bc} ± 6.96	**	**	***
Color intensity (u.a)	1.446 ^{a} ± 0.01	0.960 ^{c} ± 0.008	1.178 ^{b} ± 0.004	1.168 ^{b} ± 0.02	0.914 ^{d} ± 0.007	0.992 ^{c} ± 0.007	***	***	***
Tonality (u.a)	0.530 ^{c} ± 0.001	0.518 ^{d} ± 0.001	0.473 ^{e} ± 0.001	0.541 ^{b} ± 0.006	0.535 ^{bc} ± 0.004	0.568 ^{a} ± 0.004	***	***	***
Condensed tannins									
Monomeric Flavan-3-ols (mg/L)	43.43 ^{bc} ± 5.7	38.12 ^{c} ± 1.88	62.96 ^{a} ± 5.15	64.76 ^{a} ± 0.95	44.67 ^{bc} ± 0.55	54.94 ^{ab} ± 3.22	n.s.	***	n.s.
Oligomeric proanthocyanidins (mg/L)	131.41 ^{a} ± 3.23	34.78 ^{c} ± 10.76	59.46 ^{bc} ± 9.68	86.30 ^{b} ± 2.15	121.30 ^{a} ± 3.69	120.11 ^{a} ± 8.45	**	***	***
Polymeric proanthocyanidins (mg/L)	1405.41 ^{a} ± 38.2	993.92 ^{c} ± 8.60	817.57 ^{c} ± 28.67	1243.24 ^{b} ± 0.00	764.86 ^{e} ± 64.90	1174.32 ^{b} ± 24.84	***	***	***
Total tannins (mg/L)	1580.21 ^{a} ± 40.67	1066.82 ^{c} ± 21.24	948.99 ^{d} ± 13.83	1394.30 ^{b} ± 3.88	838.95 ^{e} ± 0.68	1349.38 ^{b} ± 36.52	***	***	***

Analytical Parameter	Syrah II								
	C b	C t	M b	M t	MB b	MB t	Position	Treatment	Position–Treatment
Color and global phenolic compounds									
Total phenols (mg/L)	823.89 ^{b} ± 10.41	812.59 ^{b} ± 15.52	865.67 ^{ab} ± 17.01	698.67 ^{c} ± 25.41	900.27 ^{a} ± 10.42	860.44 ^{ab} ± 33.31	***	***	***
Non-flavonoids (mg/L)	80.08 ^{c} ± 1.22	80.34 ^{c} ± 1.43	89.42 ^{b} ± 3.12	79.26 ^{c} ± 3.01	97.34 ^{b} ± 3.60	111.91 ^{a} ± 4.01	n.s.	***	***
Flavonoids (mg/L)	743.81 ^{b} ± 6.81	732.25 ^{b} ± 31.53	776.25 ^{ab} ± 17.64	619.45 ^{c} ± 26.21	802.93 ^{a} ± 9.62	748.53 ^{ab} ± 15.53	n.s.	***	***
Total anthocyanins (mg/L of malvidin)	485.67 ^{ab} ± 14.41	491.18 ^{ab} ± 4.53	514.46 ^{a} ± 12.31	393.74 ^{c} ± 17.61	506.05 ^{ab} ± 8.40	459.41 ^{b} ± 31.32	***	**	***
Color intensity (u.a)	0.956 ^{d} ± 0.010	0.987 ^{c} ± 0.004	1.032 ^{b} ± 0.006	0.827 ^{e} ± 0.004	1.081 ^{a} ± 0.010	1.020 ^{b} ± 0.004	***	***	***
Tonality (u.a)	0.515 ^{d} ± 0.001	0.507 ^{e} ± 0.001	0.518 ^{cd} ± 0.001	0.531 ^{b} ± 0.003	0.524 ^{bc} ± 0.002	0.572 ^{a} ± 0.004	***	***	***
Condensed tannins									
Monomeric Flavan-3-ols (mg/L)	50.84 ^{a} ± 0.91	44.53 ^{a} ± 8.83	43.51 ^{a} ± 8.62	48.62 ^{a} ± 10.91	37.34 ^{a} ± 9.02	36.31 ^{a} ± 7.23	n.s.	*	n.s.
Oligomeric proanthocyanidins (mg/L)	30.03 ^{a} ± 0.82	26.34 ^{a} ± 8.81	48.91 ^{a} ± 24.21	35.22 ^{a} ± 18.60	36.03 ^{a} ± 3.22	30.53 ^{a} ± 1.10	n.s.	n.s.	n.s.
Polymeric proanthocyanidins (mg/L)	404.51 ^{a} ± 74.90	445.54 ^{a} ± 171.14	511.81 ^{a} ± 38.92	440.73 ^{a} ± 67.12	704.42 ^{a} ± 156.01	576.21 ^{a} ± 67.02	n.s.	n.s.	n.s.
Total tannins (mg/L)	485.49 ^{a} ± 74.98	516.29 ^{a} ± 131.10	604.25 ^{a} ± 6.23	524.48 ^{a} ± 37.66	771.37 ^{a} ± 150.10	642.99 ^{a} ± 60.84	n.s.	n.s.	n.s.

Analytical Parameter	Trincadeira								
	C b	C t	M b	M t	MB b	MB t	Position	Treatment	Position–Treatment
Color and global phenolic compounds									
Total phenols (mg/L)	446.05 ^{d} ± 0.90	678.58 ^{c} ± 1.07	461.79 ^{d} ± 6.84	468.58 ^{d} ± 12.71	1110.23 ^{a} ± 5.73	928.58 ^{b} ± 4.12	**	***	***
Non-flavonoids (mg/L)	100.50 ^{ab} ± 1.65	84.74 ^{cd} ± 4.53	89.35 ^{bc} ± 0.59	67.67 ^{e} ± 8.05	72.30 ^{de} ± 1.45	106.13 ^{a} ± 6.77	n.s.	***	***
Flavonoids (mg/L)	345.53 ^{e} ± 3.22	595.53 ^{c} ± 5.97	372.44 ^{de} ± 7.12	399.98 ^{d} ± 1.56	1037.11 ^{a} ± 5.28	825.56 ^{b} ± 9.94	*	***	***
Total anthocyanins (mg/L of malvidin)	194.46 ^{cd} ± 7.19	312.77 ^{a} ± 11.48	239.59 ^{b} ± 3.80	230.62 ^{bc} ± 9.67	249.38 ^{bc} ± 3.48	172.33 ^{d} ± 5.04	*	***	***
Color intensity (u.a)	0.348 ^{e} ± 0.000	0.603 ^{a} ± 0.001	0.386 ^{d} ± 0.002	0.431 ^{c} ± 0.010	0.445 ^{b} ± 0.001	0.337 ^{f} ± 0.001	***	***	***
Tonality (u.a)	0.631 ^{c} ± 0.000	0.595 ^{e} ± 0.001	0.585 ^{f} ± 0.001	0.614 ^{d} ± 0.010	0.656 ^{b} ± 0.003	0.681 ^{a} ± 0.002	***	***	***

Table 3. *Cont.*

Condensed tannins									
Monomeric Flavan-3-ols (mg/L)	16.67 ^{ab} ± 0.68	12.65 ^{bc} ± 1.28	8.44 ^{c} ± 2.11	6.96 ^{c} ± 0.68	20.03 ^{a} ± 1.79	18.86 ^{ab} ± 5.20	n.s.	***	n.s.
Oligomeric proanthocyanidins (mg/L)	14.87 ^{ab} ± 0.46	21.20 ^{b} ± 3.23	28.20 ^{ab} ± 0.83	44.78 ^{a} ± 1.23	44.13 ^{a} ± 4.61	40.39 ^{ab} ± 2.71	**	***	n.s.
Polymeric proanthocyanidins (mg/L)	413.24 ^{cd} ± 0.96	584.46 ^{a} ± 4.78	287.16 ^{e} ± 0.96	339.19 ^{de} ± 7.64	451.35 ^{bc} ± 34.40	484.19 ^{b} ± 22.36	***	***	***
Total tannins (mg/L)	446.60 ^{c} ± 0.07	617.60 ^{a} ± 1.07	322.58 ^{d} ± 1.52	390.93 ^{c} ± 5.73	516.55 ^{b} ± 29.81	543.44 ^{b} ± 24.86	***	***	**

Analytical Parameter	**Alicante Bouschet**								
	C b	**C t**	**M b**	**M t**	**MB b**	**MB t**	**Position**	**Treatment**	**Position–Treatment**
Color and global phenolic compounds									
Total phenols (mg/L)	2006.56 ^{a} ± 32.04	1627.19 ^{b} ± 21.05	1312.13 ^{d} ± 27.03	1406.81 ^{c} ± 27.75	1656.30 ^{b} ± 10.53	1424.28 ^{c} ± 3.76	***	***	***
Non-flavonoids (mg/L)	83.53 ^{ab} ± 4.63	68.09 ^{c} ± 1.96	66.68 ^{c} ± 7.70	72.42 ^{bc} ± 7.68	96.46 ^{a} ± 1.21	75.37 ^{bc} ± 0.89	n.s.	***	**
Flavonoids (mg/L)	1872.25 ^{a} ± 9.89	1570.71 ^{b} ± 9.51	1245.39 ^{d} ± 1.24	1281.16 ^{cd} ± 6.49	1559.64 ^{b} ± 2.89	1348.43 ^{c} ± 14.93	***	***	***
Total anthocyanins (mg/L of malvidin)	1017.92 ^{a} ± 46.45	898.46 ^{b} ± 10.23	751.46 ^{c} ± 27.31	732.96 ^{cd} ± 1.47	757.90 ^{c} ± 14.48	662.77 ^{d} ± 18.06	***	***	*
Color intensity (u.a)	2.322 ^{a} ± 0.001	1.839 ^{c} ± 0.041	1.684 ^{d} ± 0.010	1.691 ^{d} ± 0.010	1.933 ^{b} ± 0.021	1.562 ^{e} ± 0.022	***	***	***
Tonality (u.a)	0.477 ^{d} ± 0.002	0.483 ^{cd} ± 0.010	0.499 ^{b} ± 0.001	0.490 ^{bc} ± 0.002	0.531 ^{a} ± 0.001	0.493 ^{bc} ± 0.011	***	***	***
Condensed tannins									
Monomeric Flavan-3-ols (mg/L)	60.19 ^{a} ± 0.29	24.56 ^{c} ± 0.10	43.25 ^{a} ± 3.99	25.76 ^{bc} ± 0.64	30.34 ^{a} ± 11.92	34.73 ^{bc} ± 0.15	***	**	n.s.
Oligomeric proanthocyanidins (mg/L)	146.38 ^{b} ± 7.17	60.51 ^{d} ± 3.59	152.76 ^{a} ± 0.71	46.52 ^{cd} ± 1.23	80.68 ^{bc} ± 3.99	45.83 ^{d} ± 5.38	***	***	***
Polymeric proanthocyanidins (mg/L)	1497.75 ^{bc} ± 47.78	1261.26 ^{c} ± 31.85	1098.65 ^{a} ± 17.20	1043.92 ^{ab} ± 4.78	1400.90 ^{c} ± 25.48	837.84 ^{d} ± 6.37	**	***	**
Total tannins (mg/L)	1704.31 ^{a} ± 40.31	1346.33 ^{c} ± 28.16	1294.66 ^{c} ± 13.91	1116.20 ^{d} ± 5.37	1512.04 ^{b} ± 7.93	918.40 ^{e} ± 11.60	**	***	**

Means followed by the same letter in the lines did not differ by Tukey's test at 5% ($p \leq 0.05$). Standard deviation of triplicate analysis. Total phenols, non-flavonoids, and flavonoids expressed as mg L^{-1} of gallic acid; u.a (absorbance unit); n.s. (not significant); * (significant differences at a 95% confidence level); ** (significant differences at a 99.9% confidence level); *** (significant differences at a 99.99% confidence level). Control (C), mulched (M), and biochar and mulched (MB). Bottom topographic position (b) and top topographic position (t).

https://doi.org/10.3390/environments13010044

The **total phenol** content exhibited significant differences between treatments and topographic positions. Control samples displayed total phenol contents of 1438.58 mg/L at C b and 984.03 mg/L at C t. In treated samples, the highest concentration was found in MB t (1265.54 mg/L), followed by M t (1225.54 mg/L), while the lowest was observed in MB b (989.09 mg/L). The **non-flavonoid** content exhibited the highest concentration in MB b (72.40 mg/L) and the lowest in MB t (59.49 mg/L). Control samples showed values of 60.83 mg/L (C b) and 55.65 (C t), while for samples with M treatment, ranged from 63.46 (M b) to 62.64 (M t). The non-flavonoid content exhibited the highest concentration in MB b (72.40 mg/L) and the lowest in MB t (59.49 mg/L). Control samples showed values of 60.83 mg/L (C b) and 55.65 (C t), while samples with M treatment were in the range from 63.46 (M b) to 62.64 (M t). **Flavonoid** content values ranged from 1378.44 mg/L in C b to 917.55 mg/L in MB b. C t and MB t samples showed values of 927.87 mg/L and of 1206.38 mg/L, respectively, while samples treated with M showed values of 930.48 mg/L (M b) and 1162.03 (M t). **Total anthocyanin** concentration displayed the highest concentration in C b (610.72 mg/L of malvidin-3-glucoside equiv.) and the lowest in MB b (460.10 mg/L of malvidin-3-glucoside equiv.). C t and MB t samples showed values of 463.23 mg/L and of 481.69 mg/L of malvidin-3-glucoside equiv., respectively, while samples treated with M showed values of 545.03 and 560.65 mg/L of malvidin-3-glucoside equiv. for M b and M t, respectively.

Color intensity displayed significant differences between treatments and positions. The highest color intensity was shown in C t (1.446), while the lowest was observed in MB b (0.914). For treated samples, the values ranged from 1.178 to 1.168 under M treatment and from 0.914 to 0.992 under MB treatment. **Tonality** values ranged from 0.473 in M b to 0.568 in MB t. No significant differences were observed for **monomeric flavan-3-ol** content across position and interaction between both factors. Control samples ranged from 43.40 (C b) to 38.12 (C t). In treated samples, the highest concentration was found in M t (64.76 mg/L), followed by M b (62.96 mg/L), while the lowest values were observed in MB b (44.67 mg/L) and MB t (54.94 mg/L). The **oligomeric proanthocyanidin** content showed moderate significant positional differences after treatment, with the lowest value observed in samples treated with wheat straw mulch (M b 59.46 and M t 86.30). Control samples ranged from 131.41 (C b) to 34.78 (C t), while samples treated with MB showing concentrations that varied from 121.30 and 120.11 mg/L for MB b and MB t, respectively. Polymeric anthocyanins were significantly affected, with the highest level observed for C b (1405.41 mg/L) and lowest shown for MB b (764.86 mg/L). C t and MB t samples showed values of 993.93 mg/L and of 1174.32 mg/L, respectively, while samples treated with M displayed values of 817.57 mg/L (Mb) and 1243.24 (Mt).

Considering the results summarized in Table 3 for **Syrah II**, the color parameters (color intensity and tonality), total phenols, and total anthocyanins were significantly affected by position and treatment, with significant position × treatment interaction in all four parameters. In contrast, non-flavonoids and flavonoids only varied by treatment. No significant differences were shown for monomeric flavan-3-ol content and oligomeric proanthocyanins.

The **total phenol** content exhibited significant differences between treatments and vineyard topographic positions. Control samples displayed total phenol contents of 823.89 mg/L at C b and 812.59 mg/L at C t. In treated samples, the highest concentration was found in MB b (900.27 mg/L), followed by M b (865.67 mg/L), while the lowest was observed in M t (698.67 mg/L). The **non-flavonoid** content displayed the highest concentration in MB t (111.91 mg/L) and the lowest in M t (79.26 mg/L). Control samples showed values of 80.08 mg/L (C b) and 80.34 (C t), while bottom samples treated with MB and Mb showed values of 97.34 and 89.42 mg/L, respectively. **Flavonoid** content showed

 https://doi.org/10.3390/environments13010044

significant differences between treatments and positions and no significant differences in the vineyard topographic positions, ranging from 802.93 mg/L in MB b to 619.45 mg/L in M t. Control samples showed values of 743.81 mg/L (C b) and 732.25 mg/L (C t), while M b and MB t showed values of 776.25 mg/L and 748.53 mg/L, respectively. **Total anthocyanin** concentration exhibited the highest concentration in MB b (506.05 mg/L of malvidin-3-glucoside equiv.) and the lowest in M t (393.74 mg/L of malvidin-3-glucoside equiv.). Control samples showed values of 485.67 mg/L (C b) and 491.18 mg/L of malvidin-3-glucoside equiv. (C t), while M b and MB t showed values of 514.46 and 459.41 mg/L of malvidin-3-glucoside equiv., respectively. The **color intensity** displayed significant differences between treatments and positions. The highest color intensity was shown in MB b (1.081), while the lowest was observed in M t (0.827). For control samples, the values ranged from 0.956 (C b) to 0.987 (C t), while the values for samples in M b and MB t were 1.032 and 1.020, respectively. **Tonality** values ranged from 0.507 in C t to 0.572 in MB t. No significant differences were observed for **monomeric flavan-3-ol** content across vineyard topographic position and interaction between both factors. Control samples ranged from 50.8 (C b) to 44.5 (C t). In treated samples, the highest concentration was found in M t (48.62 mg/L) followed by M b (43.51 mg/L), while the lowest values were observed in MB b (37.34 mg/L) and MB t (36.31 mg/L). The **oligomeric proanthocyanidin** content showed no significant positional, treatment, and position × treatment differences, with the lowest value observed in control samples (C b = 30.03 and C t = 26.34). Treated samples ranged from 48.91 (M b) to 35.22 (M t) under M treatment, while samples treated with MB had concentrations that varied from 36.03 and 30.53 mg/L for MB b and MB t, respectively. **Polymeric anthocyanin** concentrations were not significantly affected, with the highest level observed for MB b (704.42 mg/L) and lowest shown for C b (404.51 mg/L). C t and MB t samples showed values of 445.54 mg/L and of 576.21 mg/L, respectively, while samples treated with M showed values of 511.81 mg/L (M b) and 440.73 (M t).

Considering the results summarized in Table 3 for **Trincadeira**, the color parameters (color intensity and tonality), total phenols, and total anthocyanins were significantly affected by position and treatment, with significant vineyard topographic position × treatment interaction in all four parameters. In contrast, non-flavonoids only varied by treatment. No significant differences were shown for monomeric flavan-3-ol content.

Total phenol content showed significant effects of both vineyard topographic position and treatment. The highest concentration was shown for MB b (1110.23 mg/L), followed by MB t (928.58 mg/L), while the lowest value was found in C b (446.05 mg/L). **Non-flavonoid** content ranged from 106.13 mg/L (MB t) to 67.67 mg/L (M t). Control samples ranged from 84.74 (C t) to 100.50 (C b), while the values for samples in M b and MB b were 89.35 and 72.30 mg/L, respectively. **Flavonoid** content ranged from 345.53 (C b) to 1037.11 mg/L (MB b). For control and MB top samples, the values were 595.53 mg/L and 825.56 mg/L, respectively, while for M-treated samples, were 372.44 mg/L (M b) and 399.98 mg/L (M t). **Total anthocyanin concentration** exhibited the highest concentration in C t (312.77 mg/L of malvidin-3-glucoside equiv.) and the lowest in MB t (172.33 mg/L of malvidin-3-glucoside equiv). In M-treated samples, the concentrations varied from 239.59 (M b) to 230.62 mg/L of malvidin-3-glucoside equiv. (M t), while for C b and MB b, the values were 194.46 and 249.38 mg/L of malvidin-3-glucoside equiv., respectively. **Color intensity** values for Trincadeira differed significantly among topographic positions and treatments. The highest values were shown for C t (0.603), while the lowest was obtained in MB t (0.337). In treated samples, the values ranged from 0.386 in M b to 0.445 in MB b. **Tonality** values varied significantly between treatment, with the highest value observed in MB t (0.681) and the lowest in M b (0.585). Control samples ranged

 https://doi.org/10.3390/environments13010044

from 0.595 (C t) to 0.631 (C b). **Monomeric flavan-3-ols** showed significant treatment effects, with concentrations ranging from 6.96 mg/L in M t to 20.03 mg/L in MB b. The highest values were shown in samples with MB treatment at both positions (MB t 18.86 mg/L), while control samples ranged from 12.65 (C t) to 16.67 mg/L (C b). **Oligomeric proanthocyanins** varied between 14.87 mg/L (C b) and 44.78 mg/L (M t). MB-treated samples ranged from 40.39 (MB t) to 44.13 (MB b). **Polymeric proanthocyanidins** were also significantly affected by position and treatment, ranging from 287.16 mg/L in M b to 584.46 mg/L in Ct. The highest values were shown for C t and MB t (484.19 mg/L), while the lowest were correlated with M b.

Considering the results summarized in Table 3 for **Alicante Bouschet**, the color parameters (color intensity and tonality), total phenols, and total anthocyanins were significantly affected by vineyard topographic position and treatment, with significant position × treatment interaction in all four parameters. In contrast, non-flavonoids only varied by treatment. Significant differences were shown for monomeric flavan-3-ol content by position × treatment interaction.

Total phenol content showed significant effects of both factors. The highest values were shown for treated samples at the bottom position with M (M b 1312.13 mg/L) and with MB (MB b1656.30 mg/L). **Non-flavonoid** content ranged from 66.68 mg/L (M b) to 96.46 mg/L (MB b). Control samples ranged from 68.09 (C t) to 83.53 (C b), while the values for treated samples in M t and MB t were 72.42 and 75.37 mg/L, respectively. **Flavonoid** content ranged from 1245.39 (M b) to 1872.25 mg/L (C b). At the top position, the concentrations were 1570.71 mg/L in C t, 1281.16 mg/L in M t, and 1348.43 mg/L in MB t. **Total anthocyanin concentration** exhibited the highest concentration in C b (1017.92 mg/L of malvidin-3-glucoside equiv.) and the lowest in MB t (662.77 mg/L of malvidin-3-glucoside equiv.). In mulch-treated samples, the concentrations varied from 751.46 mg/L of malvidin-3-glucoside equiv. (M b) to 732.96 mg/L of malvidin-3-glucoside equiv. (M t), while for C t and MB b, the values were 898.46 and 757.90 mg/L of malvidin-3-glucoside equiv., respectively. **Monomeric flavan-3-ols** showed significant topographic position and treatment effects, with concentrations ranging from 24.56 mg/L in C t to 60.19 mg/L in C b. In MB-treated samples, the concentrations varied from 30.34 mg/L (MB b) to 34.73 mg/L (MB t), while for M b and M t, the values were 45.83 and 25.76 mg/L, respectively. **Oligomeric proanthocyanins** varied between 46.83 mg/L (MB t) and 152.76 mg/L (M b). In control samples, the concentrations varied from 60.51 mg/L (C t) to 146.38 mg/L (C b), while for M t and MB b, the values were 46.52 and 80.68 mg/L, respectively. **Polymeric proanthocyanidins** were also significantly affected by position and treatment, ranging from 837.84 mg/L in MB t to 1497.75 mg/L in C b. The highest concentrations displayed for M-treated samples were M b (1098.65 mg/L) and M t (1043.92 mg/L), while C t and MB b were 1261.26 and 1400.90 mg/L, respectively.

3.3. Individual Monomeric Anthocyanins in Skin and Seed Extracts

The HPLC data for individual monomeric anthocyanins obtained for Syrah I, Syrah II, Trincadeira, and Alicante Boushet cultivars are presented in Figure 3. In all cases, differences in monomeric anthocyanins profile were significantly affected between treatments and positions. Six of the fourteen monomeric anthocyanins identified were at higher concentration in all treated samples and controls in the following order: Malvidin 3-O-glucoside, Malvidin 3-O-acetylglucoside, Malvidin 3-O-coumarylglucoside, Peonidin 3-O-glucoside, Petunidin 3-O-glucoside, and Delfinidin 3-O-coumarylglucoside. The quantification of anthocyanins within each group (monoglucosylated, monoacetylated, monocoumarylated, and monocaffeoylated) is expressed as milligrams per liter (mg/L) of the corresponding

 https://doi.org/10.3390/environments13010044

monoglucosylated anthocyanin equivalent per liter. However, for simplicity, the results below are presented as mg/L.

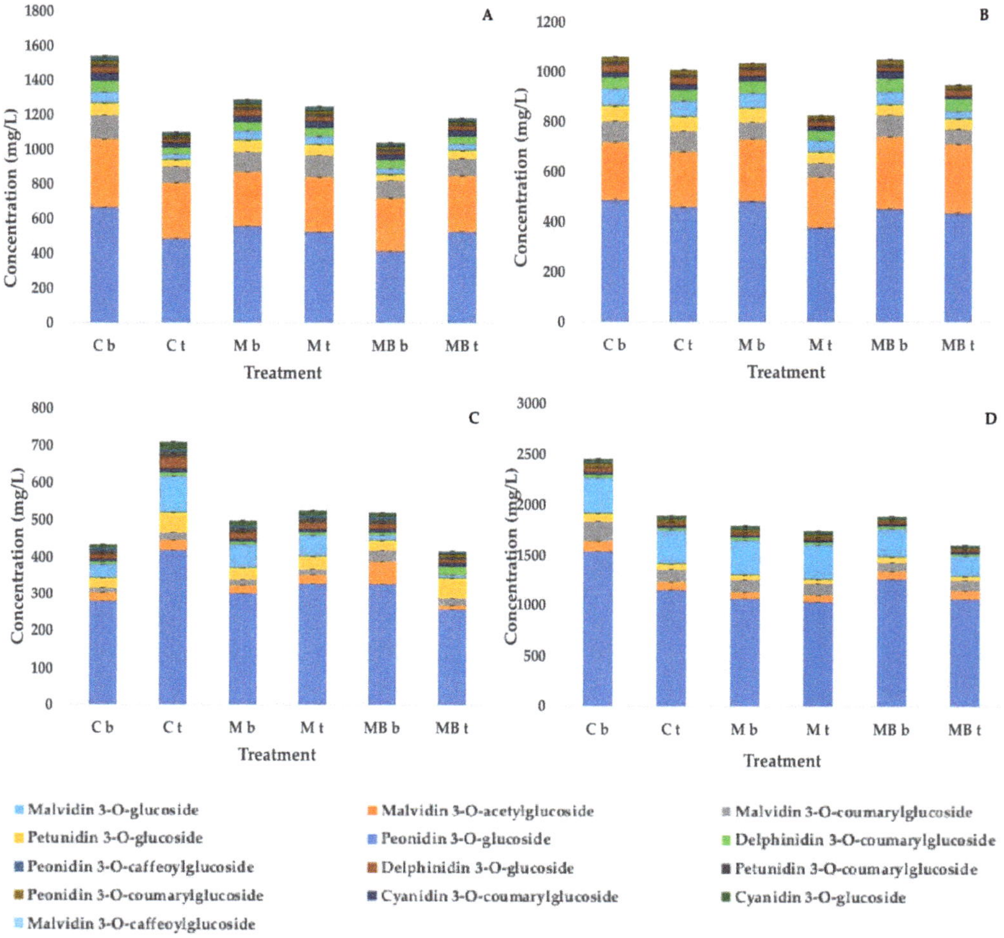

Figure 3. Concentration of individual monomeric anthocyanins (mg/L) determined with HPLC in grape samples from control (C), M treatment, and MB treatment and topographic positions (bottom—b and top—t): C b, C t, M b, Mt, MB b, and MB t. Each color represents a distinct anthocyanin compound, including delphinidin, cyanidin, peonidin, petunidin, and malvidin derivatives. Bars represent mean values for each compound and treatment. All monoglucosylated, monoacetylated, monocoumarylated, and monocaffeoylated are expressed as milligrams per liter (mg/L) of the corresponding monoglucosylated anthocyanin equivalent per liter: (**A**) Syrah I, (**B**) Syrah II, (**C**) Trincadeira, and (**D**) Alicante Bouschet.

In **Syrah I**, Malvidin 3-O-glucoside ranged from 663.8 mg/L in C b to 411.0 mg/L in MB b. Peonidin 3-O-glucoside ranged from 31.9 mg/L (C t) to 65.5 mg/L (C b), Petunidin 3-O-glucoside displayed the maximum value in C b (68.4 mg/L) and the minimum in MB b (33.0 mg/L). Malvidin 3-O-acetylglucoside exhibited significant differences among treatments. The highest concentration was found in C b (395.1 mg/L), while the lowest was obtained in MB b (307.2 mg/L). Malvidin 3-O-coumarylglucoside ranged from 138.0 mg/L in C b to 95.5 mg/L in C t and delphinidin 3-O-coumarylglucoside from 40.3 mg/L (C t) to 65.1 mg/L (C b). Total monomeric anthocyanin content ranged from 1038.6 mg/L

https://doi.org/10.3390/environments13010044

(MB b) to 1543.3 mg/L (C b). (Figure 3A) The amounts for other monomeric anthocyanins obtained for all varieties are presented in the Supporting Materials.

In **Syrah II**, Malvidin 3-O-glucoside ranged from 485.1 mg/L in C b to 376.1 mg/L in M t. Peonidin 3-O-glucoside ranged 40.8 mg/L (MB t) and 69.1 mg/L (C b), while Petunidin 3-O-glucoside ranged from 30.6 mg/L (MB t) to 59.5 mg/L (C b). Malvidin 3-O-acetylglucoside exhibited significant differences among treatments and positions. The highest concentration was shown in MB b (289.8 mg/L), while the lowest was obtained in M t (202.5 mg/L). Malvidin 3-O-coumarylglucoside ranged from 55.7 mg/L in M t to 86.6 mg/L in MB b and delphinidin 3-O-coumarylglucoside from 40.1 mg/L (C t) to 54.8 mg/L (C b). Total monomeric anthocyanin content ranged from 824.9 mg/L (M t) to 1056.2 mg/L (C b). Control bottom samples (C b) consistently showed the highest overall concentration of anthocyanins, followed by MB b (1037.5 mg/L) and M b (1035.9 mg/L) (Figure 3B).

In **Trincadeira**, Malvidin 3-O-glucoside ranged from 256.0 mg/L in MB t to 418.4 mg/L in C t. Peonidin 3-O-glucoside varied between 55.3 mg/L (MB t) and 97.4 mg/L (C t), while Petunidin 3-O-glucoside ranged from 24.6 mg/L (MB t) to 52.7 mg/L (C t). Malvidin 3-O-acetylglucoside exhibited significant differences among treatments and positions. The highest concentration was shown in to 28.3 mg/L in Ct (28.3 mg/L), while the lowest was found in MB t (20.1 mg/L). Malvidin 3-O-coumarylglucoside ranged from 8.5 mg/L (MB t) to 20.2 mg/L (C t), and Delphinidin 3-O-coumarylglucoside varied between 7.4 mg/L (C b) and 9.8 mg/L (C t). Total monomeric anthocyanin content ranged from 417.5 mg/L in MB t to 724.2 mg/L in C t. Control top samples (C t) consistently exhibited the highest total anthocyanin concentrations, and MB t (417.45 mg/L) and M b (498.90 mg/L) presented the lowest levels (Figure 3C).

In **Alicante Bouschet**, Malvidin 3-O-glucoside ranged from 1057.4 mg/L in MB t to 1542.7 mg/L in C b. Peonidin 3-O-glucoside varied between 208.1 mg/L (MB t) and 351.4 mg/L (C b), while Petunidin 3-O-glucoside ranged from 40.6 6 mg/L (MB t) to 74.6 mg/L (C b). Malvidin 3-O-acetylglucoside exhibited significant differences among treatments and positions. The highest concentration was shown in C b (99.7 mg/L), while the lowest was found in M b (71.8 mg/L). Malvidin 3-O-coumarylglucoside ranged from 87.6 mg/L (MB b) to 196.5 mg/L (C b), and Delphinidin 3-O-coumarylglucoside varied between 22.9 mg/L (MB t) and 30.0 mg/L (C b). Total monomeric anthocyanin content ranged from 1615.7 mg/L in MB t to 2466.3 mg/L in C b. Control top samples (C b) consistently exhibited the highest total anthocyanin concentrations, and MB t (1615.71 mg/L) and M t (1749.59 mg/L) presented the lowest levels (Figure 3D).

4. Discussion

4.1. Classical Chemical Parameters

4.1.1. Total Acidity and pH

The pH values observed among all cultivars appeared relatively high; however, this may also be associated with sample treatment, as the freezing and subsequent defrosting of the grapes—required due to the large number of samples analyzed—cannot be excluded as a factor affecting berry cell integrity and the distribution of organic acids, potentially influencing the measured pH values [28].

The mulch treatment significantly influenced grape must chemical composition, most notably in the Syrah I vines, where a reduction in pH was observed. This decrease in pH was generally associated with slightly higher total acidity, a relationship commonly reported in previous studies investigating the effects of organic amendments on grape must composition. Such responses have been attributed to delayed ripening and modified nutrient dynamics, as a response to higher yield and increased nitrogen availability [37].

 https://doi.org/10.3390/environments13010044

This could indicate that reduced pH observed in Syrah I is related to increased nitrogen availability. However, in this study, no consistent increase in YAN was observed under both treatments when compared with the corresponding controls at the same topographic position, with the exception of MB b. Differences in the composition of nitrogen sources between the control and treated samples may influence the buffering capacity of the must and, consequently, affect pH and total acidity.

Other studies have reported increased potassium concentration following mulch application [17,21], which can neutralize juice acids and increase pH. This may be explained by mulch's ability to regulate soil temperature, as potassium uptake is temperature-dependent [6]. Conversely, lower acidity might result from a microclimatic change in the canopy, where warmer conditions in treated vines could enhance the consumption of organic acids, reducing their concentration at harvest. These effects could be linked to changes in soil surface properties following organic amendment application.

In Alicante Bouschet, pH values decreased under M b and increased under MB b, while total acidity remained relatively stable. These findings may reflect modifications in cation concentration (K^+ and Ca^{2+}) rather than changes in acid concentration. For Trincadeira and Antão Vaz, a decrease in pH and a corresponding increase in total acidity was observed under M treatment at the top position (M t), indicating that mulch is probably capable of modifying soil conditions differently along the topographic gradient by reducing acidity at the bottom position while increasing at the top position. This bottom–top variation pattern was not evident in MB treatment, supporting previous reports that biochar may buffer or decrease the effects of mulch, likely due to its high surface area and capacity to stabilize nutrient and moisture dynamics. The reduction in positional differences in total acidity shown for M treatment suggests that this amendment might have a potential role in homogenizing vine response across the vineyard topographic position.

4.1.2. Total Soluble Solids (TSSs)

There were no significant main effects of treatments on TSS °Brix values across the vineyard, indicating that the amendments did not systematically delay sugar accumulation, although they showed significant differences for the position factor under MB treatment in all varieties. Positional differences in TSS concentrations were reduced when treatment was applied, particularly following MB treatment, which yielded higher TSSs in the bottom samples for Syrah I. Similar patterns were found in Trincadeira, Alicante Bouschet, and Antão Vaz grapes treated with mulch. These results suggest a possible interaction between amendment type and soil water distribution, which are in line with previous reports describing a general reduction in TSSs following amendment treatments [37], while others observed these effects only during dry-year conditions [19]. The reduction in positional differences might suggest improved soil water-holding capacity and structure after organic treatment, whereby the bottom topographic position is able to retain more water that would otherwise leach down in the absence of organic amendments and promote more balanced moisture conditions at both topographic positions in the vineyard.

4.1.3. Yeast Assimilable Nitrogen (YAN)

Yeast assimilable nitrogen (YAN) was the most dynamic parameter in response to the treatments. A significant treatment × topographic position interaction for the overall varieties revealed that the effect of each amendment on nitrogen availability was highly dependent on its topographic position. In most cases, grapes treated with MB reported higher YAN values than controls or grapes treated with M, suggesting a treatment-dependent improvement in nitrogen availability. The amendments applied in this study were not particularly rich in nutrients compared to other organic amendments, such as manure,

 https://doi.org/10.3390/environments13010044

sludge, or green manure. However, biochar can be conceptualized as a highly porous "ion sponge", able to absorb and release surrounding nutrients depending on soil chemical conditions. This property contributes to an increase in cation exchange capacity (CEC) and improves nutrient retention [12].

For Syrah II, a similar pattern of the classical parameters was observed as in Syrah I, except for the YAN parameter. Syrah II displayed low YAN values in control samples (61.25 mg/L) and the highest values in MB t (195.42 mg/L). The Syrah II plots were likely nitrogen-deficient, and the application of M or MB treatment may have contributed to the much higher assimilable nitrogen levels observed in the must when compared to the control plot at both topographic positions. The variation in YAN values between Syrah I and Syrah II may be related to differences in the vineyard orientation and microclimate as the two plots were placed in opposite vineyard directions [38]. Other local factors such canopy of the vines, soil temperature, and sun exposure could have contributed to yielding significant YAN variations between both Syrah plots [38]. Slight but significant changes in YAN between both positions were also found in other varieties, suggesting a different mineral composition not only between vineyards but also within the same vineyard. Given the slow nutrient release of the OA applied (due to their chemical characteristics) and the limited observation period (first year), a large increase in total nitrogen supply was not expected. However, the observed reduction in positional differences under M treatment for all grape varieties except for Alicante Bouschet suggests that mulching contributed to a more balanced distribution at both topographic positions in the vineyard.

4.1.4. General Trends of Position and Treatment Effects on Classical Parameters

Overall, the application of M and MB treatment promoted considerable shifts in the must chemical composition mainly associated with nitrogen accumulation and sugar concentration, particularly at different topographic positions, while changes in pH levels were less evident. The reduction in positional differences obtained under M treatment suggests improved soil structure and water retention at the top topographic position, potentially favoring a more balanced nutrient uptake at both topographic positions for some varieties. Although the observed treatment effects are discussed in relation to potential changes in soil water and nutrient dynamics, direct measurements of soil moisture, temperature, and nutrient availability were not collected during the growing season; therefore, future studies integrating detailed soil monitoring are necessary to better elucidate the mechanisms driving these responses. The MB treatment yielded the strongest effects, highlighting its potential as a sustainable soil management strategy for reducing bottom–top variability and enhancing must quality parameters. Further analysis of the phenolic composition and chromatic characteristics was performed to better understand how these parameters were affected by topographic position and treatment.

4.2. Color and Global Phenolic Composition

In this study, clear effects of vineyard topographic position and soil management practices were observed for the global phenolic composition and color parameters (color intensity and tonality) of the grape varieties examined. Significant differences were observed with grapes samples treated with mulch or mulch combined with biochar, affording the highest phenolic concentrations in some cultivars when compared with control samples. Despite the strong and consistent influence of treatment, the level and trend of the response varied among cultivars, given the intrinsic varietal characteristics and site-specific environmental conditions. These variations are discussed in detailed and structured according to the main analytical parameters evaluated.

 https://doi.org/10.3390/environments13010044

4.2.1. Total Phenols, Non-Flavonoids, and Flavonoids

Total phenol content was significantly influenced by both treatment and position across all grape varieties. In Syrah I and Alicante Bouschet, the highest values were observed in the control (untreated) bottom samples (1438.58 and 2006.56 mg/L, respectively), while in Syrah II and Trincadeira, the highest levels were mostly shown with MB treatment, particularly at the bottom position (900.27 and 1110.23 mg/L). These results indicated a consistent enrichment of total phenols under MB treatment, yielding the highest values. Across all cultivars, M and MB treatments reduced the positional differences when compared to controls, suggesting a homogenizing effect of the amendments in grape phenolic composition. This effect was more pronounced for all cultivars except Syrah II under M treatment, where phenolic concentrations increased at the bottom and decreased at the top (Syrah II: M b 865.67 and M t 698.67 mg/L; C b 823.89 and C t 812.59 mg/L).

Non-flavonoid concentrations showed smaller but significant differences among treatments. The highest values were consistently found in MB treatment samples of Syrah I (72.40 mg/L), Syrah II (111.91 mg/L), and Alicante Bouschet (96.46 mg/L). While these patterns varied slightly among cultivars, when biochar was used in organic treatment, non-flavonoid content seemed to increase regardless of position. These evidences underline the sensitivity of non-flavonoids to soil organic amendment type and highlights the influence of biochar on the accumulation of small phenolic compounds.

Flavonoid content followed the same overall pattern as shown for total phenols. The highest concentration was observed in treated samples for all cultivars except Alicante Bouschet (C b 1872.25 mg/L), MB t for Syrah I—1206.38 mg/L, MB b for Syrah II—802.93 mg/L, and MB b for Trincadeira—1037.11 mg/L. In contrast, control and M-treated samples displayed lower levels of flavonoid, suggesting that biochar enrichment intensified treatment effects. Across all varieties, positional differences in flavonoid content were reduced when organic treatment was applied, which was also seen for total phenols.

4.2.2. Total Anthocyanins, Color Intensity, and Tonality

Total anthocyanin concentrations demonstrated clear varietal and treatment-related significant differences. In Syrah I and Alicante Bouschet, the highest anthocyanin concentration was shown in C b (610.72 and 1017.92 mg/L of malvidin-3-glucoside equiv., respectively), whereas in Syrah II, MB treatment displayed comparable or, in some cases, higher values than controls. The positional differences were reduced under M and MB treatment, suggesting that organic treatments weakened bottom–top variability. In Trincadeira, the top control sample (C t) exhibited the highest anthocyanin content, while MB treatment increased total polyphenols and flavonoids but not anthocyanins, suggesting compositional adjustments within the phenolic profile.

Color intensity followed the same general trend as shown for total phenols and anthocyanins, with significant effects of both factors (position and treatment). These highest color intensity values were observed in control samples of Syrah I (1.446) and Alicante Bouschet (2.322), while in Syrah II and Trincadeira, MB treatments yielded higher color intensities compared to controls (1.081 and 0.445, respectively). Tonality displayed significant differences among treatment and position. The highest tonality was consistently found in top samples treated with MB across varieties, 0.568 in Syrah I, 0.572 in Syrah II, 0.681 in Trincadeira, and 0.531 in Alicante Bouschet, suggesting that treatment had a considerable impact on color balance.

4.2.3. Monomeric Flavan-3-Ols, Oligomeric, and Polymeric Proanthocyanidins

Monomeric flavan-3-ols exhibited minor variation compared with other phenolic parameters. Slight increases were found in samples treated with M for Syrah I and MB

https://doi.org/10.3390/environments13010044

for Trincadeira, while Syrah II and Alicante Bouschet showed no consistent trend. The highest concentrations were shown in M t in Syrah I (64.76 mg/L) and MB b in Trincadeira (20.03 mg/L), suggesting that organic treatment enhanced this fraction in some varieties.

Oligomeric and polymeric proanthocyanidins displayed more distinct treatment responses. In Syrah I and Trincadeira, the highest level of oligomeric proanthocyanidins were shown in MB b (121.30 and 44.13 mg/L, respectively), while polymeric forms peaked in C b for Syrah I (1405.4 mg/L) and in C t for Trincadeira (584.46 mg/L). In Syrah II and Alicante Bouschet, M b treatments afforded higher oligomeric fractions while polymeric proanthocyanidins were highest in MB b t for Syrah II (704.42 mg/L) and for Alicante Bouschet (1400.90 mg/L).

4.2.4. General Trends of Position and Treatment Effects on Global Phenolic Parameters

Across all varieties, Alicante Bouschet and Syrah I exhibited the highest overall phenolic concentrations, followed by Trincadeira and Syrah II. The MB treatment consistently improved total phenol, non-flavonoid, and flavonoid content across varieties, while positional effects were reduced compared with the control. In contrast, anthocyanins and color parameters displayed more distinct responses, suggesting varietal differences in sensitivity to soil and microclimatic conditions.

These patterns are in line with previous studies reporting that soil water availability, temperature, and nutrient status are the main factors for regulating grape polyphenol biosynthesis [24,37,39–47]. Given the limited differences in YAN content shown for both treatments within each cultivar, which can serve as an indicator of nitrogen supply, nitrogen availability played a limited role in the observed differences between positions and treatments. Water supply probably represented the leading factor influencing variation across treatments and positions. Several authors suggest that under water stress conditions, the final polyphenol concentration in grapes tends to increase [39,45]. Some studies have found reductions or no changes in polyphenol concentrations following treatment, while others reported increases [11,23]. Additionally, the same mulching treatment applied in different vineyards within the same study generated divergent responses: an increase in one site and no change in the other site [11,22,23]. However, none of these studies reported responses to the same OA mulching depending on vine topographic position within the same vineyard, as shown in this study.

Mulch and biochar applications are known to reduce runoff and enhance soil moisture retentions [8,48], thus moderating vine water stress and homogenizing berry development along the topographic gradient. However, the response varies depending on the intensity and timing of water stress, as well as variety and vintage, highlighting the complexity of the mechanism and its dependence on additional *terroir* factors [6,8,39,49,50].

Considering the cultivars studied, Syrah I and Alicante Bouschet exhibited the richest phenolic profile probably due to warmer or drier vineyard site conditions, which promoted smaller berry size and higher skin-juice ratios, and may have increased pigment concentration. In contrast, Syrah II and Trincadeira displayed lower phenolic levels, probably related to variations in the microclimate and different vineyard environments.

Differences between M and MB treatments further suggested that biochar contributed additional effects beyond those of straw mulch alone. MB treatment afforded higher total phenol and flavonoid content across most varieties, likely reflecting its enhanced capacity to improve soil physical properties and water-holding potential, although further studies are necessary to confirm soil structure and moisture. When M treatment was used, minor positional differences were observed, suggesting significant improvements in soil structure and water retention at the top position, potentially favoring more balanced nutrient uptake at both topographic positions for some grape varieties. Future studies on soil monitoring are necessary to better elucidate the

https://doi.org/10.3390/environments13010044

mechanisms driving these responses. However, local differences in soil texture, slope inclination, and microclimate probably modulated treatment effects.

Overall, the results confirm that vineyard topographic position and organic treatment jointly affect grape phenolic composition. The consistent enhancement of total phenols, flavonoids, total anthocyanins, and color characteristics under MB treatment demonstrates the potential of biochar-enriched mulches as sustainable vineyard practices for improving grape quality. Additionally, M treatment is beneficial in improving the phenolic composition at the top topographic position, thereby reducing bottom–top variability. This could be particularly important for certain cultivars planted in specific sites that lack optimal conditions for producing high-quality grapes. However, subsequent years of study are necessary to further highlight some of the trends identified here.

4.3. Individual Monomeric Anthocyanins in Skin and Seed Extract HPLC

4.3.1. General Trends in Anthocyanin Variation

The composition and concentration of anthocyanins varied significantly among treatments and topographic positions across all cultivars. Overall, control samples displayed the highest total anthocyanin content, while MB-treated samples, particularly at the bottom position, consistently presented lower levels (Figure 4). The total anthocyanin concentration were in the ranges of approximately 1000–1500 mg/L in Syrah I, 800–1100 mg/L in Syrah II, 400–700 mg/L in Trincadeira, and 1600–2400 mg/L in Alicante Bouschet. These variations might result from both the cultivar's intrinsic genetic capacity for anthocyanin biosynthesis and site-specific factors, such as microclimates, topographic exposure, and soil water availability. Despite these differences, a consistent pattern was observed among cultivars; organic treatment, especially MB treatments, reduced positional differences and probably homogenized anthocyanin distribution at both topographic positions.

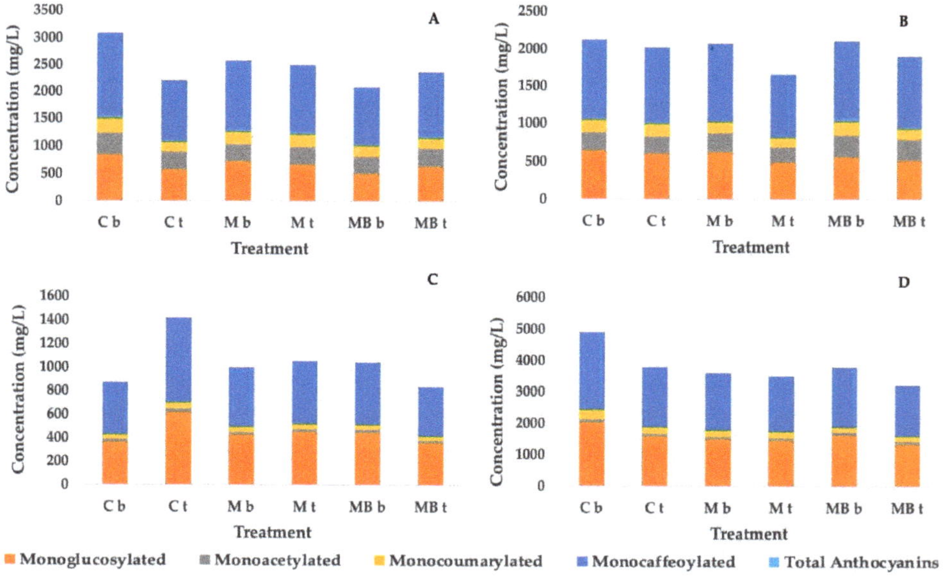

Figure 4. Total concentrations of monomeric anthocyanin groups (monoglucosylated, monoacetylated, monocoumaroylated, and monocaffeoylated) determined with HPLC in grape samples from different treatments and topographic positions (C b, C t, M b, M t, MB b, and MB t). Each bar color represents a different group of anthocyanins. Mean values of all monomeric anthocyanin groups are expressed as milligrams per liter (mg/L) of the corresponding monoglucosylated anthocyanin equivalent per liter: (**A**) Syrah I, (**B**) Syrah II, (**C**) Trincadeira, and (**D**) Alicante Bouschet.

https://doi.org/10.3390/environments13010044

In Syrah I, the total anthocyanin concentrations measured spectrophotometrically in control bottom position samples was 610.72 mg/L expressed as malvidin-3-O-glucoside equivalents, while HPLC quantifications for the same sample yielded 1543.4 mg/L, with malvidin 3-O-glucoside accounting for 663.8 mg/L. This apparent divergence results from methodological differences between the two approaches. In this study, HPLC quantification was performed using individual calibration curves for each monoglucoside anthocyanin (in total, six calibration curves), providing compound-specific response factors and accurately reflecting the contribution of each pigment. On contrary, the Ribéreau-Gayon and Stonestreet method uses a single calibration based on malvidin-3-O-glucoside and considers identical molar absorptivity for all anthocyanins. This generalization, combined with the fixed wavelength measurement at 520 nm, can reduce analytical precision and may underestimate pigments with higher molar extinction coefficients, such as delfinidin or petunidin derivatives. Although previous studies have used individual calibrations curves for anthocyanin quantification by HPLC, none have directly compared such data with Ribéreau-Gayon results. Considering that compound-specific calibration improves analytical accuracy, it is reasonable to consider that the values obtained represent the real anthocyanin concentrations in grape extracts for each cultivar studied [51–54].

4.3.2. Effects of Treatments on Anthocyanin Concentration

Across the four grape varieties, clear differences in monomeric anthocyanin composition and concentration were observed among treatments and individual monomeric anthocyanins classes (monglucosylated, monoacetylated, monocoumaroylated, and monocaffeoylated). Variations between positions (bottom vs. top) and treatments (C, M, and MB) indicated that both factors significantly influenced anthocyanin biosynthesis and accumulation.

Monoglucosylated anthocyanins consistently controlled the pigment profile; particularly, Malvidin 3-O-glucoside showed the highest concentrations across all treatments, followed by petunidin, peonidin, delphinidin, and cyanidin derivatives. The highest levels of these compounds were consistently observed in control bottom samples, while lower concentrations were displayed under M and MB treatments. In Syrah I and Alicante Bouschet, malvidin 3-glucoside accounted for more than half of total anthocyanins, which confirms the importance of this pigment in color expression. In general, the decrease in monoglucosylated forms under MB treatments suggests that mulch and biochar may influence grape metabolism due to modifications in soil moisture and temperature that can affect anthocyanin biosynthesis and degradation processes.

Monoacetylated and monocoumaroylated anthocyanins displayed similar treatment-related trends but with larger variations among cultivars. Malvidin 3-O-acetylglucoside concentrations were highest in control samples for Syrah I and Alicante Bouschet, while MB treatments showed slightly lower concentrations but with reduced positional differences. Coumaroylated anthocyanin forms, including delfinidin, peonidin, and malvidin 3-O-coumarylglucoside, exhibited clear positional and treatment effects, with higher content observed in controls and M-treated samples compared to MB. These anthocyanin derivatives are related to increased pigment and color stability. MB-treated samples reported lower levels, suggesting that soil water retention conditions might have increased due to biochar in the soil amendment, consequently promoting a shift towards less acylated anthocyanins profiles.

Monocaffeoylated anthocyanins typically represented less than 3% of total anthocyanins, although they followed a similar pattern as described previously for monoglycosylated anthocyanins. In Syrah I and Trincadeira, minor increases were observed in MB-treated top samples, possibly related to varietal differences in phenolic metabolism or microclimatic conditions between topographic positions.

https://doi.org/10.3390/environments13010044

4.3.3. Comparative Cultivar Responses

Despite strong treatment effects, the level of anthocyanin response differed considerably among cultivars. Alicant Bouschet exhibited the highest total anthocyanin content, which is consistent with its known high anthocyanin profile and teinturier nature. Syrah I displayed clear positional differences, which were reduced under organic treatment. Syrah II showed lower concentrations overall compared to Syrah I, probably due to differences in orientation and local microclimate. Trincadeira reported the lowest anthocyanin concentrations but had the most pronounced relative increase at the treated top position, likely due to its delayed ripening.

4.3.4. Implications for Vineyard Management

Across all cultivars with M and MB treatments, a decrease in anthocyanin content was consistently observed, suggesting that organic treatment may influence secondary metabolism by influencing environmental stress. While control bottom samples showed highest pigment concentrations when compared to Ct, this effect may also reflect local hydric and nutritional imbalances. Likewise, the more balanced anthocyanin pattern in treated samples may suggest improved homogeneity at both topographic vineyard positions. Therefore, the use of mulch and biochar may enhance vineyard uniformity and stability, while leading to a modest decrease in anthocyanin concentration. The level of these effects depends on the cultivar-specific physiology, topographic position, and microclimatic factors, highlighting the importance of designing soil management strategies to site conditions and production goals.

The cultivar-specific responses observed highlight the importance of evaluating grape varieties individually under semi-arid conditions. Differences among cultivars indicate that organic amendments and vineyard topographic position can differentially influence technologically relevant parameters, including acidity, sugar accumulation, nitrogen availability, and phenolic potential, supporting the need for site- and cultivar-specific soil management strategies.

5. Conclusions

The conclusions into the effects of straw mulch and mulch+ biochar treatment effects on grape composition, with particular emphasis on global phenolic compounds across vineyard topographic position (bottom vs. top), were as follows:

- Classical chemical parameters showed significant variation across all varieties, although no consistent trend was observed among them. In some varieties, mulching decreased pH and increased total acidity, while YAN was primarily driven by topographic position rather than soil treatment. The most notable observation in some cultivars concerns TSS content, which closely followed the anthocyanin trend, reinforcing that these responses may reflect shifts in ripening progression across treatments and topographic positions.
- Total phenols, flavonoids, total anthocyanins, and color characteristics were consistently enhanced under MB treatment, which demonstrates the potential of biochar-enriched mulches as sustainable vineyard practices for improving grape quality due to its capacity to improve soil physical properties and water-holding potential. Additionally, M treatment is beneficial in improving the phenolic composition at the top topographic position, thereby reducing bottom–top variability.
- Total tannins were slightly reduced by both M and MB treatments and were higher on the bottom topographic position, although no consistent pattern was observed among cultivars. In some cases, samples treated with MB showed higher levels at the bottom position, while in other cultivars tannins, concentrations increased under M treatment.

 https://doi.org/10.3390/environments13010044

- Total anthocyanins determined by HPLC were consistently reduced on M and MB treatments and were higher on bottom topographic positions. The treated samples frequently showed reduced total anthocyanin content relative to control samples, although reduced positional differences within the same organic treatment were observed.

In conclusion, even after a single season, mulch, and mulch and biochar emerge as powerful drivers of grape quality, with the potential to reshape topography-related patterns in berry composition. Long-term investigations are necessary to assess the stability of these effects over multiple vintages. Beyond its immediate impact on grape chemistry, mulching offers valuable insights into the interplay between organic soil management and vineyard landscape features, with promising implications for both wine quality enhancement and the long-term sustainability of viticulture in Mediterranean and other water-limited environments.

Supplementary Materials: The following supporting information can be downloaded at https://www.mdpi.com/article/10.3390/environments13010044/s1, Table S1: Concentration of individual monomeric anthocyanins (mg/L) identified and quantified by HPLC-DAD in Syrah I grape samples; Table S2: Concentration of individual monomeric anthocyanins (mg/L) identified and quantified by HPLC-DAD in Syrah II grape samples; Table S3: Concentration of individual monomeric anthocyanins (mg/L) identified and quantified by HPLC-DAD in Trincadeira grape samples; Table S4: Concentration of individual monomeric anthocyanins (mg/L) identified and quantified by HPLC-DAD in Alicante Bouschet grape samples; Table S5: Mean grape yield (kg plant^{-1}) and standard deviation for grapevines under different soil management treatments at the Fundação Eugénio de Almeida (FEA) and Herdade do Mouchão (MOU) vineyards. Treatments include control (U), wheat straw mulch (M), and wheat straw mulch combined with biochar (MB).

Author Contributions: Conceptualization, J.M.R.-d.-S., S.C., S.P. and M.J.C.; methodology, J.M.R.-d.-S., S.C., S.P. and M.J.C.; validation, J.M.R.-d.-S., S.C., S.P., M.J.C. and. S.G.H.; formal analysis, M.P., S.G.H. and N.S.; investigation, M.P., S.G.H. and N.S.; resources, J.M.R.-d.-S., S.P. and M.J.C.; writing—original draft preparation, M.P. and S.G.H.; writing—review and editing, J.M.R.-d.-S., S.C., S.P., M.J.C. and S.G.H.; supervision, J.M.R.-d.-S., S.C. and S.G.H.; project administration, J.M.R.-d.-S.; funding acquisition, S.P. All authors have read and agreed to the published version of the manuscript.

Funding: This research was funded by the European Union—PRR (Plano de Recuperação e Resiliência), under the project SOLVIT (Vine & Wine Portugal C644866286-00000011). This work was also supported by FCT—Fundação para a Ciência e Tecnologia, I.P. through projects UID/04129/2025 (https://doi.org/10.54499/UID/04129/2025) of LEAF-Linking Landscape, Environment, Agriculture and Food, and LA/P/0092/2020 (TERRA).

Data Availability Statement: Data are contained within this article and its Supplementary Materials.

Acknowledgments: The authors acknowledge the R&D unit MED—Mediterranean Institute for Agriculture, Environment and Development (https://doi.org/10.54499/UIDB/05183/2020) and the Associate Laboratory CHANGE—Global Change and Sustainability Institute (https://doi.org/10.54499/LA/P/0121/2020). We gratefully acknowledge Filipa Cachapa, João Canedo, Luisa Coelho, Vasilis Daimonakos, Bernardo Lameira, Hyo Ju Mendez Pinto, and Elmar Stoutjesdijk for their assistance in implementing the treatments and collecting the samples. The authors extend their sincere gratitude to the vineyard managers Iain Reynolds, Carolina Sardinha, Ana Sampaio, and André Pilirito for their support and invaluable assistance with site selection and monitoring. We thank the staff of the Laboratory of Enology of Instituto Superior de Agronomia, namely Ana Catarina Mendes and Joana Granja-Soares for their administrative and technical support. Sergio Prats was supported by the Portuguese Science Foundation (FCT) through the research contract CEECIND/01473/2020 and the Spanish Ministry of Science, Innovation and Universities through the research contract RYC2022-035489-I. The author thank Fundação Eugénio de Almeida and Herdade do Mouchão for the contribution in the present study. The authors have reviewed and edited the output and take full responsibility for the content of this publication.

Conflicts of Interest: The authors declare no conflicts of interest. The funders had no role in the design of this study; in the collection, analyses, or interpretation of data; in the writing of the manuscript; or in the decision to publish the results.

Abbreviations

The following abbreviations are used in this manuscript:

SOM	Soil organic matter
OAs	Organic amendments
TSS	Total soluble solids
YAN	Yeast assimilable nitrogen
CEC	Cation exchange capacity

References

1. World Atlas of Desertification (WAD). Aridity Projections European Commission Joint Research Centre. Available online: https://wad.jrc.ec.europa.eu/aridityprojections (accessed on 14 January 2025).
2. FAO. *The State of Food and Agriculture 2025-Addressing Land Degradation Across Landholding Scales*; FAO: Rome, Italy, 2025. [CrossRef]
3. European Court of Auditors. *Desertification in the EU Background Paper*; European Union: Luxembourg, 2018.
4. Komatsuzaki, M.; Ohta, H. Soil management practices for sustainable agro-ecosystems. *Sustain. Sci.* **2007**, *2*, 103–120. [CrossRef]
5. Tilman, D.; Cassman, G.K.; Matson, A.P.; Naylor, R.; Polasky, S. Agricultural sustainability and intensive production practices. *Nature* **2002**, *418*, 671–677. [CrossRef]
6. Steenwerth, K.; Guerra, B. Influence of floor management technique on grapevine growth, disease pressure, and juice and wine composition: A review. *Am. J. Enol. Vitic.* **2012**, *63*, 149–164.
7. Lal, R. Carbon management in agricultural soils. *Mitig. Adapt. Strateg. Glob. Change* **2007**, *12*, 303–322. [CrossRef]
8. Lippi, P.; Mattii, G.B.; Cataldo, E. Biochar, properties, and skills with a focus on implications for vineyard land and grapevine performance. *Phyton-Int. J. Exp. Bot.* **2025**, *94*, 33–64. [CrossRef]
9. Gholamahmadi, B.; Jeffery, S.; Gonzalez-Pelayo, O.; Prats, S.A.; Bastos, A.C.; Keizer, J.J.; Verheijen, F.G.A. Biochar impacts on runoff and soil erosion by water: A systematic global scale meta-analysis. *Sci. Total Environ.* **2023**, *871*, 161860. [CrossRef]
10. Canedo, J.N.G.V.; Coelho, L.; Castro, L.; Verheijen, F.G.A.; Prats, S. Biochar and mulch: Hydrologic, erosive, and phytotoxic responses across different application strategies and agricultural soils. *Agronomy* **2025**, *15*, 926. [CrossRef]
11. Mairata, A.; Valls-Fonayet, J.; Labarga, D.; Puelles, M.; Cluzet, S.; Portu, J.; Pou, A. Influence of organic mulches and soil properties on the phenolic profile of leaves, canes, and grape skins in grapevine (*Vitis vinifera* L.). *J. Sci. Food Agric.* **2025**, *105*, 4893–4904. [CrossRef]
12. Campos, P.; Miller, A.Z.; Prats, S.A.; Knicker, H.; Hagemann, N.; De la Rosa, J.M. Biochar amendment increases bacterial diversity and vegetation cover in trace element-polluted soils: A long-term field experiment. *Soil Biol. Biochem.* **2020**, *150*, 108014. [CrossRef]
13. Coelho, L.; Canedo, J.N.G.V.; Custódio, M.; Flores, D.; Mourão, P.; Palma, P.; Prats, S.A. Feedstock and pyrolysis conditions of biochars: Influence on soil phytotoxicity and water ecotoxicity. *Soil Biol. Biochem.* **2025**, *211*, 109935. [CrossRef]
14. Verheijen Frank, G.A.; Bastos, A.C.; Santos, M.; Khodaparast, Z.; Oliveira, F.J.; Silva, F.R.; Martins, M.A.S.; Almeida, C.; Rocha, S.M. Biochar had no detrimental effects on markers linked to grape and wine sensory attributes: The case study of a Portuguese Appellation. *Vitis* **2025**, *64*, 13.
15. Diacono, M.; Montemurro, F. Long-term effects of organic amendments on soil fertility. A review. *Agron. Sustain. Dev.* **2010**, *30*, 401–422. [CrossRef]
16. Botelho, M.; Cruz, A.; Ricardo-Da-Silva, J.; De Castro, R.; Ribeiro, H. Mechanical pruning and soil fertilization with distinct organic amendments in vineyards of syrah: Effects on vegetative and reproductive growth. *Agronomy* **2020**, *10*, 1090. [CrossRef]
17. Pinamonti, F. Compost mulch effects on soil fertility, nutritional status and performance of grapevine. *Nutr. Cycl. Agroecosyst.* **1998**, *51*, 239–248. [CrossRef]
18. Botelho, M.; Ribeiro, H.; Cruz, A.; Martins, M.; Khairnar, K.S.; Pardal, R.; Catarino, S.; De Castro, R.; Ricardo-Da-Silva, J. Mechanical pruning, and soil organic amendments in vineyards of Syrah: Effects on wine mineral composition. *Cienc. Tec. Vitivinic.* **2021**, *36*, 151–162. [CrossRef]
19. Varga, P.; Májer, J. The use of organic waste for soil-covering of vineyards. *Acta Hortic.* **2004**, *652*, 191–197. [CrossRef]

 https://doi.org/10.3390/environments13010044

20. Botelho, M.; Ribeiro, H.; Cruz, A.; Duarte, D.F.; Faria, D.L.; de Castro, R.; Ricardo-Da-Silva, J. Mechanical pruning, and soil organic amendments in vineyards of Syrah: Effects on grape composition. *Oeno One* **2021**, *55*, 267–277. [CrossRef]
21. Mundy, D.C.; Agnew, R.H. Effects of mulching with vineyard and winery waste on soil fungi and botrytis bunch rot in Marlborough vineyards. *N. Z. Plant Prot.* **2002**, *55*, 135–138. [CrossRef]
22. Mairata, A.; Pou, A.; Martínez, J.; Puelles, M.; Labarga, D.; Portu, J. Organic mulches slightly influence the wine phenolic profile and sensory evaluation. *Food. Chem.* **2024**, *457*, 140045. [CrossRef]
23. Buesa, I.; Mirás-Avalos, J.M.; De Paz, J.M.; Visconti, F.; Sanz, F.; Yeves, A.; Guerra, D.; Intrigliolo, D.S. Soil management in semi-arid vineyards: Combined effects of organic mulching and no-tillage under different water regimes. *Eur. J. Agron.* **2021**, *123*, 126198. [CrossRef]
24. Botelho, M.; Ribeiro, H.; Cruz, A.; Duarte, D.F.; Faria, D.L.; Khairnar, K.S.; Pardal, R.; Susini, M.; Correia, C.; Catarino, S.; et al. Mechanical pruning, and soil organic amending in two terroirs. effects on wine chemical composition and sensory profile. *Am. J. Enol. Vitic.* **2022**, *73*, 26–38. [CrossRef]
25. Cataldo, E.; Salvi, L.; Sbraci, S.; Storchi, P.; Mattii, G.B. Sustainable viticulture: Effects of soil management in *Vitis vinifera*. *Agronomy* **2020**, *10*, 1949. [CrossRef]
26. Prats, S.A.; Merino, A.; Gonzalez-Perez, J.A.; Verheijen, F.G.A.; De la Rosa, J.M. Can straw-biochar mulching mitigate erosion of wildfire-degraded soils under extreme rainfall? *Sci. Total Environ.* **2021**, *761*, 143219. [CrossRef]
27. Carbonneau, A.; Champagnol, F. Nouveaux Systèmes de Culture Integré du Vignoble. Programme AIR-3-CT 93. 1993, *Unpublished Protocol*.
28. Gump, H.B.; Zoecklein, W.B.; Fugelsang, C.K.; Whiton, S.R. Comparison of analytical methods for prediction of prefermentation nutritional status of grape juice. *Am. Soc. Enol. Vitic.* **2002**, *53*, 325–329. [CrossRef]
29. Ribéreau-Gayon, P. Le dosage de composes phenoliques totaux dans les vin rouges. *Anal. Chim. Acta* **1970**, *52*, 627–631.
30. Kramling, T.E.; Singleton, V.L. An estimate of the nonflavonoid phenols in wines. *Am. J. Enol. Vitic.* **1969**, *20*, 86–92. [CrossRef]
31. Ribéreau-Gayon, P.; Stonestreet, E. Le dosage des anthocyanes dans le vin rouge. *Bull. Soc. Chim. Fr.* **1965**, *9*, 2649–2652. [PubMed]
32. Somers, T.C.; Evans, M.E. Spectral evaluation of young red wines: Anthocyanin equilibria, total phenolics, free and molecular SO_2, "chemical age". *J. Sci. Food Agric.* **1977**, *28*, 279–287. [CrossRef]
33. Roggero, J.P.; Coen, S.; Ragonnet, B. High performance liquid chromatography survey on changes in pigment content in ripening grapes of Syrah. an approach to anthocyanin metabolism. *Am. J. Enol. Vitic.* **1986**, *37*, 77–83. [CrossRef]
34. Sun, B.; Leandro, C.; Ricardo da Silva, J.M.; Spranger, I. Separation of grape and wine proanthocyanidins according to their degree of polymerization. *J. Agric. Food Chem.* **1998**, *46*, 1390–1396. [CrossRef]
35. Sun, B.; Ricardo-da-Silva, J.M.; Spranger, I. Critical factors of vanillin assay for catechins and proanthocyanidins. *J. Agric. Food Chem.* **1998**, *46*, 4267–4274. [CrossRef]
36. Sun, B.; Ricardo-Da-Silva, J.M.; Spranger, M.I. Quantification of catechins and proanthocyanidins in several portuguese grapevine varieties and red wines. *Cienc. Tec. Vitivinic.* **2001**, *16*, 23–34.
37. Mondini, C.; Fornasier, F.; Sinicco, T.; Sivilotti, P.; Gaiotti, F.; Mosetti, D. Organic amendment effectively recovers soil functionality in degraded vineyards. *Eur. J. Agron.* **2018**, *101*, 210–221. [CrossRef]
38. Bell, S.J.; Henschke, P.A. Implications of nitrogen nutrition for grapes, fermentation, and wine. *Aust. J. Grape Wine Res.* **2005**, *11*, 242–295. [CrossRef]
39. Van Leeuwen, C.; Roby, J.P.; De Rességuier, L. Soil-related terroir factors: A review. *Oeno One* **2018**, *52*, 173–188. [CrossRef]
40. Keller, M. *The Science of Grapevines*, 3rd ed.; Academic Press: London, UK, 2015.
41. Hilbert, G.; Soyer, J.P.; Molot, C.; Giraudon, J.; Milin, S.; Gaudillere, J.P. Effects of nitrogen supply on must quality and anthocyanin accumulation in berries of cv. Merlot. *Am. J. Enol. Vitic.* **2003**, *54*, 150–154.
42. Delgado, R.; Martín, P.; Del Álamo, M.; González, M.R. Changes in the phenolic composition of grape berries during ripening in relation to vineyard nitrogen and potassium fertilisation rates. *J. Sci. Food Agric.* **2004**, *84*, 623–630. [CrossRef]
43. Heimler, D.; Romani, A.; Ieri, F. Plant polyphenol content, soil fertilization and agricultural management: A review. *Eur. Food Res. Technol.* **2017**, *243*, 1107–1115. [CrossRef]
44. Pérez-Álvarez, E.P.; Martínez-Vidaurre, J.M.; Garde-Cerdán, T. Anthocyanin composition of grapes from three different soil types in cv. Tempranillo A.O.C. Rioja vineyards. *J. Sci. Food Agric.* **2019**, *99*, 4833–4841. [CrossRef]
45. Teixeira, A.; Eiras-Dias, J.; Castellarin, S.D.; Gerós, H. Berry phenolics of grapevine under challenging environments. *Int. J. Mol. Sci.* **2013**, *14*, 18711–18739. [CrossRef]
46. Cortell, J.M.; Halbleib, M.; Gallagher, A.V.; Righetti, T.L.; Kennedy, J.A. Influence of vine vigor on grape (*Vitis vinifera* L. Cv. Pinot Noir) anthocyanins. 2. Anthocyanins and pigmented polymers in wine. *J. Agric. Food. Chem.* **2007**, *55*, 6585–6595. [CrossRef]
47. Downey, M.O.; Dokoozlian, N.K.; Krstic, M.P. Cultural practice and environmental impacts on the flavonoid composition of grapes and wine: A review of recent research. *Am. J. Enol. Vitic.* **2006**, *57*, 257–268. [CrossRef]
48. Seguin, G. "Terroirs" and pedology of wine growing. *Experientia* **1986**, *42*, 861–873. [CrossRef]
49. Davis, J.G.; Wilson, C.R. *Choosing a Soil Amendment*; Colorado State University Extension: Fort Collins, CO, USA, 2005.

https://doi.org/10.3390/environments13010044

50. Asadullayev, R.; Abasova, K.; Mammadova, K.; Guliyeva, A. Assessment of grape cultivar traits and environmental adaptation in the Absheron Peninsula, Azerbaijan. *Dysona Appl. Sci.* **2025**, *6*, 51–59.
51. Gonzalo-Rivas, J.C.; Gutierrez, E.; Hebrero, E.; Santos-Buelga, C. Comparisons of methods for determination of anthocyanins in red wines. *Am. J. Enol. Vitic.* **1992**, *43*, 210–214. [CrossRef]
52. Hohnová, B.; Šťavíková, L.; Karásek, P. Determination of anthocyanins in red grape skin by pressurised fluid extraction and HPLC. *Czech J. Food Sci.* **2008**, *26*, 39–44. [CrossRef]
53. Custodio-Mendoza, E.; Bonilla-Cruz, J.; González-Hernández, M.D.; Sánchez-Pérez, Y.; Hernández-Guerrero, C. A review of quantitative and topical analysis of anthocyanins in food. *Molecules* **2024**, *29*, 1735. [CrossRef]
54. Mattioli, R.; Francioso, A.; Mosca, L.; Silva, P. Anthocyanins: A comprehensive review of their chemical structure, occurrence and quantitative analysis by hplc and spectrophotometry. *Molecules* **2020**, *25*, 5141.

https://doi.org/10.3390/environments13010044

environments

Article

Influence of Vermicompost Tea on Metabolic Profile of *Diplotaxis muralis*: An NMR Spectroscopic Analysis

Sami ur Rehman †, Federica De Castro †, Alessio Aprile, Michele Benedetti * and Francesco Paolo Fanizzi

Department of Biological and Environmental Sciences and Technologies (DiSTeBA), University of Salento, Via Monteroni, I-73100 Lecce, Italy; sami.urrehman@unisalento.it (S.u.R.); federica.decastro@unisalento.it (F.D.C.); alessio.aprile@unisalento.it (A.A.); fp.fanizzi@unisalento.it (F.P.F.)

* Correspondence: michele.benedetti@unisalento.it

† These authors contributed equally to this work.

Abstract: Recently, we reported the efficacy of organic nutrient solutions in supporting the hydroponic cultivation of *Diplotaxis muralis.* The aim of this study was to elucidate the influence of standard and enhanced vermitea formulations, compared to the conventional Hoagland solution, on phytochemical and metabolomic changes in *D. muralis.* Using NMR-based metabolomics and multivariate analysis, we observed significant metabolite variation among treatments. Both vermitea formulations increased the levels of acetate, alanine, and 2-oxoglutarate, and boosted the biosynthesis of key secondary metabolites, including methoxy flavonoids and glucosinolates. The standard vermitea treatment further resulted in a higher accumulation of leucine and citrate, while the Hoagland solution induced higher glucose concentrations. Enhanced vermitea improved copper and zinc uptake, positively correlating with methoxy flavonoid production. In contrast, the higher phosphorus and potassium content of the Hoagland solution correlated with increased glucose levels in *D. muralis*. Metabolite profiling coupled with multivariate analysis identified the enhanced vermitea as the best alternative to chemical nutrient solution for improving the nutritional and phytochemical quality of *D. muralis* leaves.

Keywords: metabolites; nuclear magnetic resonance; vermicompost tea; *Diplotaxis muralis*; organic fertilizer

1. Introduction

Salad crops from the Brassicaceae family, collectively known as rocket, rucola, or arugula, are native to Mediterranean countries, specifically those in the western region. This area is considered a center of diversity due to its unique climatic conditions [1]. Rocket crops comprise various species from the *Eruca* and *Diplotaxis* genera. The *Diplotaxis* genus includes a group of herbaceous annuals, typically diploid, with rare instances of polyploidy and dysploidy. *D. muralis*, also known as the annual wall rocket, is an allotetraploid produced by the hybridization of *D. viminea* and *D. tenuifolia*. The *D. muralis* is used in Mediterranean cuisine as a salad ingredient and garnish due to its distinctive pungent and peppery flavor [2]. The organoleptic properties of *D. muralis*, such as its spicy flavor, are attributed to a rich array of bioactive compounds, including glucosinolates (GSLs) and flavonoids. The hydrolysis of GSLs by myrosinase results in the production of isothiocyanates, thiocyanates, and nitriles [3]. These compounds are believed to possess numerous health-promoting potentials, such as antioxidant and anticarcinogenic properties [4]. Rockets are effective in the prevention of various diseases, including coronary heart disease and inflammatory disorders [5].

 https://doi.org/10.3390/environments12100366

Regarding the agricultural production of *Diplotaxis* species, these are commonly grown in open fields but can also be cultivated in soilless systems. Soilless culture systems produce clean, fresh vegetables in a significantly shorter time compared to traditional cultivation and facilitate postharvest processes in the industry. However, the environmental impact resulting from the disposal of hydroponic solutions after the crop growth cycle cannot be overlooked [6]. In hydroponics, chemical fertilizer salts are dissolved in water to create nutrient solutions essential for plant growth. These nutrient solutions need to be replaced periodically, resulting in the production of hydroponic wastewater, which is often disposed of directly into the environment, causing contamination [7]. Therefore, the use of organic nutrient solutions as a potential alternative to inorganic nutrient solutions in hydroponic farming is crucial [6]. The ability to transform diverse organic waste into organic fertilizers enriched with all essential elements for plant growth has made vermicompost a viable option for sustainable agriculture [8]. The aqueous extract of vermicompost, known as vermicompost tea or vermitea, contains nutrients, microbes, and phytohormones beneficial to plant growth when applied as a foliar application or in the soil [9]. The symbiotic relationship between plants and microbes enhances the production of phytohormones and nutrient acquisition [10]. Microbes have the potential to chemically communicate with their host plants, stimulating the immune system and affecting the plant's physiology and health [11], thereby influencing the plant's metabolism. Moreover, the application of vermicompost influences plant metabolites such as phenols, flavonoids, and various amino acids [12,13].

In our previous study, *D. muralis* plants were cultivated using three different hydroponic solutions: Hoagland solution, standard vermitea, and enhanced vermitea [14]. The variations in pH, electrical conductivity (EC), and the nature of the nutrient solutions (organic vs. chemical) significantly influenced the plant growth parameters and the uptake of macro and micronutrients in the leaves of *D. muralis*. These variations in plant growth parameters and nutrient acquisition can affect the metabolic profile of the plant. Primary metabolites serve nutritional purposes, whereas secondary metabolites are used by plants as a defense against various biotic and abiotic stresses [15]. The effect of different agronomic practices, such as water supply and nitrogen nutrition, on quality traits and metabolism of different rocket species has been confirmed by several studies [16–18]. A recent study by Buitrago-Villanueva et al. [19] assessed the impact of the culture system and harvest time on metabolite composition in *Eruca sativa* leaves, reporting that some compounds were significantly altered by the culture system. These studies explored the metabolite composition, mainly focusing on GC-MS and LC-MS metabolomic studies. NMR-based metabolomics is a powerful approach to studying plant metabolism, providing comprehensive insights into the chemical composition of plant systems. Several studies have employed an NMR-based metabolomic approach to evaluate changes in the metabolic profiles of edible plant parts [20–22]. NMR spectroscopy provides comprehensive information on metabolite composition in complex plant samples and is widely applied in food science for traceability, authenticity, and safety studies [23].

In this context, to investigate sustainable farming approaches for *D. muralis* production and its significance as an important food source, three different nutrient solutions were used in hydroponic cultivation: Hoagland solution (HS), standard vermitea (SVT), and enhanced vermitea (EVT). The impact of different growth conditions on the metabolite pattern of rockets, specifically *D. muralis*, using nuclear magnetic resonance (NMR) spectroscopy, has not been previously characterized. The present investigation employs ^1H NMR spectroscopy to determine the metabolic composition of *D. muralis* to understand the bioactive chemicals associated with the type of plant growth medium. These findings will enhance our understanding of the biochemical composition of *Diplotaxis* species and contribute valuable data to the broader field of plant metabolomics. The insights gained could pave the way for new therapeutic applications and enhance the nutritional profile of *D. muralis*.

2. Materials and Methods

2.1. Experimental Design

The preparation of SVT and EVT and their physicochemical properties are reported in our previous study [14]. SVT was prepared by mixing green waste vermicompost (VC) with double-distilled water (DDW) in a 1:10 ratio. The pH of SVT ranged between 7.5 and 8.0, and no further pH alteration was made. The electrical conductivity (EC) of SVT was 1.292 mS/cm. EVT, on the other hand, was prepared by mixing VC with DDW in a 1:5 ratio. The pH of EVT was maintained within the range of 5.80–6.30 by gradual addition of a few drops of diluted sulfuric acid (10% H_2SO_4). The EC of EVT was approximately 2.0 mS/cm. The higher EC value can be interpreted as indicating higher nutrient content in the growth media, due to the presence of more dissolved nutrients, specifically ions like NO_3^-, K^+, and PO_4^{3-}, in solution with a high EC value. Therefore, SVT and EVT are differentiated by pH value and nutrient concentration. Both organic nutrient solutions (SVT and EVT) can contain various phytohormones and beneficial microbes (which are absent in chemical nutrient solutions) that influence plant growth. The pH and EC of EVT and Hoagland solution (HS) are comparable; however, the nutrient composition varies between the two solutions due to the differences in their organic and chemical sources, respectively.

D. muralis seeds were germinated in plastic Petri dishes containing a double layer of moist filter paper. Following six days of germination, five sprouted seeds were carefully transferred into each plastic pot filled with perlite. These pots were then placed into a hydroponic system. Fresh hydroponic solutions, such as SVT, EVT, and Hoagland solution, were circulated through the pots to ensure a consistent supply. An aquarium pump was used to maintain an adequate supply of oxygen. The pH of EVT and Hoagland solution was adjusted to a range of 5.80–6.30 using diluted sulfuric acid and was continuously monitored every two days. To prevent nutrient depletion, the hydroponic solutions were replaced with fresh nutrient solutions every two weeks. A completely randomized design (CRD) experiment consisting of three treatments and six replications was employed. The plants were grown in a growth chamber at the University of Salento, Lecce, Italy (40°20′12″ N; 18°07′24″ E) under controlled conditions (25/18 °C day/night temperature, 14 h of daylight, and 65% humidity). Leaves of *D. muralis* were harvested after 45 days, at the onset of flowering.

2.2. Sample Preparation for NMR Metabolomics Analysis

Samples were prepared according to a previously reported experimental method [24]. Specifically, *D. muralis* leaf samples were rapidly frozen in liquid nitrogen, pulverized to a fine powder using a stainless-steel blender, and subsequently lyophilized for 48 h. For metabolite extraction, 0.75 mL of CD_3OD and 0.75 mL of KH_2PO_4 buffer in D_2O (pH 5.9) containing 0.05% w/v TSP-d4 (sodium salt of trimethylsilylpropionic acid) were combined with plant material (100 mg). The mixtures were carefully vortexed at room temperature for 1 min, followed by 10 min of sonication under the same conditions. Samples were centrifuged at 17,000× *g* for 20 min, and 700 μL of the resulting supernatant was transferred into a NMR tube (5 mm).

2.3. NMR Measurements and Metabolite Identification

All measurements were conducted using a Bruker AVANCE III 600 Ascend NMR spectrometer (Bruker, Ettlingen, Germany), operating at 600.13 MHz for proton detection. The spectrometer was fitted with a TCI cryoprobe that included a z-axis gradient coil and automatic tuning/matching (ATM). The experiments were conducted at 300 K under automation, with samples introduced via a Bruker Automatic Sample Changer integrated with IconNMR software "Version 5" (Bruker) for sample loading. Each spectrum was acquired using a 1D sequence with pre-saturation, and a composite pulse for selection (zgcppr, a standard pulse sequence from Bruker) was executed, comprising 16 transients and 16 dummy scans, with a relaxation delay of 5 s. The free induction decay (FID) data consisted of 64 K data points, with a spectral width of 12,019.230 Hz (20.0276 ppm) and an acquisition time of 2.73 s. Prior to Fourier transformation, the resulting FIDs were processed with an exponential weighting function corresponding to a line broadening of 0.3 Hz, followed by automated phasing and baseline correction [25].

Metabolites were identified through ^{1}H and ^{13}C assignments by 1D and 2D homo- and heteronuclear experiments (Figures S1–S3), supported by comparison with other reported literature [26–29]. ^{1}H NMR data processing was performed by using TopSpin 3.6.1 (Bruker) and Amix 3.9.15 (Bruker, BioSpin, Italy) for visual inspection and the bucketing process for further multivariate statistical analyses. All spectra were referenced to the TSP signal (δ = 0.00 ppm).

2.4. ^{1}H NMR Data Processing and Multivariate Statistical Analysis

The ^{1}H NMR chemical shifts corresponding to the identified metabolites are presented in Table S1. The bucketing preprocessing procedure was utilized for the ZGCPPR spectra across the 10.0 to 0.5 ppm region. Each NMR spectrum was systematically divided into rectangular buckets of fixed width, measuring 0.04 ppm, and integrated using Bruker Amix 3.9.15 software (Bruker, BioSpin).

The remaining buckets were normalized to the total area to reduce minor discrepancies and then mean-centered. Multivariate statistical analyses, including unsupervised principal component analysis (PCA) and supervised orthogonal partial least squares discriminant analysis (OPLS-DA), were conducted to investigate the inherent variation within the data, utilizing SIMCA 14 software (Sartorius Stedim Biotech, Umeå, Sweden) [30,31]. The Pareto scaling method was applied by dividing the mean-centered data by the square root of the standard deviation. The robustness of the statistical models was tested by the cross-validation default method (7-fold) and further evaluated with a permutation test (100 permutations). The total variations in the data and the internal cross-validation, hence the quality of the statistical models, were described with R^2(cum) and Q^2(cum) parameters and p values (p[CV-ANOVA]) (a p-value of < 0.05, confidence level of 95%, was considered statistically significant) obtained from analysis of variance testing of cross-validated predictive residuals (CV-ANOVA).

2.5. Chemicals

Deuterium oxide (99.9 atom%D) containing 0.05% wt 3-(trimethylsilyl)propionic-2,2,3,3 d4 acid sodium salt (TSP), and potassium phosphate monobasic was purchased from Armar Chemicals (Döttingen, Switzerland). Methanol-d4 (99.9 atom %D) and potassium phosphate monobasic were purchased from CARLO ERBA Reagents (Milan, Italy).

3. Results

3.1. ^{1}H-NMR Characterization of *D. muralis* Metabolic Extract

The metabolic variation induced by the different nutrient solutions was analyzed using NMR-based metabolomic approaches. The ^{1}H NMR spectra of the leaves' hydroalcoholic extracts, for each treatment condition (HS, SVT, EVT), were collected, and a multivariate statistical analysis was subsequently applied. The representative ^{1}H NMR spectrum, Figure 1, and the primary resonances were assigned to individual metabolites and are reported in Table S1.

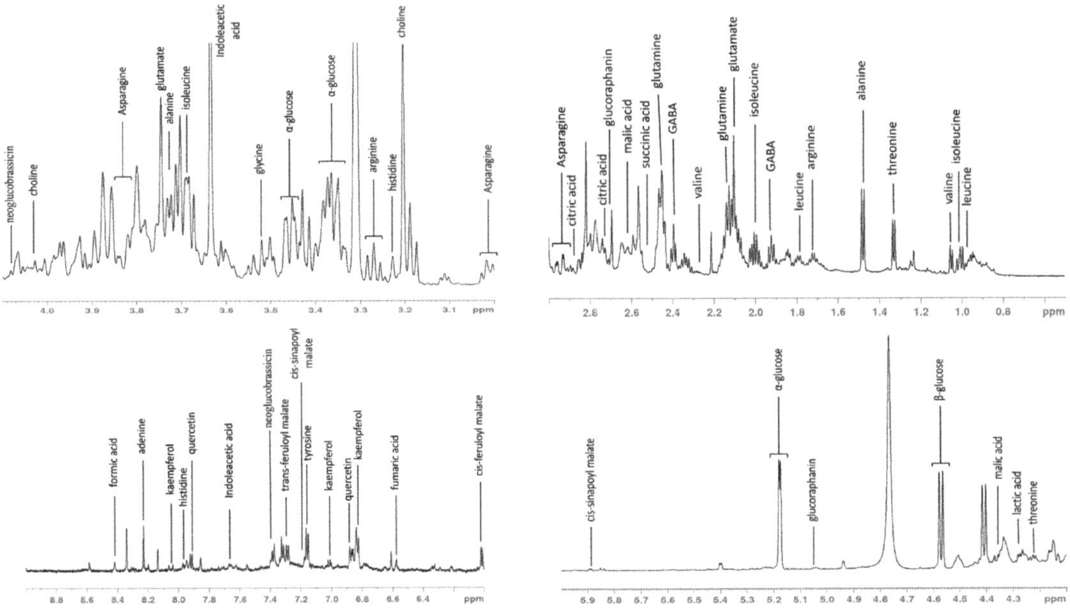

Figure 1. Expansions of a representative 600 MHz ^{1}H ZGCPPR NMR spectrum of hydroalcoholic extract from *Diplotaxis muralis* leaves.

A total of 28 compounds were identified through 1D and 2D NMR spectral analysis and compared with previously reported data [24]. The spectrum is characterized by a complex pattern of signals attributable to sugars (α and β glucose), amino acids (alanine, glutamate, glutamine, histidine, isoleucine, leucine, threonine, asparagine, arginine, phenylalanine, tyrosine, glycine, γ-amino-butyrate, and valine), organic acids (citric acid, formic acid, fumaric acid, malic acid, and succinic acid), flavonoids (Kaempferol analogues, quercetin analogues), glucosinolates (glucoraphanin, neoglucobrassicin) and others (adenine, choline). Notably, the spectra of *D. muralis* exhibit pronounced peaks corresponding to flavonoids and glucosinolates. The antioxidant activity of Brassicaceae leaves is closely correlated with their flavonoid content, whereas glucosinolates play multifaceted roles, including key functions in plant defense [32,33].

3.2. Multivariate Statistical Analysis

A preliminary unsupervised analysis (PCA) was performed on the entire dataset to observe a general trend of data clustering (Figure 2). Bucket tables obtained from the aqueous leaves' extracts (^{1}H ZGCPPR NMR spectra) were used as input variables. A two-component PCA model describes 68.9% of the total variance of the dataset. Samples grouping along the t[1] component of the PCA score plot model was observed (Figure 2), particularly evident among HS and SVT samples, indicating a significant difference in the metabolic profile of *D. muralis* depending on the different growth conditions.

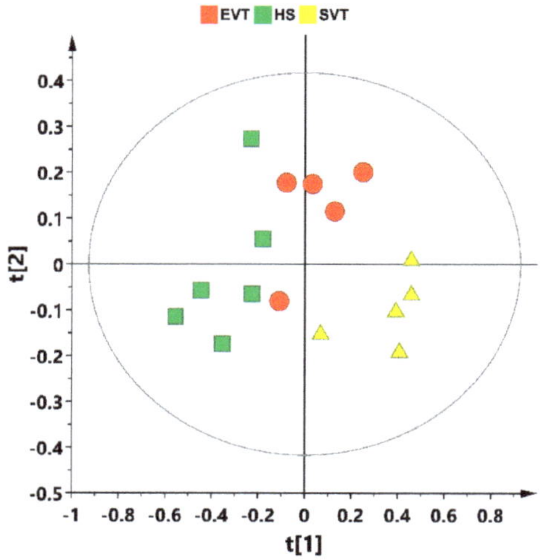

Figure 2. Two-component PCA t[1]/t[2] scores plot model for Hoagland solution (HS), standard vermitea (SVT), and enhanced vermitea (EVT). t[1] and t[2] account for 57.3% and 11.6% of the total variance, respectively.

Further supervised OPLS-DA analysis was applied to the NMR bucketed data to discriminate between the predefined groups of samples by separating predictive (correlated with class separation) from orthogonal components (unrelated to the class separation). This approach aimed to improve the interpretability of the model and simplify the identification of key variables contributing to class differences. Pairwise supervised OPLS-DA analysis performed on leaf samples' extracts according to the variable "growth conditions", i.e., Hoagland solution vs. standard vermitea and Hoagland solution vs. enhanced vermitea, yielded good models (1 + 1 + 0) with significant predictive capability $Q^2 = 0.87$ and $Q^2 = 0.69$, respectively (Figure 3B,D), thus reflecting more differentiation among the variables.

The score scatter plots of the two OPLS-DA models (Figure 3A,C) demonstrated a clear separation in each pairwise comparison (HS vs. SVT and HS vs. EVT). The corresponding S-line plots identified the metabolites responsible for the observed class separation between HS vs. SVT and HS vs. EVT (Figure 3B,D). From our analysis, the SVT and EVT growth conditions induce higher levels of acetate, alanine, 2-oxoglutarate, flavonoids (especially methoxy flavonoids comparable to quercetin analogues in which the hydroxy group is replaced by a methoxy group), and glucosinolates (glucoraphanin, neoglucobrassicin), and lower levels of glucose compared to HS. Furthermore, SVT also enhances the content of leucine and citrate in comparison to HS.

The Box-and-Whisker plots, Figure 4, illustrate the alteration trends of significant metabolites, identified in the OPLS-DA pairwise comparison models, concerning the different treatment conditions (SVT, EVT, HS). Further, one-way ANOVA with Tukey's post hoc test revealed leucine, glucose, 2-oxoglutarate, methoxy flavonoids, citrate, acetate, and alanine as highly significant metabolites ($p < 0.05$). Glucosinolates did not show highly significant changes ($p > 0.05$).

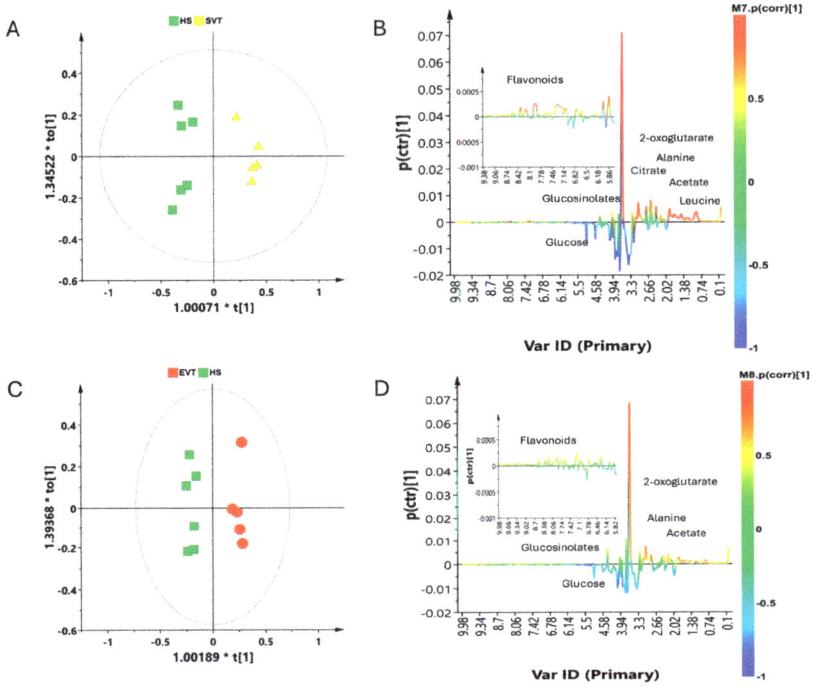

Figure 3. Pairwise comparison of OPLS-DA models and corresponding S-line plots: (**A**,**B**) comparison between Hoagland solution (HS) and standard vermitea (SVT); (**C**,**D**) comparison between Hoagland solution (HS) and enhanced vermitea (EVT).

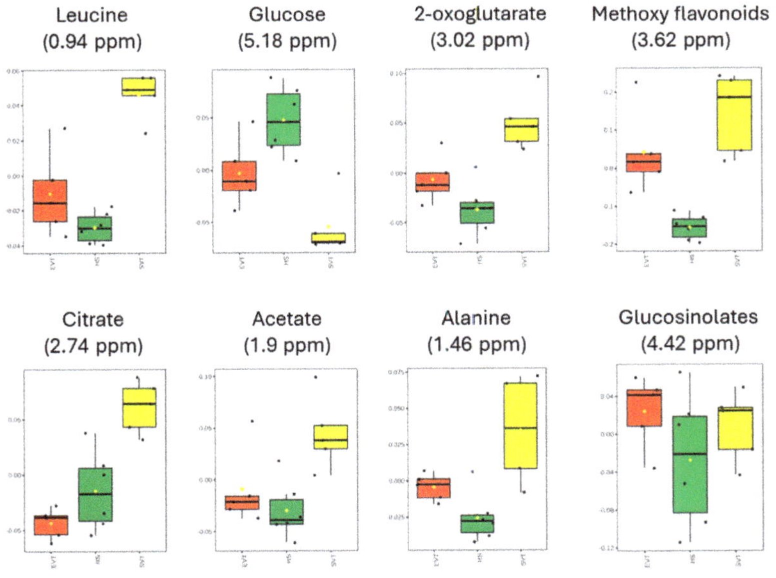

Figure 4. Box-and-Whisker plots summarizing the normalized Bucket values of the discriminant identified metabolites distribution among *Diplotaxis muralis* leaf samples treated with enhanced vermitea (EVT), standard vermitea (SVT), and Hoagland solution (HS). The yellow diamond indicates the mean value for each group, the notch represents the 95% confidence interval around the median of each group, and the dots beyond the line edges indicate outliers.

To correlate the content of discriminant metabolites in *D. muralis* grown under different hydroponic conditions and the accumulation of metals, as previously studied [14], an additional partial least squares (PLS) regression model was built (4 components, R2X(cum) = 0.788, R2Y (cum) = 0.926, Q2 (cum) = 0.595), as shown in Figure 5. The predictor variables (loadings) were related to the metal content (Cu, Zn, Fe, Mn, P, Ca, and K), quantified by ICP-AES [14]. The loading plot, Figure 5B, and its expansion, Figure 5C, display how the predictor variables contribute to the PLS components. The axes (w*c [1] and w*c [2]) represent how strongly each variable is correlated with the first two PLS components, highlighting that the accumulation of Fe, Mn, P, Ca, and K is associated with a higher amount of glucose in HS samples compared to the SVT, which was characterized by a higher content of methoxy flavonoids (quercetin analogues) and lower content of Fe, Mn, P, Ca, K, Cu, and Zn. EVT samples showed an intermediate profile between SVT and HS. The elevated concentration of methoxy flavonoids observed in EVT samples is associated with increased levels of Cu and Zn. Additionally, higher concentrations of Fe and Mn were also detected in EVT. The HS samples are characterized by lower content of Cu and Zn, which, from our analysis, is related to the higher content of glucose and K compared to SVT.

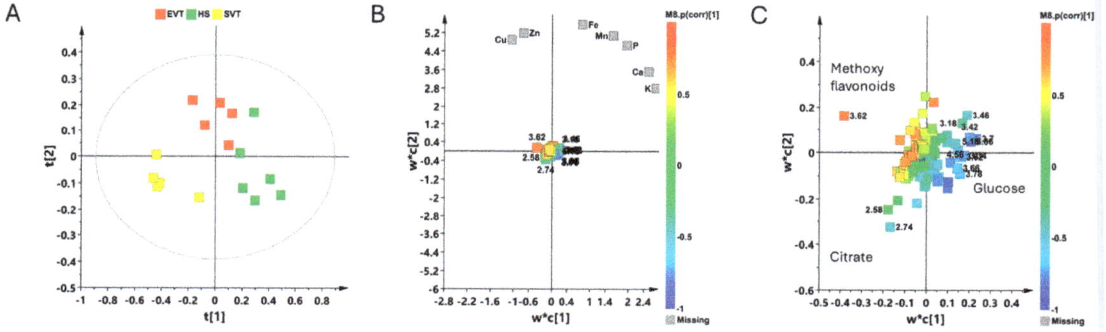

Figure 5. (**A**) PLS regression model score plot (4 components, R2X (cum) = 0.788, R2Y (cum) = 0.926, Q2 (cum) = 0.595) obtained from ^{1}H NMR Extracts of *Diplotaxis muralis* leaves; (**B**) relative loading plot; (**C**) expansion of the loading plot with highlighted NMR variables. Y variables = copper (Cu), zinc (Zn), iron (Fe), manganese (Mn), phosphorus (P), calcium (Ca), and potassium (K) content in enhanced vermitea (EVT), standard vermitea (SVT), and Hoagland solution (HS) samples.

4. Discussion

In the present study, the variations in the metabolic profile of *D. muralis*, influenced by different growth media, HS, SVT, and EVT, were evaluated. The characterization of *D. muralis* leaves grown in organic (vermiteas) and chemical nutrient solutions using nuclear magnetic resonance (NMR) spectroscopy, along with multivariate data analysis, was provided. A preliminary unsupervised principal component analysis (PCA) performed on the entire dataset described the variability due to different growth conditions, resulting in a clear differentiation in *D. muralis* primary and secondary metabolites. The type and concentration of plant metabolites are greatly affected by crop type, cultivar, culture environment, and fertilization level [34]. The PCA described a higher degree of separation between HS and SVT as compared to HS and EVT. This effect may be due to the significant differences in pH and nutrient contents between the HS and SVT hydroponic solutions. The differences in the pH of growth media can alter the accumulation of secondary metabolites [35]. Besides, SVT contains low levels of essential plant nutrients such as nitrogen, phosphorus, potassium, and other micronutrients. Such deficiencies may disrupt the regulation of metabolic pathways, leading to alterations in the metabolite profile [19]. In contrast, EVT contains an adequate concentration of nutrients to meet the nutritional demands of rocket plants. Additionally, the pH values of both EVT and HS were maintained at similar levels as required for optimal plant uptake [14].

To further explore the variation in metabolites, pairwise OPLS-DA comparisons (HS vs. SVT and HS vs. EVT) with their corresponding S-line models were performed (Figure 3). As described by significant predictive capability (Q2) values, clear discrimination was observed in HS vs. SVT and HS vs. EVT comparisons, suggesting the influence of vermitea on the metabolite patterns in *D. muralis*. Other authors have also confirmed the impact of vermicompost on the metabolic profiles of *Moringa oleifera* [36] and *Ocimum basilicuum* leaves [37]. However, the alterations in the metabolite profile of soilless-grown *D. muralis* in response to vermicompost tea (vermitea) have not been studied previously.

Vermicompost possesses plant biostimulatory activity due to the presence of phytohormones, organic acids, fulvic acids, humic acids, and beneficial microbes [14,38]. These plant biostimulants may influence the production of secondary metabolites [39]. Türkay and Öztürk [40] investigated the effects of solid vermicompost and vermitea on phenylpropene biosynthesis in basil (*Ocimum basilicum* L.) plants. Their findings revealed that the application of vermitea significantly enhanced the levels of phenylpropenes, which are key compounds that contribute to the characteristic aroma and antimicrobial activities of basil. The positive influence of vermicompost on the flavonoid content and antioxidant capacity and antioxidant capacity of various plant species has also been reported [41,42].

The overall composition of plant metabolites reflects the different metabolic pathways. Metabolites such as acetate, alanine, 2-oxoglutarate, flavonoids (especially methoxy flavonoids comparable to quercetin analogues in which the hydroxy group is replaced by a methoxy group), and glucosinolates (glucoraphanin, neoglucobrassicin) are present in higher amounts in SVT and EVT compared to HS, which, on the contrary, showed a higher content of glucose. In plants, acetate primarily functions in its activated form as acetyl-coenzyme A (Acetyl-CoA) [43]. It plays a central role in energy metabolism, storage, and utilization through the citric acid cycle (also known as the Krebs cycle or TCA cycle). Furthermore, acetyl-CoA serves as a central hub linking lipid and carbohydrate metabolism, and secondary metabolites biosynthesis, including flavonoids [44]. The accumulation of glucosinolates and flavonoids in *D. muralis* leaves improved its organoleptic properties, such as spicy flavor [3]. Additionally, 2-oxoglutarate, an intermediate of the Krebs cycle, plays a key role in carbon and nitrogen metabolism in plants. It also acts as a cofactor for a diverse range of enzymes involved in amino acid, glucosinolate, flavonoid, and alkaloid metabolism [45].

A significant aspect was the elevated flavonoid content observed in both types of vermitea (EVT and SVT), with particular emphasis on methoxy-modified quercetin analogues. This was evidenced by a highly significant signal at 3.62 ppm in the NMR spectrum, corresponding to the methoxy group. The application of vermitea upregulated the O-methyltransferase gene expression in plants [40]. O-methyltransferase plays a crucial role in the biosynthesis and diversification of flavonoids by methylation of hydroxyl groups on flavonoid precursors [46]. Methylation is a modification that flavonoids undergo, which can alter their stability, reactivity, and solubility. Flavonoid methylation requires methyl donors (SAM: S-adenosylmethionine), which are linked to nitrogen metabolism [47]. Moderate stress, as the presence of organic inputs, can shift flavonoid metabolism from hydroxy (-OH) to methoxy ($-OCH_3$) derivatives to increase the antioxidant capacity or the stability against oxidation.

Organic fertilizers stimulate the production of shikimic acid, which leads to an increased production of phenolic compounds and flavonoids [48]. Vermitea provides bioactive compounds or induces stress responses that trigger the synthesis of protective metabolites such as flavonoids and glucosinolates, which are involved in plant defense and antioxidant activity. This is also confirmed by the higher amount of glucose found in Hoagland solution growth conditions, where there is a lack of stress that induces a metabolic shift, and the more stable carbohydrate metabolism is preferred. Indeed, HS provides a well-balanced supply of nutrients, leading to normal growth in *D. muralis* without the need for increased secondary metabolism, such as flavonoids and glucosinolates. Moreover, flavonoids are powerful antioxidants that contribute to human health by reducing oxidative stress and inflammation. Glucosinolates, including glucoraphanin and neoglucobrassicin, are known for their potential anticancer properties and their role in detoxification processes in the human body [49]. Since these compounds are more abundant in SVT and EVT, plants grown under these conditions could offer greater health benefits.

From a nutritional perspective, SVT and EVT improve the nutrient profile of *D. muralis* compared to HS [14]. Brassica vegetables are a rich source of phenolic compounds, characterized by the presence of flavonoids and glucosinolates, which impact the flavor and acceptance of food and have potential health benefits [50]. Flavonoids, low molecular weight secondary metabolites located in cell vacuoles, play key roles in plant defense mechanisms. Several studies have explored the chemoprotective characteristics of rocket plants and have found that their flavonoid contents are induced by different light levels during the growth period [51–53]. In the present study, the environmental conditions were kept consistent for all treatments, so the observed variation in metabolites is not influenced by light factors. Therefore, the accumulation of phenolic compounds results also greatly influenced by biotic and abiotic stress, as well as nutrient availability [54].

Vermitea is rich in nitrogen, phosphorus, and potassium (NPK), as well as trace elements such as iron (Fe), zinc (Zn), and copper (Cu). The PLS-DA analysis performed here showed an interesting correlation between specific metabolites and metal contents, including differences in the absorption of these metals as previously discussed [14]. The higher levels of copper (Cu) and zinc (Zn) significantly modulated the phenol and flavonoid contents [55–57]. The enhanced content of Cu and Zn is closely related to higher methoxy flavonoids in EVT, whereas the P and K contents are associated with higher glucose in HS growth conditions. This effect may be due to the role of both P and K in various metabolic processes that influence sugar production and utilization [58]. Furthermore, an adequate phosphorus supply enhanced the activity of Ribulose-1,5-bisphosphate carboxylase/oxygenase (Rubisco), a central enzyme in the Calvin–Benson cycle, thereby promoting photosynthetic efficiency and subsequently increasing glucose accumulation [59].

Another modification in the metabolism of plants grown in vermitea (EVT and SVT) is related to certain amino acids, particularly phenylalanine, which is converted into 4-coumaroyl-CoA by the phenylpropanoid pathway, producing flavonoids [60]. Amino acids contribute to the accumulation of osmolytes, thus playing a key role in plant stress tolerance mechanisms [61]. The amino acids present in plant physiological structures serve as fundamental precursors for protein synthesis, which supports plants in metabolism regulation and stress adaptation [62]. Additionally, various metabolites differentiating EVT and SVT treatments were also identified. The plants grown in SVT accumulated higher levels of amino acids, including leucine and citrate, compared to those grown in HS. This is because the vermitea contains organic nitrogen released by organic matter decomposition, which contributes to the increased amino acid profile.

Focusing on amino acids such as leucine, which is a branched-chain amino acid (BCAA), it plays a significant role in plant energy metabolism. Leucine was found in higher content in vermitea compared to HS, particularly in SVT samples. The carbon skeleton of BCAA is converted to the intermediates of the Krebs cycle for ATP production [63]. Additionally, citrate is the 1st intermediate of the Krebs cycle and is critical for various metabolic networks involved in the production of amino acids and phytohormones. Citrate also induces antioxidant defense systems, increases chlorophyll content and influences secondary metabolism to improve plant growth even under stress [64].

These findings are also supported by the referenced literature [65–67], indicating a higher accumulation of amino acids in plants treated with vermicompost.

5. Conclusions

The present study elucidated the influence of organic nutrient solutions, specifically SVT and EVT, in comparison to the conventional HS, on metabolite differentiation in *D. muralis* leaves. Plants cultivated in SVT and EVT systems accumulated higher concentrations of phenolic compounds compared to those grown in HS. Both organic nutrient solutions (SVT and EVT) enhanced the accumulation of acetate, alanine, and 2-oxoglutarate, along with key secondary metabolites including flavonoids and glucosinolates. Furthermore, SVT exhibited a higher level of leucine and citrate and a lower accumulation of glucose in comparison to HS. Notably, the increased accumulation of phenolic compounds in plants treated with EVT and SVT not only enriched the nutritional and nutraceutical value of *D. muralis* leaves but also improved sensory attributes such as aroma and taste. However, higher crop yield was observed in EVT and HS treatments relative to SVT. Therefore, the use of EVT as an organic hydroponic solution appears to be more beneficial for achieving higher yields along with improved nutritional quality and antioxidant contents. This approach may contribute to enhancing farmers' economic returns while promoting better health among consumers. Moreover, the variations in the metabolite profile constitute a criterion for selecting the appropriate growth media to produce rocket plants with improved nutraceutical characteristics. In this context, the EVT system is recommended, as it supports improved plant growth alongside elevated levels of nutrients and flavonoids, key antioxidant compounds.

Future research should aim to identify the microbial communities present in vermicompost tea to uncover the key plant–microbe interactions responsible for the observed metabolic shifts in treated plants. Integrating metagenomic and metatranscriptomic approaches will be essential to elucidate these complex mechanisms. In parallel, transcriptomic analyses are recommended to clarify the molecular pathways driving the enhanced flavonoid production induced by organic nutrient solutions.

Supplementary Materials: The following supporting information can be downloaded at: https://www.mdpi.com/article/10.3390/environments12100366/s1, Table S1: Chemical shifts (δ) and assignments of metabolite resonances in the ^{1}H ZGCPPR NMR spectra of *Diplotaxis muralis* leaves; Figure S1: 2D ^{1}H *J*-resolved (JRESQC) spectrum for *Diplotaxis muralis* leaves extract; Figure S2: 2D ^{1}H-^{1}H COSY spectrum for *Diplotaxis muralis* leaves extract; Figure S3: The representative 2D ^{1}H–^{13}C HSQC spectrum of *Diplotaxis muralis* leaves extract.

Author Contributions: Conceptualization, S.u.R., F.D.C., A.A. and M.B.; methodology, S.u.R., F.D.C. and M.B.; formal analysis and data curation, S.u.R. and F.D.C.; writing—original draft preparation, S.u.R. and F.D.C.; supervision, F.D.C., A.A. and M.B.; writing—review and editing, M.B. and F.P.F. All authors have read and agreed to the published version of the manuscript.

Funding: This research was funded by the Ph.D. program titled: "Development of New Aerobic Composting Processes to Reduce the Environmental Impact Generated by Waste Disposal" (project code: CUP:F85F21005750001) of "Dottorati su tematiche Green" del Programma Operativo Nazionale (PON) R&I 2014–2020 and the project "National Biodiversity Future Center—NBFC" financed by the Italian National Recovery and Resilience Plan (PNRR) (project code: CN00000033).

Data Availability Statement: The data presented in this study are available upon request from the corresponding author.

Acknowledgments: The authors would like to express sincere gratitude to Paolo Marini of Compost Natura s.r.l., located at Via Mallacca Zummari 32, I-73010 Arnesano, Italy, for providing the platform for preparing the vermicompost tea used in this study.

Conflicts of Interest: The authors declare no conflicts of interest.

Abbreviations

The following abbreviations are used in this manuscript:

NMR	Nuclear magnetic resonance
EVT	Enhanced vermitea
SVT	Standard vermitea
HS	Hoagland solution
PCA	Principal component analysis
OPLS-DA	Orthogonal partial least squares discriminant analysis
PLS-DA	Partial least squares discriminant analysis

References

1. Tripodi, P.; Francese, G.; Mennella, G. Rocket salad: Crop description, bioactive compounds and breeding perspectives. *Adv. Hortic. Sci.* **2017**, *31*, 107–114.
2. Pasini, F.; Verardo, V.; Cerretani, L.; Caboni, M.F.; D'Antuono, L.F. Rocket salad (*Diplotaxis* and *Eruca* spp.) sensory analysis and relation with glucosinolate and phenolic content. *J. Sci. Food Agric.* **2011**, *91*, 2858–2864. [CrossRef]
3. Nguyen, V.T.; Stewart, J.; Lopez, M.; Ioannou, I.; Allais, F. Glucosinolates: Natural occurrence, biosynthesis, accessibility, isolation, structures, and biological activities. *Molecules* **2020**, *25*, 4537. [CrossRef]
4. Zhang, Y. Cancer-preventive isothiocyanates: Measurement of human exposure and mechanism of action. *Mutat. Res./Fundam. Mol. Mech. Mutagen.* **2004**, *555*, 173–190. [CrossRef] [PubMed]
5. Ramos-Bueno, R.; Rincón-Cervera, M.; González-Fernández, M.; Guil-Guerrero, J. Phytochemical composition and antitumor activities of new salad greens: Rucola (*Diplotaxis tenuifolia*) and corn salad (*Valerianella locusta*). *Plant Foods Hum. Nutr.* **2016**, *71*, 197–203. [CrossRef] [PubMed]
6. Jiang, X.; Lu, C.; Hu, R.; Shi, W.; Zhou, L.; Wen, P.; Jiang, Y.; Lo, Y.M. Nutritional and microbiological effects of vermicompost tea in hydroponic cultivation of maple peas (*Pisum sativum* var. *arvense* L.). *Food Sci. Nutr.* **2023**, *11*, 3184–3202. [CrossRef]
7. Bertoldi, F.; Sant'Anna, E.; Barcelos-Oliveira, J. *Chlorella vulgaris* cultivated in hydroponic wastewater. In Proceedings of the International Symposium on Soilless Culture and Hydroponics 843, Lima, Peru, 25–28 August 2008; pp. 203–210.
8. Hajam, Y.A.; Kumar, R.; Kumar, A. Environmental waste management strategies and vermi transformation for sustainable development. *Environ. Chall.* **2023**, *13*, 100747. [CrossRef]
9. Abdel Salam, M.M.; Roshdy, N.M. The influence of different applications of vermicompost tea on the quality of pomegranate fruits. *SVU-Int. J. Agric. Sci.* **2022**, *4*, 107–118. [CrossRef]
10. Pearson, C.; Gawel, R.; Maki, D. Effects of vermicompost and beneficial microbes on biomass and nutrient density in purple lady bok choy (*Brassica rapa* var. *chinensis*) in a vertical hydroponic grow tower system. *Fine Focus* **2023**, *9*, 46–64.
11. Berg, G. Plant–microbe interactions promoting plant growth and health: Perspectives for controlled use of microorganisms in agriculture. *Appl. Microbiol. Biotechnol.* **2009**, *84*, 11–18. [CrossRef]
12. Manning, W.J.; Paoletti, E.; Sandermann, H., Jr.; Ernst, D. Ethylenediurea (EDU): A research tool for assessment and verification of the effects of ground level ozone on plants under natural conditions. *Environ. Pollut.* **2011**, *159*, 3283–3293. [CrossRef]
13. Rehman, S.u.; De Castro, F.; Marini, P.; Aprile, A.; Benedetti, M.; Fanizzi, F.P. Vermibiochar: A novel approach for reducing the environmental impact of heavy metals contamination in agricultural land. *Sustainability* **2023**, *15*, 9380. [CrossRef]
14. Rehman, S.u.; Aprile, A.; De Castro, F.; Negro, C.; Migoni, D.; Benedetti, M.; Sabella, E.; Fanizzi, F.P. Assessing the Effectiveness of Vermi-Liquids as a Sustainable Alternative to Inorganic Nutrient Solutions in Hydroponic Agriculture: A Study on *Diplotaxis muralis*. *Agronomy* **2024**, *14*, 1310. [CrossRef]
15. Kumar, D. Nuclear magnetic resonance (NMR) spectroscopy for metabolic profiling of medicinal plants and their products. *Crit. Rev. Anal. Chem.* **2016**, *46*, 400–412. [CrossRef] [PubMed]
16. Schiattone, M.I.; Boari, F.; Cantore, V.; Castronuovo, D.; Denora, M.; Di Venere, D.; Perniola, M.; Sergio, L.; Todorovic, M.; Candido, V. Effect of water regime, nitrogen level and biostimulants application on yield and quality traits of wild rocket [*Diplotaxis tenuifolia* (L.) DC.]. *Agric. Water Manag.* **2023**, *277*, 108078. [CrossRef]
17. Paradiso, R.; Di Mola, I.; Cozzolino, E.; Ottaiano, L.; El-Nakhel, C.; Rouphael, Y.; Mori, M. Nutrient and nutraceutical quality of rocket as a function of greenhouse cover film, nitrogen dose and biostimulant application. *Agronomy* **2023**, *13*, 638. [CrossRef]

18. Di Mola, I.; Ottaiano, L.; Cozzolino, E.; Senatore, M.; Giordano, M.; El-Nakhel, C.; Sacco, A.; Rouphael, Y.; Colla, G.; Mori, M. Plant-based biostimulants influence the agronomical, physiological, and qualitative responses of baby rocket leaves under diverse nitrogen conditions. *Plants* **2019**, *8*, 522. [CrossRef] [PubMed]
19. Buitrago-Villanueva, I.; Barbosa-Cornelio, R.; Coy-Barrera, E. Influence of the Culture System and Harvest Time on the Specialized Metabolite Composition of Rocket Salad (*Eruca sativa*) Leaves. *Horticulturae* **2023**, *9*, 235. [CrossRef]
20. Burdziej, A.; Da Costa, G.; Gougeon, L.; Le Mao, I.; Bellée, A.; Corio-Costet, M.-F.; Mérillon, J.-M.; Richard, T.; Szakiel, A.; Cluzet, S. Impact of different elicitors on grapevine leaf metabolism monitored by ^{1}H NMR spectroscopy. *Metabolomics* **2019**, *15*, 67. [CrossRef]
21. Colì, C.S.; Girelli, C.R.; Cesari, G.; Hussain, M.; Fanizzi, F.P. Biodynamic, organic and integrated agriculture effects on cv. Italia table grapes juice, over a 3-year period experiment: An ^{1}H NMR spectroscopy-based metabolomics study. *Chem. Biol. Technol. Agric.* **2024**, *11*, 35. [CrossRef]
22. Lanzotti, V.; Anzano, A.; Grauso, L.; Zotti, M.; Sacco, A.; Senatore, M.; Moreno, M.; Diano, M.; Parente, M.; Esposito, S. NMR metabolomics and chemometrics of lettuce, *Lactuca sativa* L., under different foliar organic fertilization treatments. *Plants* **2022**, *11*, 2164. [CrossRef]
23. De Castro, F.; Benedetti, M.; Del Coco, L.; Fanizzi, F.P. NMR-based metabolomics in metal-based drug research. *Molecules* **2019**, *24*, 2240. [CrossRef]
24. Kim, H.K.; Choi, Y.H.; Verpoorte, R. NMR-based metabolomic analysis of plants. *Nat. Protoc.* **2010**, *5*, 536–549. [CrossRef]
25. De Castro, F.; Stefano, E.; De Luca, E.; Muscella, A.; Marsigliante, S.; Benedetti, M.; Fanizzi, F.P. A NMR-Based Metabolomic Approach to Investigate the Antitumor Effects of the Novel [Pt (η^1-C_2H_4OMe)(DMSO)(phen)]$^+$ (phen = 1,10-Phenanthroline) Compound on Neuroblastoma Cancer Cells. *Bioinorg. Chem. Appl.* **2022**, *2022*, 8932137. [CrossRef]
26. Abdel-Farid, I.B.; Kim, H.K.; Choi, Y.H.; Verpoorte, R. Metabolic characterization of *Brassica rapa* leaves by NMR spectroscopy. *J. Agric. Food Chem.* **2007**, *55*, 7936–7943. [CrossRef]
27. Abdel-Farid, I.; Jahangir, M.; Van Den Hondel, C.; Kim, H.; Choi, Y.; Verpoorte, R. Fungal infection-induced metabolites in *Brassica rapa*. *Plant Sci.* **2009**, *176*, 608–615. [CrossRef]
28. Jahangir, M.; Kim, H.K.; Choi, Y.H.; Verpoorte, R. Metabolomic response of *Brassica rapa* submitted to pre-harvest bacterial contamination. *Food Chem.* **2008**, *107*, 362–368. [CrossRef]
29. Liang, Y.-S.; Choi, Y.H.; Kim, H.K.; Linthorst, H.J.; Verpoorte, R. Metabolomic analysis of methyl jasmonate treated *Brassica rapa* leaves by 2-dimensional NMR spectroscopy. *Phytochemistry* **2006**, *67*, 2503–2511. [CrossRef]
30. Trygg, J.; Wold, S. Orthogonal projections to latent structures (O-PLS). *J. Chemom.* **2002**, *16*, 119–128. [CrossRef]
31. Veskovski, L.; Andersson, P.-O.; Turesson, I.; Malmodin, D.; Pedersen, A.; Mellqvist, U.-H. Serum Metabolomic Profiling Correlated with ISS and Clinical Outcome for Multiple Myeloma Patients Treated with High Dose Melphalan and Autologous Stem Cell Transplantation. *Blood* **2019**, *134*, 1771. [CrossRef]
32. Martínez-Sánchez, A.; Llorach, R.; Gil, M.I.; Ferreres, F. Identification of new flavonoid glycosides and flavonoid profiles to characterize rocket leafy salads (*Eruca vesicaria* and *Diplotaxis tenuifolia*). *J. Agric. Food Chem.* **2007**, *55*, 1356–1363. [CrossRef]
33. Martínez-Sánchez, A.; Gil-Izquierdo, A.; Gil, M.I.; Ferreres, F. A comparative study of flavonoid compounds, vitamin C, and antioxidant properties of baby leaf Brassicaceae species. *J. Agric. Food Chem.* **2008**, *56*, 2330–2340. [CrossRef]
34. Lim, J.-E.; Cho, M.-J.; Yun, H.-J.; Ha, S.-K.; Lee, D.-B.; Sung, J.-K. The relation between fertilization practices and functional metabolites of crops: A review. *Korean J. Soil Sci. Fertil.* **2016**, *49*, 168–180. [CrossRef]
35. Radić, S.; Vujčić, V.; Glogoški, M.; Radić-Stojković, M. Influence of pH and plant growth regulators on secondary metabolite production and antioxidant activity of *Stevia rebaudiana* (Bert). *Period. Biol.* **2016**, *118*, 9–19. [CrossRef]
36. Manzano-Gómez, L.A.; Guzmán-Albores, J.M.; Rincón-Rosales, R.; Winkler, R.; Rincón-Molina, C.I.; Castañón-González, J.H.; Ruiz-Lau, N.; Gutiérrez-Miceli, F.A.; Rincón-Molina, F.A.; Ruíz-Valdiviezo, V.M. Evaluation of Metabolomic Profile and Growth of *Moringa oleifera* L. Cultivated with Vermicompost under Different Soil Types. *Agronomy* **2021**, *11*, 2061. [CrossRef]
37. Kosem, H.; Kocak, M.Z.; Kaysim, M.G.; Celikcan, F.; Kulak, M. Liquid leachate produced from vermicompost effects on some agronomic attributes and secondary metabolites of sweet basil (*Ocimum basilicum* L.) exposed to severe water stress conditions. *Horticulturae* **2022**, *8*, 1190. [CrossRef]
38. Wong, W.S.; Zhong, H.T.; Cross, A.T.; Yong, J.W.H. Plant biostimulants in vermicomposts: Characteristics and plausible mechanisms. In *The Chemical Biology of Plant Biostimulants*; Wiley: Hoboken, NJ, USA, 2020; pp. 155–180.
39. Posmyk, M.M.; Szafrańska, K. Biostimulators: A new trend towards solving an old problem. *Front. Plant Sci.* **2016**, *7*, 748. [CrossRef]
40. Türkay, I.; Öztürk, L. The form, dose, and method of application of vermicompost differentiate the phenylpropene biosynthesis in the peltate glandular trichomes of methylchavicol chemotype of *Ocimum basilicum* L. *Ind. Crops Prod.* **2023**, *198*, 116688. [CrossRef]
41. Latifi, M.; Kheiry, A.; Razavi, F.; Sanikhani, M. Influence of vermicompost and nitrogen treatments on the enhancement of morphophysiological traits and essential oil composition of spearmint (*Mentha spicata* L.) for double harvesting seasons. *J. Med. Plants By-Prod.* **2024**, *13*, 1016–1025.

42. Souffront, D.K.S.; Salazar-Amoretti, D.; Jayachandran, K. Influence of vermicompost tea on secondary metabolite production in tomato crop. *Sci. Hortic.* **2022**, *301*, 111135. [CrossRef]
43. Moffett, J.R.; Puthillathu, N.; Vengilote, R.; Jaworski, D.M.; Namboodiri, A.M. Acetate revisited: A key biomolecule at the nexus of metabolism, epigenetics and oncogenesis—Part 1: Acetyl-CoA, acetogenesis and acyl-CoA short-chain synthetases. *Front. Physiol.* **2020**, *11*, 580167. [CrossRef]
44. Li, H.; Xu, Y.; Lin, J.; Feng, B.; Zhu, A.; Zhao, X.; Wang, D.; Zeng, Y.; Yang, H.; Wang, S.; et al. Acetate prevents pistil dysfunction in rice under heat stress by inducing methyl jasmonate and quercetin synthesis. *J. Adv. Res.* 2025; *in press*.
45. Araújo, W.L.; Martins, A.O.; Fernie, A.R.; Tohge, T. 2-Oxoglutarate: Linking TCA cycle function with amino acid, glucosinolate, flavonoid, alkaloid, and gibberellin biosynthesis. *Front. Plant Sci.* **2014**, *5*, 552. [CrossRef] [PubMed]
46. Koirala, N.; Thuan, N.H.; Ghimire, G.P.; Van Thang, D.; Sohng, J.K. Methylation of flavonoids: Chemical structures, bioactivities, progress and perspectives for biotechnological production. *Enzym. Microb. Technol.* **2016**, *86*, 103–116. [CrossRef] [PubMed]
47. Wang, Y.; Huang, R.; Gao, S.; Yue, M.; Zhang, X.; Zeng, W.; Tang, B.; Zhou, J.; Huang, D.; Xu, S. Identification of two new flavone 4′-O-methyltransferases and their application in de novo biosynthesis of (2S)-hesperetin in *Yarrowia lipolytica*. *Synth. Syst. Biotechnol.* **2025**, *10*, 728–736. [CrossRef]
48. Rostaei, M.; Fallah, S.; Carrubba, A.; Lorigooini, Z. Organic manures enhance biomass and improve content, chemical compounds of essential oil and antioxidant capacity of medicinal plants: A review. *Heliyon* **2024**, *10*, e36693. [CrossRef]
49. Baldelli, S.; Lombardo, M.; D'Amato, A.; Karav, S.; Tripodi, G.; Aiello, G. Glucosinolates in Human Health: Metabolic Pathways, Bioavailability, and Potential in Chronic Disease Prevention. *Foods* **2025**, *14*, 912. [CrossRef]
50. D'Antuono, L.F.; Elementi, S.; Neri, R. Glucosinolates in *Diplotaxis* and *Eruca* leaves: Diversity, taxonomic relations and applied aspects. *Phytochemistry* **2008**, *69*, 187–199. [CrossRef]
51. Jin, J.; Koroleva, O.A.; Gibson, T.; Swanston, J.; Magan, J.; Zhang, Y.; Rowland, I.R.; Wagstaff, C. Analysis of phytochemical composition and chemoprotective capacity of rocket (*Eruca sativa* and *Diplotaxis tenuifolia*) leafy salad following cultivation in different environments. *J. Agric. Food Chem.* **2009**, *57*, 5227–5234. [CrossRef]
52. Ali, A.; Franzoni, G.; Petrini, A.; Santoro, P.; Mori, J.; Ferrante, A.; Cocetta, G. Investigating physiological responses of Wild Rocket subjected to artificial Ultraviolet B irradiation. *Sci. Hortic.* **2023**, *322*, 112415. [CrossRef]
53. Neugart, S.; Krumbein, A.; Zrenner, R. Influence of light and temperature on gene expression leading to accumulation of specific flavonol glycosides and hydroxycinnamic acid derivatives in kale (*Brassica oleracea* var. sabellica). *Front. Plant Sci.* **2016**, *7*, 326. [CrossRef] [PubMed]
54. Larbat, R.; Paris, C.; Le Bot, J.; Adamowicz, S. Phenolic characterization and variability in leaves, stems and roots of Micro-Tom and patio tomatoes, in response to nitrogen limitation. *Plant Sci.* **2014**, *224*, 62–73. [CrossRef]
55. Ahmad, P.; Ahanger, M.A.; Alyemeni, M.N.; Wijaya, L.; Egamberdieva, D.; Bhardwaj, R.; Ashraf, M. Zinc application mitigates the adverse effects of NaCl stress on mustard [*Brassica juncea* (L.) Czern & Coss] through modulating compatible organic solutes, antioxidant enzymes, and flavonoid content. *J. Plant Interact.* **2017**, *12*, 429–437.
56. Mamat, D.D.; Chong, C.-S.; Abd Samad, A.; Chai, T.-T.; Abd Manan, F. Effects of copper on total phenolics, flavonoids and mitochondrial properties of *Orthosiphon stamineus* callus culture. *Int. J. Agric. Biol.* **2015**, *17*, 1243–1248. [CrossRef]
57. Son, Y.-J.; Park, J.-E.; Lee, N.; Ju, Y.-W.; Pyo, S.-H.; Oh, C.; Yoo, G.; Nho, C.W. Copper-or Zinc-Fortified Nutrient Solution in Vertical Farming System Enriches Copper or Zinc and Elevates Phenolic Acid and Flavonoid Contents in *Artemisia annua* L. *Agronomy* **2024**, *14*, 135. *Agronomy* **2024**, *14*, 135. [CrossRef]
58. Shen, C.; Wang, J.; Jin, X.; Liu, N.; Fan, X.; Dong, C.; Shen, Q.; Xu, Y. Potassium enhances the sugar assimilation in leaves and fruit by regulating the expression of key genes involved in sugar metabolism of Asian pears. *Plant Growth Regul.* **2017**, *83*, 287–300. [CrossRef]
59. Yan, J.; Ye, X.; Song, Y.; Ren, T.; Wang, C.; Li, X.; Cong, R.; Lu, Z.; Lu, J. Sufficient potassium improves inorganic phosphate-limited photosynthesis in Brassica napus by enhancing metabolic phosphorus fractions and Rubisco activity. *Plant J.* **2023**, *113*, 416–429. [CrossRef]
60. Nabil-Adam, A.; Elnosary, M.E.; Ashour, M.L.; Abd El-Moneam, N.M.; Shreadah, M.A. Flavonoids biosynthesis in plants as a defense mechanism: Role and function concerning pharmacodynamics and pharmacokinetic properties. In *Flavonoid Metabolism-Recent Advances and Applications in Crop Breeding*; IntechOpen: Rijeka, Croatia, 2023.
61. Jogawat, A. Osmolytes and their role in abiotic stress tolerance in plants. In *Molecular Plant Abiotic Stress: Biology and Biotechnology*; Wiley: Hoboken, NJ, USA, 2019; pp. 91–104.
62. Kocaman, A.; İnci, Y.; Kıtır, N.; Turan, M.; Argın, S.; Yıldırım, E.; Giray, G.; Ersoy, N.; Güneş, A.; Katırcıoğlu, H.; et al. The effect of novel biotechnological vermicompost on tea yield, plant nutrient content, antioxidants, amino acids, and organic acids as an alternative to chemical fertilizers for sustainability. *BMC Plant Biol.* **2024**, *24*, 868. [CrossRef] [PubMed]
63. Hildebrandt, T.M.; Nesi, A.N.; Araújo, W.L.; Braun, H.-P. Amino acid catabolism in plants. *Mol. Plant* **2015**, *8*, 1563–1579. [CrossRef]

64. Tahjib-Ul-Arif, M.; Zahan, M.I.; Karim, M.M.; Imran, S.; Hunter, C.T.; Islam, M.S.; Mia, M.A.; Hannan, M.A.; Rhaman, M.S.; Hossain, M.A.; et al. Citric acid-mediated abiotic stress tolerance in plants. *Int. J. Mol. Sci.* **2021**, *22*, 7235. [CrossRef] [PubMed]
65. Wang, D.; Shi, Q.; Wang, X.; Wei, M.; Hu, J.; Liu, J.; Yang, F. Influence of cow manure vermicompost on the growth, metabolite contents, and antioxidant activities of Chinese cabbage (*Brassica campestris* ssp. chinensis). *Biol. Fertil. Soils* **2010**, *46*, 689–696. [CrossRef]
66. Turan, M.; Kocaman, A.; Tüfenkçi, Ş.; Katırcıoğlu, H.; Güneş, A.; Kıtır, N.; Giray, G.; Gürkan, B.; Ersoy, N.; Yildirim, E. Development of organic phosphorus vermicompost from raw phosphate rock using microorganisms and enzymes and its effect on tomato yield. *Sci. Hortic.* **2023**, *321*, 112323. [CrossRef]
67. Abbey, L.; Annan, N.; Asiedu, S.K.; Esan, E.O.; Iheshiulo, E.M.-A. Amino acids, mineral nutrients, and efficacy of vermicompost and seafood and municipal solid wastes composts. *Int. J. Agron.* **2018**, *2018*, 6419467. [CrossRef]

environments

MDPI

Article

Sustainable Soybean Production Using Residual Vermicompost Inputs in Corn-Soybean Rotation

Ivan Oyege [1,2] and Maruthi Sridhar Balaji Bhaskar [1,*]

1 Department of Earth and Environment, Florida International University, Miami, FL 33199, USA; ioyeg002@fiu.edu
2 Department of Chemistry, Busitema University, Tororo P.O. Box 236, Uganda
* Correspondence: mbalajib@fiu.edu

Abstract: Soybeans (*Glycine max* L.), a globally significant crop, play a critical role in economic, nutritional, and ecological systems, particularly in rotational farming due to their nitrogen-fixing capacity. This study investigated the residual effects of vermicompost (VC) and vermicompost tea (VCT) applied during a preceding corn cycle on subsequent soybean growth and productivity in an organic corn–soybean rotation. Soybeans were grown in raised beds previously treated with different VCT concentrations and combinations of VC+VCT, without additional fertilization during the soybean phase. Physiological traits, including leaf chlorophyll content (SPAD values) and stomatal conductance, were measured alongside biomass, yield, and plant leaves nutrient concentrations. VC+VCT treatments significantly increased biomass and yield, with VC1+VCT20 achieving the highest biomass (3.02 tons/ha) and yield (1.68 tons/ha). Leaf nutrient analysis revealed increased uptake of both macro- and micronutrients in amended treatments, while SPAD and stomatal conductance values remained consistently higher than in the control. Soil analyses confirmed improved nutrient retention and cation exchange capacity in amended plots, demonstrating the legacy benefits of organic inputs. Therefore, residual VCT and VC+VCT applications improved soybean productivity, nutrient acquisition, and physiological performance in rotational systems. By reducing reliance on synthetic fertilizers and enhancing soil fertility, this strategy supports climate-smart agriculture principles and contributes to SDG 2 (Zero Hunger), SDG 12 (Responsible Consumption and Production), and SDG 13 (Climate Action).

Keywords: soybean; vermicompost; vermicompost tea; crop rotation; nutrient uptake; sustainable agriculture; organic amendments; stomatal conductance; SPAD

1. Introduction

Soybeans is a cornerstone of global agriculture, recognized as one of the most economically and nutritionally important crops worldwide. Its seeds are a rich source of amino acids, high protein content, oil, and essential amino acids [1,2] making it indispensable for human diets and livestock feed formulations. Beyond its role in food and feed, soybean-derived products contribute to diverse industrial applications, including biofuels, bioplastics, and pharmaceuticals, thereby reinforcing its strategic importance in sustainable development and the bioeconomy [2,3]. Global production is highly concentrated, with Argentina, Brazil, and the United States collectively accounting for over 80% of the total output [4]. In the United States alone, soybeans generated revenues exceeding USD 56 billion in 2023, and this demonstrates their critical role in rural economies, food security, and biofuel supply chains [4,5].

Beyond their economic value, soybeans are integral to sustainable agriculture due to their unique biological and ecological functions. Soybeans form symbiotic associations with nitrogen-fixing bacteria, thereby enriching the soil's nitrogen content and reducing dependence on synthetic fertilizers [6,7]. This biological nitrogen fixation not only supports soybean growth but also benefits subsequent crops in rotations, enhancing soil fertility and productivity. As a result, soybeans serve as a cornerstone in crop rotations, where they can improve soil health and productivity for subsequent crops [8,9]. By reducing reliance on synthetic fertilizers, soybean-based systems contribute to climate-smart agriculture goals of lowering greenhouse gas emissions and building soil carbon stocks [10,11], which directly support SDG 2 (Zero Hunger) and SDG 13 (Climate Action).

Despite these advantages, optimizing yields of soybean in low-input or organic systems remains a challenge, primarily due to the need for sustained nutrient availability throughout the growing season without synthetic agrochemical inputs [12]. VC and VCT are promising organic amendments for sustainable nutrient management [13]. VC is a nutrient-dense, slow-release organic material that enhances soil structure, microbial activity, and long-term fertility [14,15]. VCT, a liquid extract of VC, delivers readily available nutrients and bioactive compounds that can improve the uptake efficiency of nutrients and enhance plant growth [16,17].

While the direct benefits of VC and VCT on crop performance are well documented, their residual effects, particularly in rotational systems where no further amendments are applied, remain poorly understood [18]. Soybeans, with their nitrogen-fixing ability and tolerance to moderate nutrient stress, are ideal for assessing these legacy effects. Understanding how residual VC and VCT influence soybean physiology, nutrient uptake, and yield can guide more efficient organic nutrient management. Such strategies align with climate-smart agriculture principles by improving nutrient use efficiency, maintaining productivity under input constraints, and enhancing agroecosystem resilience [19,20], which are key targets under SDG 12 (Responsible Consumption and Production) and SDG 15 (Life on Land) [21]. This is particularly relevant for organic systems, where nutrient availability is largely governed by the mineralization dynamics of previously applied organic matter.

The novelty of this study lies in its focus on the residual effects of VC and VCT in rotational systems, specifically evaluating their impact on soybean performance following a preceding corn cycle without additional fertilization. Therefore, this study evaluates the residual effects of VC and VCT applied in a preceding corn cycle on soybean growth, physiology, and productivity. Specifically, we examined their effects on chlorophyll content, stomatal conductance, biomass, and yield. We hypothesize that (1) residual VC and VCT would enhance soybean performance compared to untreated controls, and (2) combined VC+VCT treatments would have the most potent effects because of their complementary nutrient release patterns. The findings aim to advance our understanding of organic nutrient cycling in crop rotations and support the development of resilient, low-input production systems that are both environmentally sustainable and agronomically viable.

2. Materials and Methods

2.1. Site Description and Experimental Design

The study was conducted at the Agroecology Program's organic garden (coordinates: 25.754906, −80.380032) at Florida International University, Miami, Florida, which is managed exclusively under organic practices and has calcareous sandy loam soils typical of South Florida agroecosystems. This trial followed our preceding corn experiment [22], utilizing the same raised beds without any additional fertilizer or amendments to assess the residual effects of previous applications.

Soybean seeds (black variety) were obtained from Hudson Valley Seed Company, located in Accord, NY, United States, and were sown in the first week of September 2024, two days post-corn harvest, and harvested in December 2024. Weather data from the official NOAA climate portal (https://www.weather.gov/wrh/Climate?wfo=mfl accessed on 23 July 2025) was used to obtain data on weather for the study area during this period. Monthly mean precipitation was 7.62 mm (September), 6.10 mm (October), 0.51 mm (November), and 1.02 mm (December). Average temperatures ranged from 29.2 °C (September) to 22.1 °C (December).

Seven galvanized steel planting beds (L × W × H: 243.84 × 60.96 × 30.48 cm), sourced from a commercial supplier (Vego Garden) in Houston, TX, USA, were installed over a layer of wood mulch and weed-suppressing fabric (Agfabric, Gardenport, Corona, CA, USA) in an open-air field exposed to natural environmental conditions. These beds, previously used for a corn cultivation trial, were filled with a potting substrate provided by RGS Nursery (Miami, FL, USA) and retained the same soil composition for the subsequent soybean experiment.

The treatments applied during the corn phase were carried over into the soybean cycle and included the following: V0: untreated control with no VC or VCT; VCT10: 10% VCT; VCT20: 20% VCT; VCT40: 40% VCT; VC1+VCT10: VC at 1 ton/acre combined with 10% VCT; VC1+VCT20: VC (1 ton/acre) and 20% VCT; VC1+VCT40: VC (1 ton/acre) and 40% VCT. In the corn phase, VCT was applied weekly for six weeks (100 mL per plant) via foliar and drench spray starting at the V2 corn stage, while VC was applied once before planting. These same treatments were carried over into the soybean phase without reapplication, allowing assessment of residual effects. The treatment rates were based on prior optimization studies [23] and represent agronomically and economically realistic levels for organic systems. Each raised bed was divided into three replicates, and irrigation was provided daily, except during rainfall events.

The VC used in this study was produced from mushroom waste blocks sourced from Lions Fruit Farm (Miami, FL, USA). The composting process involved *Eisenia fetida* earthworms over a four-month period, during which organic matter was decomposed through mechanical fragmentation, enzymatic activity, and microbial processes [13]. VCT was prepared by aerating steeped VC in water for 24 h, then diluting to the desired concentrations (10%, 20%, and 40%) for weekly foliar and drench applications.

2.2. Plant Growth and Physiology Measurements

Leaf chlorophyll content was measured weekly between 9:00 am and 10:30 am from 32 to 60 days after sowing (DAS) using a SPAD 502 Plus chlorophyll meter from Spectrum Technologies Inc., located in Aurora, IL, USA, as per the established protocol by [23]. Stomatal conductance was measured during the same period using a LI-600 fluorometer/porometer manufactured by LI-COR Biosciences, based in Lincoln, NE, USA. These physiological parameters were selected to assess plant health and photosynthetic efficiency under residual nutrient conditions.

2.3. Plant and Soil Sampling Procedures

At 77 days after sowing (DAS), entire soybean plants, including pods and roots, were collected and airdried at a temperature range of 25–30 °C until a constant weight was achieved, and the total biomass was recorded. Grain yield was determined by weighing the dried grains (excluding pods) and calculating yield per hectare based on the raised bed using the following formula:

$$\text{Soybean grain yield in tons/ha} = \frac{\text{Total dry grain weight in tons}}{\text{Area of the raised bed in m}^2} \times 10{,}000$$

Dried leaves from each replicate were grinded into powder using Hamilton Beach coffee grinder (Glen Allen, VA, USA), and were analyzed for macronutrients (N, K, C, Mg, P, S, Ca) and micronutrients (B, Mn, Na, Zn, Fe, Cu) using a C/N analyzer and using inductively coupled plasma mass spectrometry (ICP-MS), in accordance with the methodology outlined by [22]. Detection limits for ICP-MS were N (0.01%), P (0.005%), K (0.005%), Ca (1 mg/kg), Mg (1 mg/kg), Zn (0.1 mg/kg), Cu (0.1 mg/kg), Fe (0.1 mg/kg), Mn (0.1 mg/kg), B (0.1 mg/kg), and Na (0.01%).

Soil samples were collected on the same day from two depths in each replicate: surface (0–10 cm) and subsurface (10–20 cm). Samples were air-dried and sieved (2 mm) prior to analysis. Soil pH was assessed using a 0.01 M calcium chloride solution. The concentrations of carbon and nitrogen were measured with a C/N analyzer. Macro- and micronutrient levels were determined via inductively coupled plasma mass spectrometry (ICP-MS). Organic matter (OM) content was estimated using the loss-on-ignition technique, and cation exchange capacity (CEC) was calculated based on the total exchangeable charges of H, Zn, Na, Ca, and K, following the method outlined by [22]. Detection limits for ICP-MS in soil samples were approximately P (0.005%), K (0.005%), Ca and Mg (1 mg/kg), Na and S (0.01%), Fe (0.3 mg/kg), Zn (0.4 mg/kg), Cu (0.6 mg/kg), Mn (0.3 mg/kg), and B (2.2 mg/kg).

2.4. Statistical Analysis

Normality of the data was assessed using the Shapiro–Wilk test, and variance homogeneity was evaluated with Levene's test. In cases where these assumptions were not met, suitable data transformations or non-parametric approaches were employed. All statistical analyses were performed using Minitab version 22 (Minitab LLC, State College, PA, USA). A one-way ANOVA was conducted at a 5% significance level. When the overall F-test indicated significance and ANOVA assumptions were satisfied, Tukey's HSD post hoc test was applied to compare treatment means, ensuring control of the family-wise error rate during multiple pairwise comparisons.

3. Results

3.1. Soil Physicochemical Properties

To evaluate the carryover effects of the corn-phase treatments, baseline soil nutrient data were collected immediately after the corn harvest [22] (Table 1). Soils at both sampling depths exhibited slightly alkaline conditions, with pH values ranging between 7.77 and 7.97. Nutrient levels of N, P, and K were generally consistent across treatments, except for the control group (V0), which consistently showed significantly reduced concentrations. Comparable patterns were noted for OM, CEC, and the majority of macro- and micronutrients.

For post-soybean harvest soil data (Table 2), nutrient patterns in surface soils (0–10 cm) largely mirrored baseline trends: all VCT and VC+VCT treatments had significantly higher N, P, C, and OM compared to V0. The control consistently showed the lowest nutrient concentrations and CEC values. Statistical comparisons revealed no statistically significant differences between VC-only and VC+VCT treatments for most nutrients. However, VCT40 had significantly lower P than VC1+VCT40, higher Mg than VCT20, and lower Zn than most other treatments (except VCT20). For CEC, VCT10 and VCT20 were significantly lower than VC1+VCT10.

Table 1. Physicochemical properties of soils following corn harvest and subsequent application of vermicompost (VC) and vermicompost tea (VCT). The data adapted from [22] reflect residual soil fertility status prior to soybean cultivation.

Treatment	pH	N (%)	P (%)	K (%)	Ca (mg kg^{-1})	Mg (mg kg^{-1})	Zn (mg kg^{-1})	Na (mg kg^{-1})	C (%)	OM (%)	CEC (cmol kg^{-1})
VC [a]	7.5	1.55	0.176	0.502	1831	1391	30.5	296	22.8	39.3	33.6
Surface soils (0–10 cm)											
V0	7.83 a	0.09 b	0.002 b	0.001 b	267 b	43 b	1.6 b	2.0 b	1.80 c	3.1 c	1.7 b
VCT10	7.77 a	0.69 a	0.018 a	0.010 a	2517 a	385 a	17.0 a	19.5 a	12.4 ab	21.3 ab	16.0 a
VCT20	7.83 a	0.80 a	0.017 a	0.008 a	2578 a	368 a	17.0 a	18.7 a	14.2 ab	24.4 ab	16.2 a
VCT40	7.77 a	0.86 a	0.016 a	0.009 a	2694 a	373 a	14.1 a	17.2 a	17.6 a	30.2 a	16.8 a
VC1+VCT10	7.77 a	0.72 a	0.018 a	0.009 a	2676 a	372 a	16.3 a	15.2 a	13.1 ab	22.6 ab	16.7 a
VC1+VCT20	7.9 a	0.62 a	0.017 a	0.008 a	2579 a	356 a	15.1 a	21.2 a	11.0 b	18.9 b	16.1 a
VC1+VCT40	7.83 a	0.85 a	0.019 a	0.010 a	2641 a	397 a	15.7 a	21.8 a	15.0 ab	25.8 ab	16.8 a
Subsurface soils (10–20 cm)											
V0	7.87 ab	0.08 b	0.002 c	0.002 c	237 b	55 b	1.5 b	3.3 b	1.5 c	2.6 c	1.7 c
VCT10	7.87 ab	0.88 a	0.018 ab	0.009 b	2409 a	432 a	13.8 a	17.6 a	15.4 ab	26.4 ab	15.9 ab
VCT20	7.80 ab	0.77 a	0.018 ab	0.009 b	2505 a	457 a	14.6 a	19.3 a	13.5 ab	23.1 ab	16.6 ab
VCT40	7.73 b	0.83 a	0.017 b	0.010 b	2416 a	503 a	15.4 a	21.7 a	16.1 a	27.8 a	16.5 a
VC1+VCT10	7.93 ab	0.69 a	0.019 ab	0.011 b	2512 a	488 a	15.1 b	20.9 a	12.0 b	20.6 b	16.9 b
VC1+VCT20	7.93 ab	0.78 a	0.024 a	0.018 a	2609 a	590 a	16.6 a	25.0 a	13.7 ab	23.6 ab	18.4 ab
VC1+VCT40	7.97 a	0.70 a	0.019 ab	0.012 b	2442 a	474 a	20.0 a	21.8 a	12.2 b	21.0 b	16.5 b

Values represent mean (n = 3 per treatment). Treatments assigned different letters are significantly different from each other at the 0.05 probability level. [a] The VC data provided reflect applications made during the corn experiment.

In subsurface soils (10–20 cm), the control treatment V0 again had the lowest nutrient concentrations, OM, and CEC. Differences between VCT and VC+VCT treatments were generally non-significant for most elements. While nutrient levels were broadly similar between surface and subsurface soils, K and Na exhibited notable vertical variation: surface soils contained 20–50% more K and 40–130% more Na than subsurface soils.

Table 2. Nutrient content of soils from two depths post soybean harvest.

Treatment	pH	N (%)	P (%)	K (%)	Ca (mg kg^{-1})	Mg (mg kg^{-1})	Zn (mg kg^{-1})	Na (mg kg^{-1})	C (%)	OM (%)	CEC (cmol kg^{-1})
Surface soil (0–10 cm)											
V0	7.57 a	0.41 c	0.011 c	0.006 b	1761 c	283 c	10.84 c	11.16 b	8.13 b	13.98 b	11.32 c
VCT10	7.40 a	0.65 b	0.020 ab	0.010 ab	2816 b	377 ab	19.37 a	19.52 ab	12.41 ab	21.34 ab	17.48 b
VCT20	7.50 a	0.69 b	0.019 ab	0.012 ab	2856 ab	369 b	18.45 ab	24.62 a	12.74 ab	21.91 ab	17.66 b
VCT40	7.43 a	0.73 b	0.018 b	0.012 ab	2857 ab	450 a	16.53 b	26.00 a	14.92 b	25.66 b	18.33 ab
VC1+VCT10	7.07 a	0.70 b	0.024 a	0.014 a	3095 a	425 ab	19.84 a	28.34 a	12.41 ab	21.34 ab	19.38 a
VC1+VCT20	7.33 a	0.65 b	0.022 ab	0.012 a	2990 ab	384 ab	19.59 a	25.30 a	11.70 ab	20.13 ab	18.46 ab
VC1+VCT40	7.37 a	0.62 ab	0.022 ab	0.015 a	2974 ab	394 ab	19.21 a	25.53 a	11.04 ab	18.98 ab	18.53 ab
Subsurface soils (10–20 cm)											
V0	7.50 a	0.43 b	0.012 b	0.005 b	1749 b	320 b	10.63 b	8.09 b	8.22 b	14.14 b	11.54 b
VCT10	7.40 a	0.70 a	0.024 a	0.008 a	2910 a	481 a	19.14 a	11.91 ab	12.57 a	21.63 a	18.78 a
VCT20	7.47 a	0.73 a	0.020 a	0.009 a	3086 a	498 a	17.98 a	13.60 a	13.73 a	23.61 a	19.80 a
VCT40	7.57 a	0.71 a	0.019 a	0.008 a	2991 a	534 a	17.08 a	13.81 a	13.38 a	23.02 a	19.60 a
VC1+VCT10	7.47 a	0.66 a	0.023 a	0.010 a	3032 a	504 a	18.39 a	12.97 a	12.16 a	20.91 a	19.63 a
VC1+VCT20	7.37 a	0.72 a	0.024 a	0.009 a	2919 a	499 a	18.71 a	10.79 ab	13.28 a	22.84 a	18.97 a
VC1+VCT40	7.40 a	0.69 a	0.023 a	0.010 a	2973 a	484 a	19.03 a	11.87 ab	12.33 a	21.21 a	19.15 a

Note: Means marked with different letters indicate statistically significant differences at $p < 0.05$.

Comparing baseline data (Table 1) and post-soybean harvest data (Table 2) revealed substantial increases in nutrients, especially in the control bed. In surface soils, pH shifted from slightly alkaline to near neutral (7.07–7.57). In V0 treatment, N, C, and OM increased nearly fivefold, while Na increased sixfold, and P, K, Ca, Mg, Zn, and CEC increased about sevenfold relative to baseline. Specific treatments also showed notable increases: K doubled in VCT20, VC1+VCT20, VC1+VCT40, and VC1+VCT10, while Na doubled in VCT40 and

VC1+VCT10 compared to their baseline levels. Other nutrients showed minimal changes between baseline and post-harvest values.

Similar patterns were observed in subsurface soils. Compared to baseline, post-harvest V0 soils had about five times more N, C, and OM; about six times more P and Mg; about seven times more Ca, Zn, and CEC; about three times more K; and about two times more Na. An exception was VC1+VCT40, where Na levels dropped to less than half of baseline. For other nutrients, changes between baseline and post-harvest values were negligible.

3.2. *Plant Nutrient Content*

The soybean leaves' nutrient content shows enhanced macro- and micronutrient uptake in soils previously amended with VC and VCT compared to the control (V0) (Table 3). All measured nutrients, except Na, varied significantly across treatments. Na concentrations remained statistically unchanged among treatments, including the control.

For every nutrient except Na, V0 consistently had the lowest leaf concentrations. The highest N (1.95%), P (1.79%), and K (0.88%) contents were recorded in VCT20, VCT40, and VC1+VCT40 treatments, respectively. VC1+VCT20 had the highest Mg, B, Ca, and Mn contents, while Fe and Cu were highest in VC1+VCT40, and Zn was highest in VCT40. There was no statistically significant difference in sulfur levels between the VCT and VC+VCT treatments. However, V0 had significantly lower S levels than VCT10, VC1+VCT10, and VC1+VCT20.

Table 3. Nutrient composition of soybean leaves under VCT and VC+VCT treatments.

Sample	N (%)	P (%)	K (%)	Ca (%)	Mg (%)	S (%)	Na (%)	B (mg kg^{-1})	Fe (mg kg^{-1})	Mn (mg kg^{-1})	Zn (mg kg^{-1})	Cu (mg kg^{-1})
V0	1.00 f	0.74 d	0.40 e	1.90 e	0.36 e	0.07 b	0.04 a	42.85 g	43.06 g	22.55 f	165.85 f	7.22 d
VCT10	1.56 d	1.71 b	0.82 b	2.97 d	0.63 bc	0.14 a	0.05 a	67.46 e	101.78 b	37.56 b	259.43 e	7.32 d
VCT20	1.95 a	1.58 c	0.58 c	2.99 d	0.56 d	0.13 ab	0.05 a	65.23 f	86.25 d	35.64 d	284.42 b	9.71 b
VCT40	1.87 b	1.79 a	0.80 b	3.01 cd	0.57 d	0.11 ab	0.05 a	70.80 c	81.30 e	32.55 e	315.64 a	7.39 cd
VC1+VCT10	1.62 c	1.56 c	0.60 c	3.29 b	0.67 b	0.15 a	0.05 a	78.50 b	77.25 f	36.77 c	262.41 d	7.90 c
VC1+VCT20	1.39 e	1.70 b	0.51 d	3.74 a	0.76 a	0.14 a	0.04 a	82.22 a	88.25 c	43.61 a	259.51 e	6.99 d
VC1+VCT40	1.53 d	1.58 c	0.88 a	3.05 c	0.59 cd	0.12 ab	0.06 a	68.19 d	109.80 a	37.87 b	271.44 c	10.71 a

Mean values followed by different letters differ significantly at $p < 0.05$.

When the ratio of soybean leaf nutrient concentration to residual soil nutrient concentration was calculated (Figure 1), distinct patterns in nutrient uptake efficiency were observed. In V0 treatment, K and P uptake efficiencies were relatively high despite low absolute nutrient content. Increasing VCT concentration enhanced P and K uptake efficiency, with VCT40 showing the highest P ratio (96.8) and a high K ratio (80.0). N uptake ratios remained relatively stable across VCT-only treatments (2.3–2.7) while VC+VCT treatments generally reduced K uptake efficiency, with the lowest values in VC1+VCT20 (48.6) and VC1+VCT10 (50.0). P uptake ratios followed a similar trend, remaining high in VC1+VCT20 but not surpassing those in VCT-only treatments. Mg uptake efficiency peaked in VC1+VCT20, while Na uptake efficiency was highest in V0 and VC1+VCT40 (Figure 1).

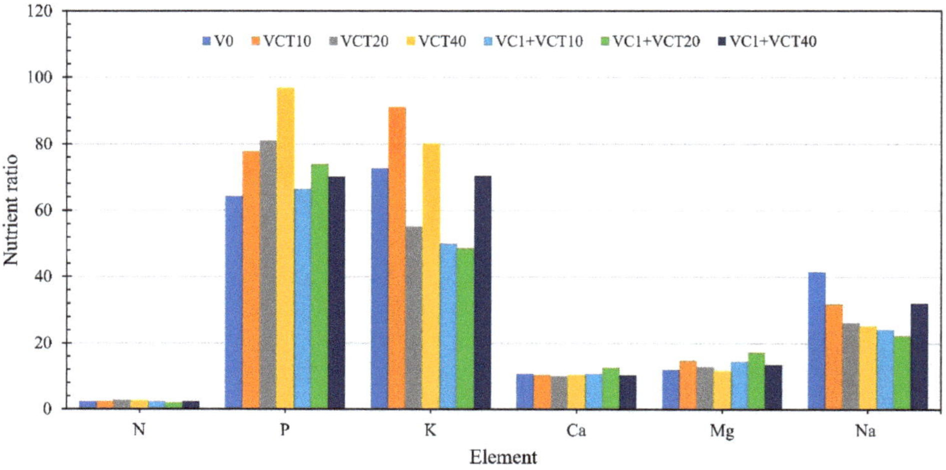

Figure 1. Ratio of soybean leaves nutrient concentration to residual soil nutrient concentration (average of surface and subsurface soils in Table 1) across different treatments. VCT-only treatments, especially VCT40, showed the highest P uptake efficiency, while VC+VCT treatments generally had lower K uptake ratios despite higher absolute K levels. V0 had low nutrient content, but relatively high uptake efficiency, indicating compensatory uptake under nutrient-limited conditions.

3.3. Relative Chlorophyll Content

SPAD measurements showed a general increase in leaf chlorophyll content with advancing plant maturity across all treatments (Figure 2). Across all sampling dates, the control treatment (V0) consistently recorded significantly lower SPAD values compared to both VCT-only and VC+VCT treatments. SPAD readings in V0 ranged from 11 to 15 throughout the experiment, showing minimal fluctuation. No significant differences were observed between VCT-only and VC+VCT treatments, except in the case of VCT40, which showed marginally higher readings from 32 to 60 DAS.

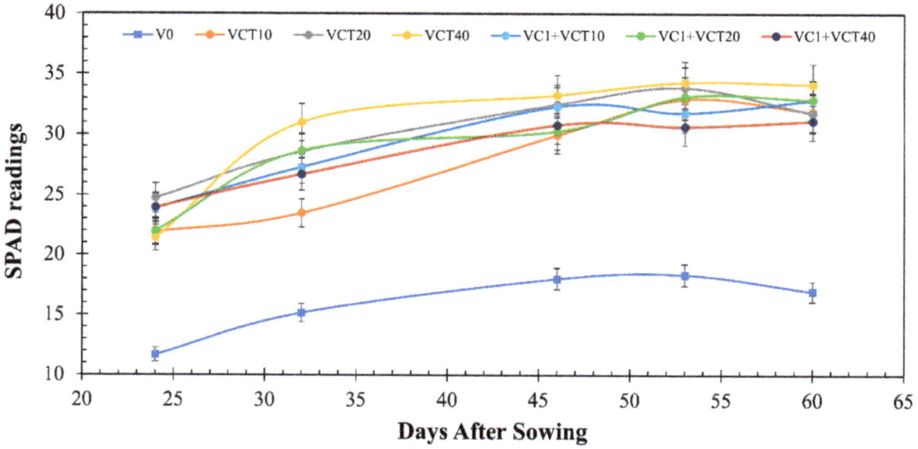

Figure 2. Variation in soybean leaf relative chlorophyll content (SPAD readings) over time, influenced by the residual nutrients from organic amendments. SPAD values increased with plant maturity and were consistently lowest in the control (V0) treatment. VCT40 and VC1+VCT20 maintained higher SPAD readings throughout, indicating enhanced nitrogen status and chlorophyll accumulation from residual organic amendments.

3.4. Stomatal Conductance

Stomatal conductance increased with plant maturity across all treatments (Figure 3). VC+VCT treatments consistently exhibited higher conductance than VCT-only and control (V0) treatments, indicating enhanced water regulation and physiological activity. Among these, VC1+VCT20 and VC1+VCT40 frequently recorded the highest conductance throughout the experiment. At 24 DAS, VC+VCT treatments already showed significantly greater conductance than both VCT-only and V0 treatments. VCT-only treatments increased gradually between 24 and 32 DAS, then remained relatively stable until 60 DAS, with the exception of VCT10, which showed a sharp increase between 53 and 60 DAS. The control (V0) maintained the lowest stomatal conductance throughout the experimental period, suggesting limited physiological responsiveness under nutrient-deficient conditions.

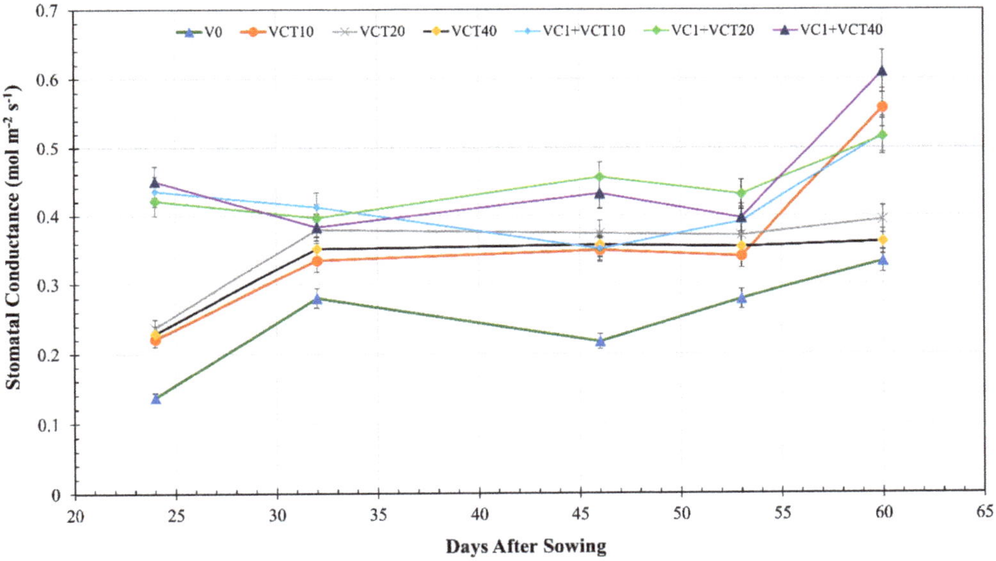

Figure 3. Variation in soybean leaf stomatal conductance with increasing days after sowing (DAS), influenced by residual nutrients from organic amendments. VC+VCT treatments, particularly VC1+VCT20 and VC1+VCT40, maintained higher conductance than VCT-only treatments, while V0 had the lowest values. These results highlight the positive legacy effect of combined organic amendments on plant water regulation and gas exchange.

3.5. Biomass and Yield

Nutrient carryover from the organic amendments had a significant influence on soybean biomass and yield (Figure 4). In terms of total biomass, VC+VCT treatments generally outperformed both VCT-only and V0, demonstrating the synergistic benefits of combined organic inputs. An exception was VCT20, which produced more biomass (2.36 tons/ha) than VC1+VCT10 (2.20 tons/ha). The highest biomass was recorded in VC1+VCT20 (3.02 tons/ha), whereas VCT40 yielded significantly less biomass than all other VCT-only and combined VC and VCT treatments. V0 had the least biomass (1.02 tons/ha), reflecting the absence of residual nutrient support.

Yield trends closely mirrored biomass patterns. VC1+VCT20 and VC1+VCT10 achieved the highest yields at 1.68 tons/ha and 1.37 tons/ha, respectively. VCT20 and VCT10 also performed well, producing higher yields than VC1+VCT10 and VCT40. VCT40 had the lowest yield among the VCT and VC+VCT treatments (0.97 tons/ha), while V0 remained the lowest overall.

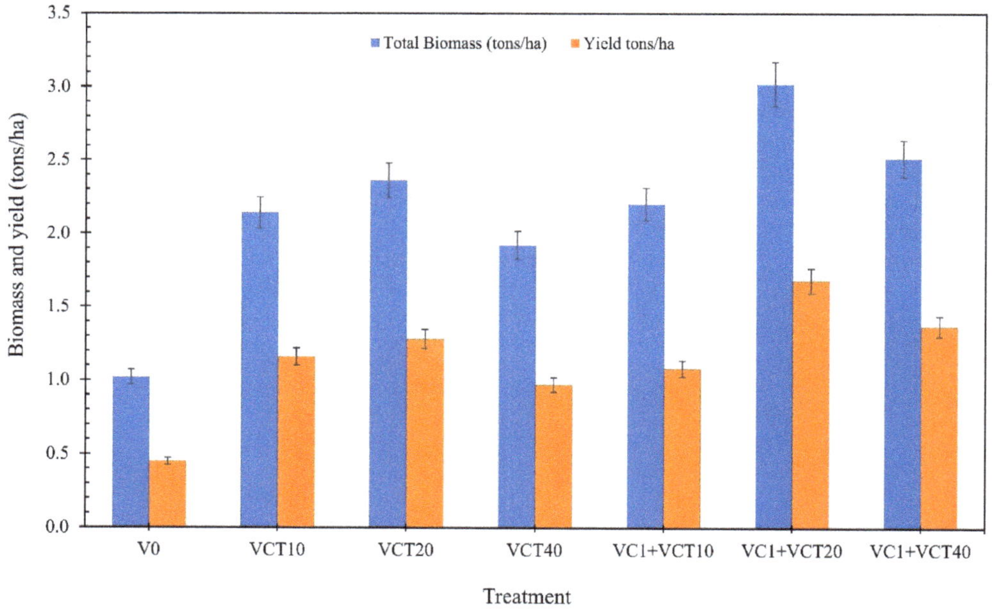

Figure 4. Total soybean dry biomass and grain yield under different organic amendment treatments using nutrient carryover from organic amendments. VC+VCT treatments outperformed both VCT-only and control treatments, with VC1+VCT20 producing the highest biomass (3.02 tons/ha) and yield (1.68 tons/ha). V0 had the lowest biomass and yield, confirming the strong carryover benefits of combined VC and VCT on crop productivity.

4. Discussion

4.1. Soil Physicochemical Properties

Baseline nutrient content (Table 1) showed a slightly alkaline pH (7.77–7.97), consistent with the calcareous sandy loam soils of South Florida [23]. The lack of significant differences in N, P, K, and most other nutrients across treatments, except in the control (V0), indicates that nutrient contributions from VC and VCT amendments applied during the corn phase persisted into the soybean cycle. This persistence suggests that organic amendments can maintain soil fertility across crop cycles, even without reapplication. Because the experiment was conducted in raised beds under controlled daily irrigation in Miami's hot and humid climate, rainfall variability was unlikely to influence soil moisture or nutrient leaching. Therefore, the observed differences are attributable to residual amendments rather than environmental factors. The findings of this study align with findings by [24], who reported that organic amendments and crop residue returns enhance nutrient retention and aggregate stability in rotational systems.

Post-harvest analysis (Table 2) showed that surface soils (0–10 cm) under organic input treatments retained significantly higher levels of N, P, C, organic matter, and CEC compared to V0, which remained lowest in all parameters. These results confirm that residual fertility from the corn phase carried over into soybean production, reinforcing the long-term benefits of organic inputs. The consistently warm conditions in Miami may also have supported microbial activity and organic matter mineralization, enhancing nutrient release from residual VC and VCT [25]. Higher nutrient levels in amended treatments may also reflect synergistic effects from soybean root exudates and rhizosphere microbial processes, which are known to stimulate nutrient cycling and retention. This is consistent with the way VC improves nutrient retention and CEC through its high OM and humic content [16,26].

In some cases, nutrient levels were similar between VCT-only and VC+VCT treatments, suggesting that VCT alone can sustain nutrient availability. However, VCT40 showed significantly lower P and Zn compared to VC1+VCT40, indicating that higher VCT concentrations alone may not improve nutrient retention and could even lead to nutrient imbalances. Under the controlled moisture conditions of this study, such differences are unlikely to result from rainfall-driven leaching. Instead, they may reflect treatment-specific nutrient dynamics such as microbial immobilization, plant uptake, or chemical interactions [27]. For example, lower P in VCT40 may result from precipitation with Ca in high-pH soils or increased plant uptake. Reduced Zn in VCT40 could be due to greater removal in biomass, given the mobility of Zn in legumes. Conversely, higher Mg in VCT40 compared to VCT20 may be attributed to solubilization from organic matter, highlighting the complex interactions between amendment concentration and nutrient availability.

Nutrient stratification was evident, with surface soils containing 20–50% more K and 40–130% more Na than subsurface layers. This likely reflects the carryover effect from foliar and drench VCT applications during corn growth, which concentrated nutrients near the soil surface. Soybean root systems, known for active nutrient cycling in the upper soil profile under organic management [28,29], likely reinforced this stratification. Our findings are consistent with [30], who reported that legume integration in crop rotations significantly boosts soil organic carbon and nutrient cycling, particularly in organic systems where synthetic inputs are absent.

When comparing baseline to post-soybean soils, even the control (V0) showed nearly fivefold increases in N, C, and OM in surface soils, despite no added inputs. Although V0 showed significant relative increases compared to its own baseline, these values remained substantially lower than all amended treatments. The observed increases in V0 reflect biological nitrogen fixation by soybeans and enhanced microbial activity stimulated by root exudates during the growing period [31,32]. These findings support the concept that legume crops can improve soil fertility through symbiotic nitrogen fixation and rhizosphere-driven nutrient cycling, even in nutrient-poor soils [31–33]. This is also consistent with the "soybean nitrogen credit" concept, where legumes enhance N availability for subsequent crops [27,34].

Notably, treatments such as VCT20 and VC1+VCT40 doubled K and Na post-harvest, demonstrating synergistic effects between residual corn-phase amendments and soybean-driven nutrient cycling. The slight pH drop from alkaline to near-neutral likely resulted from organic acid release via soybean root exudates and microbial activity, a well-documented phenomenon in organically managed legume systems [35–37].

4.2. Plant Chemical Analysis

Soybean leaf nutrient content and uptake efficiency patterns indicated apparent legacy effects of VCT and VC applications from the corn phase (Table 3). Since no inputs were applied during soybean growth, the observed differences in nutrient profiles are attributable to residual soil fertility and chemical transformations from prior treatments.

Post-harvest soil data confirmed that VCT and VC+VCT treatments maintained significantly higher nutrient reserves than V0, and these reserves translated into higher leaf macro- and micronutrient content. This supports previous findings that VC-based amendments enhance nutrient retention and mineralization over time, thereby increasing nutrient availability for subsequent crops [38,39].

The sodium content showed no significant treatment effect, likely reflecting its minor role in soybean nutrition and limited variability in soil Na. However, higher Na uptake ratios in V0 and VC1+VCT40 may indicate osmotic adjustment under nutrient-limited or ion-rich conditions, where Na accumulation helps maintain cell turgor or reflects elevated soluble salt levels [40–42]. Furthermore, nutrient-specific patterns also revealed interesting findings. The high N content in VCT20 suggests optimal mineralization without excessive loss. In contrast, the peak K content in VC1+VCT40 likely reflects contributions from both the mineral fraction of VC and the soluble K in VCT. The elevated Ca, Mg, B, and Mn content in VC1+VCT20 (Table 3) points to synergistic improvements in CEC and divalent cation availability. Furthermore, Fe, Zn, and Cu were highest in VC1+VCT40 and VCT40, consistent with the ability of VC to chelate micronutrients and enhance plant uptake [39,43].

Uptake efficiency ratios, such as high K and P ratios in V0 (Figure 1), despite low absolute nutrient levels, suggest compensatory uptake mechanisms, where plants in nutrient-limited soils enhance root allocation and scavenging capacity [44,45]. However, yield in V0 remained constrained by the small total nutrient pool. Increasing VCT rates boosted P and K uptake efficiencies, with VCT40 reaching the highest P ratio (96.8) and maintaining strong K uptake (80.0). This suggests that VCT promotes more bioavailable nutrient forms, likely through the release of organic acids and microbial solubilization processes [46]. By contrast, VC+VCT treatments often had lower K uptake ratios, possibly due to a dilution effect from higher total K availability or cation competition in soils enriched with Ca^{2+} and Mg^{2+} [43,47]. The highest Mg uptake ratios in VC1+VCT20 aligned with leaf Mg results, reinforcing the advantage of combining solid and liquid organic amendments for sustained nutrient supply.

4.3. *Relative Chlorophyll Content*

SPAD readings increased across all treatments from 32 to 60 DAS (Figure 2), indicating normal canopy development and chlorophyll accumulation during soybean vegetative growth. Since SPAD values serve as a proxy for leaf chlorophyll and nitrogen status [48], the consistently lower readings in the control (V0) indicate limited nitrogen availability, which is consistent with soil and leaf analyses showing significantly lower N content in V0 compared to the VCT and VC+VCT treatments. This highlights the importance of residual organic inputs in sustaining early-season nitrogen availability and chlorophyll synthesis.

Among the amended treatments, VCT40 and VC1+VCT20 maintained higher SPAD values throughout the growth period, indicating a nutrient environment that supported prolonged chlorophyll production. VCT likely contributed soluble nutrients and microbial stimulants from the corn phase, while VC provided slow-release nutrients that continued to nourish the soybean crop [13,23]. The absence of substantial differences among amended treatments (Figure 2) suggests a nutrient threshold effect, where once adequate N is available, additional inputs do not further enhance chlorophyll levels. This plateau may reflect nutrient saturation or uniform root uptake efficiency as plants mature.

By 60 DAS, SPAD readings converged across treatments, likely due to nutrient uptake stabilizing and internal redistribution as the crop transitioned to reproductive growth. At this stage, chlorophyll levels typically plateau or decline as nutrients are remobilized to developing pods. Decomposition of organic residues and microbial activity may also have evened out nutrient availability across treatments. These patterns are consistent with previous studies showing that organic amendments boost early chlorophyll development but have diminishing effects as plants enter reproductive stages [43,49].

4.4. *Stomatal Conductance*

Stomatal conductance increased over time across all treatments (Figure 3), reflecting progressive canopy development and rising photosynthetic demand. The highest conductance values were observed in VC+VCT treatments, particularly VC1+VCT20 and

VC1+VCT40, suggesting that the combination of slow-release nutrients from VC and bioactive compounds in VCT created an optimal soil environment for root development and water uptake. This likely enhanced the plants' ability to regulate stomatal opening, a process closely linked to water availability, nutrient status, and overall physiological vigor [46,50].

Bioactive components in VCT, such as humic substances, cytokinins, and auxins, may have stimulated stomatal opening by enhancing photosynthetic activity and internal water transport [13,23]. However, without a steady nutrient supply from VC, the effects of VCT alone appeared less durable. VCT-only treatments showed moderate early increases in stomatal conductance but plateaued from 32 to 60 DAS, except for VCT10, which exhibited a late-season increase. This pattern suggests that VCT can provide a short-term physiological boost, but VC is necessary to maintain the effect throughout the growth cycle [22]. The consistently low stomatal conductance in V0 shows how nutrient limitations can restrict root development, water uptake, and stomatal activity, ultimately affecting photosynthetic performance.

4.5. Biomass and Yield

Soybean biomass and yield were strongly influenced by residual amendments from the corn phase (Figure 4). V0, which received no prior inputs and recorded the lowest biomass and yield due to nutrient-deficient conditions. In contrast, VC+VCT treatments consistently outperformed both VCT-only and control treatments, with VC1+VCT20 producing the highest biomass and yield. These results highlight the synergistic effects of combining sustained nutrient release of VC with VCT's fast-acting nutrients and bioactive compounds, which together enhance early growth, microbial activity, and nutrient availability [22,26]. Soybean yields in our VC1+VCT20 treatment (1.68 tons/ha) were comparable to those reported by [51] in short-term organic rotation using biological fertilizers, suggesting that residual organic inputs can match the performance of actively fertilized organic systems.

Notably, VCT20 outperformed VC1+VCT10 in biomass, indicating that optimal VCT rates can rival combined treatments under certain conditions. However, VCT40 yielded significantly less biomass and grain yield than other amended treatments, suggesting that high concentrations may lead to nutrient imbalances, reduced uptake efficiency, or mild phytotoxic effects [22]. These findings align with prior research, which shows that the benefits of VCT are concentration-dependent and may plateau or decline at an excessive rate [16].

While VCT10 and VCT20 delivered respectable yields without VC, the absence of a sustained nutrient source likely limited total biomass accumulation. As expected for a no-fertilizer soybean phase, overall yields were lower than the commercial average obtained by farmers in the state of Georgia, USA (10.63 tons/ha) [52]. Nevertheless, the results demonstrate that residual organic amendments from a previous crop can provide meaningful yield gains in low-input systems. Our findings support the broader environmental benefits of organic soybean systems, consistent with [53], who reported that organic rotations reduced environmental impact by over 30% compared to conventional systems.

5. Conclusions

This study demonstrated that nutrient effects from prior applications of compost-based inputs during the preceding corn cycle can substantially improve soybean performance in an organic rotation system. By tracking plant physiology, nutrient uptake, and yield, we found that combining VC and VCT, especially at moderate rates, provided both immediate and sustained nutrient supply. This translated into higher chlorophyll content, stronger stomatal conductance, greater biomass, and improved yields.

VC1+VCT20 treatment emerged as the most effective treatment, demonstrating that a balanced combination of solid and liquid organic amendments can optimize nutrient use efficiency and promote vigorous plant growth. In contrast, higher VCT rates, such as VCT40, were less effective, highlighting the importance of avoiding nutrient imbalances. Even without amendments, soybeans maintained a baseline yield, demonstrating their resilience and nitrogen-fixing ability.

These results demonstrate the value of incorporating residual organic amendments into crop rotation systems to reduce dependence on chemical inputs, improve soil quality, and promote environmentally friendly farming practices. Future work should evaluate these approaches across multiple rotation cycles, diverse agroecological settings, and long-term multi-season trials to assess their consistency, resilience, and scalability under real-world farming conditions. Additionally, future work should include the quantification of biologically active compounds in VC and VCT to understand better their role in plant physiology, nutrient dynamics, and pest resistance. Such data would strengthen the mechanistic understanding of how these amendments function and guide more targeted application strategies.

Author Contributions: Conceptualization, I.O.; methodology, I.O. and M.S.B.B.; software, I.O.; validation, I.O.; formal analysis, I.O.; investigation, I.O.; resources, M.S.B.B.; data curation, I.O.; writing—original draft preparation, I.O.; writing—review and editing, I.O. and M.S.B.B.; visualization, I.O. and M.S.B.B.; supervision, M.S.B.B.; project administration, M.S.B.B.; funding acquisition, M.S.B.B.; All authors have read and agreed to the published version of the manuscript.

Funding: This research was partially supported by the Fulbright Foreign Students Program in Uganda, with primary funding provided through USDA-NRCS and USDA-NIFA grants under designated award numbers NR224209 XXXXG001; AWD13273, NR233 A750011 G026; AWD15131; 2023-70001-40999, 2023-77040-41154, 25010-84552-1; PO 4500167774.

Data Availability Statement: The original contributions presented in this study are included in the article. Further inquiries can be directed to the corresponding author.

Acknowledgments: We thank Fiona Nakabuye and Addy Carl for their valuable assistance with the experiment.

Conflicts of Interest: The authors declare no conflicts of interest.

References

1. Kudełka, W.; Kowalska, M.; Popis, M. Quality of Soybean Products in Terms of Essential Amino Acids Composition. *Molecules* **2021**, *26*, 5071. [CrossRef]
2. Guo, B.; Sun, L.; Jiang, S.; Ren, H.; Sun, R.; Wei, Z.; Hong, H.; Luan, X.; Wang, J.; Wang, X.; et al. Soybean genetic resources contributing to sustainable protein production. *Theor. Appl. Genet.* **2022**, *135*, 4095–4121. [CrossRef]
3. Messina, M. Perspective: Soybeans Can Help Address the Caloric and Protein Needs of a Growing Global Population. *Front. Nutr.* **2022**, *9*, 909464. [CrossRef]
4. USDA. Production—Soybeans. Available online: https://www.fas.usda.gov/data/production/commodity/2222000 (accessed on 10 January 2025).
5. Xie, W.; Wang, Q.; Guo, L.; Zhang, Q. Improved Biodiesel Production from Soybean Oil Using Molybdenum-Zirconium Doped Aluminosilicates as Heterogeneous Catalysts. *BioEnergy Res.* **2024**, *17*, 532–546. [CrossRef]
6. Abd-Alla, M.H.; Al-Amri, S.M.; El-Enany, A.-W.E. Enhancing Rhizobium–Legume Symbiosis and Reducing Nitrogen Fertilizer Use Are Potential Options for Mitigating Climate Change. *Agriculture* **2023**, *13*, 2092. [CrossRef]
7. Mahmud, K.; Makaju, S.; Ibrahim, R.; Missaoui, A. Current Progress in Nitrogen Fixing Plants and Microbiome Research. *Plants* **2020**, *9*, 97. [CrossRef] [PubMed]
8. Neupane, A.; Bulbul, I.; Wang, Z.; Lehman, R.M.; Nafziger, E.; Marzano, S.L. Long term crop rotation effect on subsequent soybean yield explained by soil and root-associated microbiomes and soil health indicators. *Sci. Rep.* **2021**, *11*, 9200. [CrossRef]

9. Yang, Y.; Zou, J.; Huang, W.; Olesen, J.E.; Li, W.; Rees, R.M.; Harrison, M.T.; Feng, B.; Feng, Y.; Chen, F. Drivers of soybean-based rotations synergistically increase crop productivity and reduce GHG emissions. *Agric. Ecosyst. Environ.* **2024**, *372*, 109094. [CrossRef]
10. e Silva, J.A.G.; de Pinho Costa, K.A.; da Silva, L.M.; da Costa Severiano, E.; Silva, F.G.; Habermann, E.; Martinez, C.A.; Vilela, L.; da Silva, A.G.; Costa, A.C.; et al. Integrated systems improve the sustainability of soybean cultivation in the tropical region. *Front. Sustain. Food Syst.* **2023**, *7*, 1224530. [CrossRef]
11. Simon-Miquel, G.; Reckling, M.; Lampurlanés, J.; Plaza-Bonilla, D. A win-win situation—Increasing protein production and reducing synthetic N fertilizer use by integrating soybean into irrigated Mediterranean cropping systems. *Eur. J. Agron.* **2023**, *146*, 126817. [CrossRef]
12. Kravchenko, A.N.; Snapp, S.S.; Robertson, G.P. Field-scale experiments reveal persistent yield gaps in low-input and organic cropping systems. *Proc. Natl. Acad. Sci. USA* **2017**, *114*, 926–931. [CrossRef]
13. Oyege, I.; Bhaskar, M.S.B. Effects of vermicompost on soil and plant health and promoting sustainable agriculture. *Soil Syst.* **2023**, *7*, 101. [CrossRef]
14. Pathma, J.; Sakthivel, N. Microbial diversity of vermicompost bacteria that exhibit useful agricultural traits and waste management potential. *Springerplus* **2012**, *1*, 26. [CrossRef] [PubMed]
15. Belbase, P.; Jayachandran, K.; Balaji Bhaskar, M.S. Assessment of Soil and Plant Nutrient Status, Spectral Reflectance, and Growth Performance of Various Dragon Fruit (Pitaya) Species Cultivated Under High Tunnel Systems. *Soil Syst.* **2025**, *9*, 75. [CrossRef]
16. Yatoo, A.M.; Ali, M.N.; Baba, Z.A.; Hassan, B. Sustainable management of diseases and pests in crops by vermicompost and vermicompost tea. A review. *Agron. Sustain. Dev.* **2021**, *41*, 7. [CrossRef]
17. Rehman, S.U.; De Castro, F.; Aprile, A.; Benedetti, M.; Fanizzi, F.P. Vermicompost: Enhancing Plant Growth and Combating Abiotic and Biotic Stress. *Agronomy* **2023**, *13*, 1134. [CrossRef]
18. Oyege, I.; Balaji Bhaskar, M.S. Residual Impacts of Vermicompost-Derived Nutrients on a Strawberry–Corn Double Cropping System Under Plasticulture in South Florida. *Environments* **2025**, *12*, 171. [CrossRef]
19. Deb, P.; Mandal, A.; Harendra; Choudhury, S.R.; Das, A.; Hazra, S.; Mukherjee, A.; Ghosh, D.; Choudhury, S.; Santra, S.C.; et al. Improving Plant Nutrient Use Efficiency for Climate-Resilient Agriculture. In *Climate-Resilient Agriculture, Volume 2: Agro-Biotechnological Advancement for Crop Production*; Hasanuzzaman, M., Ed.; Springer International Publishing: Cham, Switzerland, 2023; pp. 209–243.
20. Kabato, W.; Getnet, G.T.; Sinore, T.; Nemeth, A.; Molnár, Z. Towards Climate-Smart Agriculture: Strategies for Sustainable Agricultural Production, Food Security, and Greenhouse Gas Reduction. *Agronomy* **2025**, *15*, 565. [CrossRef]
21. Lipper, L.; Thornton, P.; Campbell, B.M.; Baedeker, T.; Braimoh, A.; Bwalya, M.; Caron, P.; Cattaneo, A.; Garrity, D.; Henry, K. Climate-smart agriculture for food security. *Nat. Clim. Change* **2014**, *4*, 1068–1072. [CrossRef]
22. Oyege, I.; Balaji Bhaskar, M.S. The Role of Vermicompost and Vermicompost Tea in Sustainable Corn Production and Fall Armyworm Suppression. *Agriculture* **2025**, *15*, 1433. [CrossRef]
23. Oyege, I.; Bhaskar, M.S.B. Evaluation of vermicompost and vermicompost tea application on corn (*Zea mays*) growth and physiology using optical plant sensors. *J. Plant Nutr.* **2024**, *48*, 1275–1293. [CrossRef]
24. Xiao, Y.; Zhou, M.; Li, Y.; Zhang, X.; Wang, G.; Jin, J.; Ding, G.; Zeng, X.; Liu, X. Crop Residue Return Rather Than Organic Manure Increases Soil Aggregate Stability under Corn–Soybean Rotation in Surface Mollisols. *Agriculture* **2022**, *12*, 265. [CrossRef]
25. Lei, J.; Su, Y.; Jian, S.; Guo, X.; Yuan, M.; Bates, C.T.; Shi, Z.J.; Li, J.; Su, Y.; Ning, D.; et al. Warming effects on grassland soil microbial communities are amplified in cool months. *ISME J.* **2024**, *18*, wrae088. [CrossRef]
26. Souffront, D.K.S.; Salazar-Amoretti, D.; Jayachandran, K. Influence of vermicompost tea on secondary metabolite production in tomato crop. *Sci. Hortic.* **2022**, *301*, 111135. [CrossRef]
27. Hall, S.J.; Russell, A.E.; Moore, A.R. Do corn-soybean rotations enhance decomposition of soil organic matter? *Plant Soil* **2019**, *444*, 427–442. [CrossRef]
28. Huang, X.; Yuan, J.; Chen, Y.; Yang, X.; Lu, W.; Ding, S.; Jiang, Y.; Zhou, X.; Mi, G.; Xu, J.; et al. Long-term cropping rotation with soybean enhances soil health as evidenced by improved nutrient cycles through keystone phylotypes interaction. *Soil Ecol. Lett.* **2024**, *6*, 240251. [CrossRef]
29. Wang, M.; Ge, A.-H.; Ma, X.; Wang, X.; Xie, Q.; Wang, L.; Song, X.; Jiang, M.; Yang, W.; Murray, J.D.; et al. Dynamic root microbiome sustains soybean productivity under unbalanced fertilization. *Nat. Commun.* **2024**, *15*, 1668. [CrossRef]
30. Al-Musawi, Z.K.; Vona, V.; Kulmány, I.M. Utilizing Different Crop Rotation Systems for Agricultural and Environmental Sustainability: A Review. *Agronomy* **2025**, *15*, 1966. [CrossRef]
31. Kebede, E. Contribution, Utilization, and Improvement of Legumes-Driven Biological Nitrogen Fixation in Agricultural Systems. *Front. Sustain. Food Syst.* **2021**, *5*, 767998. [CrossRef]
32. Coskun, D.; Britto, D.T.; Shi, W.; Kronzucker, H.J. How plant root exudates shape the nitrogen cycle. *Trends Plant Sci.* **2017**, *22*, 661–673. [CrossRef]

33. Blesh, J. Feedbacks between nitrogen fixation and soil organic matter increase ecosystem functions in diversified agroecosystems. *Ecol. Appl.* **2019**, *29*, e01986. [CrossRef]
34. Gelfand, I.; Philip Robertson, G. A reassessment of the contribution of soybean biological nitrogen fixation to reactive N in the environment. *Biogeochemistry* **2015**, *123*, 175–184. [CrossRef]
35. Sun, H.; Jiang, S.; Jiang, C.; Wu, C.; Gao, M.; Wang, Q. A review of root exudates and rhizosphere microbiome for crop production. *Environ. Sci. Pollut. Res.* **2021**, *28*, 54497–54510. [CrossRef]
36. Fan, X.; Ge, A.-H.; Qi, S.; Guan, Y.; Wang, R.; Yu, N.; Wang, E. Root exudates and microbial metabolites: Signals and nutrients in plant-microbe interactions. *Sci. China Life Sci.* **2025**, *68*, 2290–2302. [CrossRef] [PubMed]
37. Vives-Peris, V.; de Ollas, C.; Gómez-Cadenas, A.; Pérez-Clemente, R.M. Root exudates: From plant to rhizosphere and beyond. *Plant Cell Rep.* **2020**, *39*, 3–17. [CrossRef] [PubMed]
38. Domínguez, J.; Aira, M.; Kolbe, A.R.; Gómez-Brandón, M.; Pérez-Losada, M. Changes in the composition and function of bacterial communities during vermicomposting may explain beneficial properties of vermicompost. *Sci. Rep.* **2019**, *9*, 9657. [CrossRef]
39. García-Pérez, P.; De Gregorio, M.; Capri, E.; Lucini, L. Coordinated Metabolic and Phenotypic Reprogramming in Response to Vermicompost Priming Supports a Promising Sustainable Alternative for the Fortification of Microgreens. *J. Soil Sci. Plant Nutr.* **2025**, *25*, 7782–7794. [CrossRef]
40. Choudhary, S.; Wani, K.I.; Naeem, M.; Khan, M.M.A.; Aftab, T. Cellular Responses, Osmotic Adjustments, and Role of Osmolytes in Providing Salt Stress Resilience in Higher Plants: Polyamines and Nitric Oxide Crosstalk. *J. Plant Growth Regul.* **2023**, *42*, 539–553. [CrossRef]
41. Leung, H.-S.; Chan, L.-Y.; Law, C.-H.; Li, M.-W.; Lam, H.-M. Twenty years of mining salt tolerance genes in soybean. *Mol. Breed.* **2023**, *43*, 45. [CrossRef]
42. Chen, H.; Chen, X.; Gu, H.; Wu, B.; Zhang, H.; Yuan, X.; Cui, X. GmHKT1;4, a novel soybean gene regulating Na+/K+ ratio in roots enhances salt tolerance in transgenic plants. *Plant Growth Regul.* **2014**, *73*, 299–308. [CrossRef]
43. Cruz, G.S.J.; Hermes, P.H.; Miriam, S.V.; López, A.M. Benefits of Vermicompost in Agriculture and Factors Affecting its Nutrient Content. *J. Soil Sci. Plant Nutr.* **2024**, *24*, 4898–4917. [CrossRef]
44. Thomas, A.; Yadav, B.K.; Šimůnek, J. Root water uptake under heterogeneous soil moisture conditions: An experimental study for unraveling compensatory root water uptake and hydraulic redistribution. *Plant Soil* **2020**, *457*, 421–435. [CrossRef]
45. Zhang, W.-P.; Surigaoge, S.; Yang, H.; Yu, R.-P.; Wu, J.-P.; Xing, Y.; Chen, Y.; Li, L. Diversified cropping systems with complementary root growth strategies improve crop adaptation to and remediation of hostile soils. *Plant Soil* **2024**, *502*, 7–30. [CrossRef]
46. Patnaik, P.; Artala, A.; Abbasi, S.A.; Tabassum-Abbasi; Abbasi, T. *An Emerging Organic Fertilizer-Cum-Pest Repellant: Vermicompost Tea*; Springer: Singapore, 2022; pp. 225–231.
47. Suhani, I.; de Araujo, A.S.F.; Kothari, R.; Vaish, B.; Singh, R.P. Investigating the Effect of Compost and Vermicompost on Wheat Plants and Soil Health Under Natural Salt Affected Field Conditions. *Int. J. Environ. Res.* **2024**, *19*, 42. [CrossRef]
48. Mehrabi, F.; Sepaskhah, A.R. Leaf Nitrogen, Based on SPAD Chlorophyll Reading Can Determine Agronomic Parameters of Winter Wheat. *Int. J. Plant Prod.* **2022**, *16*, 77–91. [CrossRef]
49. Ievinsh, G. Review on Physiological Effects of Vermicomposts on Plants. In *Biology of Composts*; Meghvansi, M.K., Varma, A., Eds.; Springer International Publishing: Cham, Switzerland, 2020; pp. 63–86.
50. Dalorima, T.L.; Zakaria, A.J.; Majrashi, A.; Mahmud, K.; Mohd, K.S.; Muhammad, H.; Khandaker, M.M. Impacts of vermicomposting rates on growth, yield and qualities of red seedless watermelon. *Aust. J. Crop Sci.* **2018**, *12*, 1765–1773. [CrossRef]
51. Savchenko, E.D. Soybean cultivation in short rotation organic crop rotation under different fertilisation systems. *Agric. Plant Sci. Theory Pract.* **2024**, *2*, 16–19. [CrossRef]
52. USDA. 2023 State Agriculture Overview Georgia. Available online: https://www.nass.usda.gov/Quick_Stats/Ag_Overview/stateOverview.php?state=GEORGIA (accessed on 21 November 2024).
53. Hokazono, S.; Hayashi, K. Comparative Life Cycle Assessment of Organic and Conventional Soybean Production in Paddy Fields under Rotational Cropping. *J. Life Cycle Assess. Jpn.* **2012**, *8*, 2–13. [CrossRef]

environments

MDPI

Article

Residual Impacts of Vermicompost-Derived Nutrients on a Strawberry–Corn Double Cropping System Under Plasticulture in South Florida

Ivan Oyege [1,2] and Maruthi Sridhar Balaji Bhaskar [1,*]

1 Department of Earth and Environment, Florida International University, Miami, FL 33199, USA; ioyeg002@fiu.edu
2 Department of Chemistry, Busitema University, Tororo P.O. Box 236, Uganda
* Correspondence: mbalajib@fiu.edu

Abstract: The increasing demand for sustainable agricultural practices necessitates strategies such as organic fertilizer alternatives and residual nutrient use to enhance crop productivity while maintaining soil health. This study investigates the residual effects of vermicompost on strawberry growth and physiology after a corn cropping cycle. The objectives were to assess how different vermicompost application rates impact strawberry yield, biomass, chlorophyll content, and fruit quality. The experiment was conducted over six months, using raised beds previously cultivated with corn and treated with six nutrient management strategies, namely, V0 (control), VC1, VCT100, VC1+VCT50, VC3, and VC3+VCT50. Metrics such as SPAD values, Brix sugar content, and stomatal conductance were measured throughout the growing season to assess physiological responses. Soil and plant chemical concentrations were determined at the end of the study to evaluate nutrient status. Results showed that the VC1 treatment produced the highest yield (11,573 kg/acre) and biomass (38,364 kg/acre), with significantly improved fruit quality (Brix sugar content of 8.3%) compared to the control (6.8%). SPAD values declined over time and showed no statistically significant differences among treatments. In the surface soil, VC3+VCT50 exhibited the highest N, P, Mg, Na, organic matter, and cation exchange capacity (CEC), and it also resulted in the highest leaf N. Leaves had higher N, P, K, and Mg concentrations, while Fe, Mn, and Cu were more concentrated in roots. Spectral reflectance data indicated reduced chlorophyll content in the VC3+VCT50-treated plants. These findings suggest that moderate vermicompost applications, such as VC1, can significantly contribute to sustainable agriculture by enhancing strawberry productivity and reducing dependence on synthetic fertilizers. However, high-rate applications, especially VC3 and VC3+VCT50, reduced plant vigor and yield, possibly due to salinity stress and the high sodium content in the vermicompost used in this study. Such outcomes may vary depending on feedstock composition, highlighting the importance of salinity screening when using organic amendments in salt-sensitive crops like strawberries.

Keywords: sustainable agriculture; normalized difference vegetative index; sugar content; salinity stress; stomatal conductance; residual soil fertility; organic amendments; strawberry–maize rotation

1. Introduction

Strawberry (*Fragaria ananassa*) is one of the most economically significant fruit crops globally, with production exceeding 9 million metric tons in 2022 [1]. Major producers include China, the United States, Mexico, and Turkey, with the U.S. contributing over 1.3 million tons annually, generating approximately USD 3.42 billion in economic value [2,3]. California dominates national production in the U.S., but Florida ranks second, cultivating strawberries on 10,000 to 12,000 acres annually, with a market value exceeding USD 500 million [4]. The fruit's high market demand stems from its nutritional benefits, including high antioxidant content, making it a consumer favorite worldwide [5]. While most production is concentrated in Central Florida, South Florida, including Miami-Dade County, is emerging as a valuable region due to its subtropical climate and unique winter growing window [6]. However, production here remains small-scale and understudied despite increasing interest in local markets and organic production systems.

Despite its economic value, strawberry cultivation is highly input-intensive and vulnerable to various pests and diseases. In Florida, strawberries are predominantly grown using raised-bed plasticulture with drip irrigation, requiring frequent pesticide applications to combat pests such as the two-spotted spider mite (*Tetranychus urticae*), fall armyworm (*Spodoptera frugiperda*), southern armyworm (*Spodoptera eridania*), aphids, and thrips [7,8]. The phase-out of methyl bromide soil fumigant has increased reliance on chemical pesticides, leading to resistance development and environmental concerns [9]. Over 63 pesticide products are registered for strawberry cultivation [10], and nearly 70% of market strawberries have been found to carry over 20 different pesticide residues [11], placing them consistently at the top of the "Dirty Dozen" list for several years [12]. These challenges have driven interest in organic management strategies, including using organic soil amendments that enhance fertility and improve pest and disease resilience. In South Florida, the challenge is compounded by unique agronomic conditions such as calcareous, shallow soils with low organic matter, high rainfall variability, and elevated temperatures, which affect nutrient availability and pest dynamics. These site-specific challenges show the need for regionally tailored nutrient and pest management approaches that support long-term sustainability.

Vermicompost (VC), often referred to as "black gold," is a nutrient-rich organic amendment produced through the microbial decomposition of organic waste by earthworms [13]. Its aqueous derivative, vermicompost tea (VCT), is prepared by steeping VC in water under aerated or non-aerated conditions to extract soluble nutrients and beneficial microbes [14]. VC is rich in essential nutrients, beneficial microbes, phytohormones, and phenolic acids that can enhance plant growth and suppress soilborne pests and pathogens [14,15]. Recent studies have also highlighted the role of vermicompost in disease suppression and plant health promotion in various crops [16]. VCT, commonly applied as a foliar spray, provides similar benefits in a liquid form. The VC functions as a slow-release nutrient source, while VCT provides readily available plant nutrients, improving soil structure and biological activity while minimizing leaching [13,17]. While several studies have documented the short-term benefits of VC and VCT on strawberries [18,19], little is known about their effectiveness when applied during a preceding crop cycle, such as corn, a heavy nutrient feeder. This raises important questions about whether residual nutrients from organic amendments such as VC and VCT, which are applied to a preceding crop like corn, can support strawberry yield, fruit quality, and nutrient availability in a subsequent rotation in regions like South Florida, where input efficiency and pest suppression are critical.

Incorporating strawberries after corn can improve soil health and promote pest management in crop rotation. Corn is typically rotated with crops like soybeans, peas, wheat, and beans [20,21], replenishing nitrogen levels in the soil, while strawberries are usually rotated with peas, beans, and leafy greens [22,23]. Introducing strawberries into corn rotation can help break pest and disease cycles, as strawberries are not closely related to corn and share one or two common pests or diseases. However, the residual nutrient effects of VC and VCT on strawberry yield and physiology following corn cultivation have not been quantified in subtropical cropping systems.

This study addresses these gaps by evaluating the residual impacts of VC and VCT applied during a corn-growing season on the subsequent growth and nutrient performance of strawberries grown without additional fertilization within a real field rotation context. Specifically, this study aimed to (1) assess the effects of residual VC and VCT treatments on strawberry morphological traits and physiological responses; (2) quantify macro- and micronutrient concentrations in soil and plant tissues to determine the extent of nutrient carryover from the previous crop season; and (3) evaluate treatment-induced differences in fruit quality, particularly Brix sugar content, as a proxy for marketable quality under organic management.

While strawberry–corn rotation is not widely practiced, the experimental design allowed for assessing the residual impacts of organic inputs across growing seasons. These findings offer broader implications for agricultural systems incorporating organic nutrient sources, managing salinity risks, and reducing dependence on synthetic fertilizers in high-value horticultural crops.

2. Methodology

2.1. Experimental Setup

This study was conducted at the organic garden (1.5 acres) of the Agroecology Program, Department of Earth and Environment, Florida International University, Miami, FL, USA (25.7539281, −80.3800388). The site is characterized by calcareous sandy loam soils with a history of organic management under the USDA People's Garden Initiative [24]. This research served as a follow-up to a previous corn experiment [25], exactly one month after the experiment concluded, using the same six raised beds (457 × 30 × 91 cm, L × W × H). The aim was to evaluate the residual effects of prior VC treatments on strawberry growth without any additional fertilizer inputs. Treatments applied during the preceding corn experiment were as follows: V0 (control—The Old Farmer's Almanac organic fertilizer with NPK percentage of 8-4-8, applied twice at a rate of 0.3 ton/acre); VC1 (vermicompost at 1 ton/acre); VCT100 (vermicompost tea, 100 mL per plant); VC1+VCT50 (vermicompost 1 ton/acre and vermicompost tea, 50 mL per plant); VC3 (vermicompost 3 ton/acre), and VC3+VCT50 (vermicompost 3 ton/acre and vermicompost tea, 50 mL per plant). The VC and VCT were sourced from Lions Fruit Farm, Redlands, Miami, FL, using rabbit manure, fruit waste (e.g., avocado), and cocoplum as feedstock. VC and VCT preparation details are described in [25]. Soil nutrient content following the corn harvest and before strawberry planting is summarized in Table 1, derived from the same study (Table 2 in [25]), to contextualize nutrient availability at the start of the strawberry trial. The nutrient composition of the VC used in the preceding corn study, whose residues were evaluated here, is detailed in [25], including total N, P, K, and Na levels.

Beds were covered with cloth mulch, and each was planted with 12 Chandler bare-root strawberry plants (sourced from Willis Orchards, Cartersville, GA, USA) in November 2022. To ensure consistent replication across treatments, the strawberry plants in each raised bed were grouped into three replicates by dividing the 12 plants into three sets of four. These plant-based replicates were used for biomass and nutrient analysis within each treatment. The experiment lasted six months, ending in April 2023. Throughout the strawberry growing season, beds were irrigated manually using a hand-held sprinkler every evening to ensure uniform moisture coverage across each raised bed. Watering was skipped on days when rainfall occurred. The weather data of the study area during the strawberry growing season (November 2022 to April 2023) were obtained from the National Oceanic and Atmospheric Administration (NOAA, https://www.weather.gov/wrh/Climate?wfo=mfl

accessed on 15 May 2025). Monthly average temperatures ranged from 22.3 °C in January to 26.3 °C in April. The precipitation was low with values of 7.1 mm (November 2022), 1.5 mm (December 2022), 0 mm (January 2023), 5.1 mm (February 2023), 3.8 mm (March 2023), and 8.4 mm (April 2023).

2.2. *Strawberries Physiology*

Physiological measurements were taken weekly between weeks 10 and 24 after transplanting, corresponding to Chandler strawberry's vegetative stage, flowering, and early fruit development stages in subtropical climates. These time points were selected to capture changes in chlorophyll content, water use efficiency, and canopy development during peak physiological activity. The relative chlorophyll content was determined using a Soil-Plant Analysis Development chlorophyll meter (SPAD 502 plus, Spectrum Technologies Inc., Aurora, IL, USA) from weeks 10 to 24 after planting. The stomatal conductivity of the strawberries was measured using a porometer and fluorometer (Li-600, LI-COR Environmental, Lincoln, NE, USA) from week 10 to week 24. The SPAD readings and stomatal conductance were taken from five mature leaves (at the mid-height of the plant) per plant between 8:30 am and 10:30 am to reduce diurnal variations. Leaf-level (five leaves per plant, four plants per replicate) and canopy-level (three plants per replicate) spectral reflectance of the strawberries was taken in week 17 at 11:00 am using a hand-held spectral radiometer (Spectral Evolution Inc., Haverhill, MA, USA) across the 350–2500 nm.

2.3. *Strawberry Harvesting and Yield Determination*

The strawberry fruits were harvested continuously throughout the experimental period from November 2022 to April 2023, coinciding with the typical fruiting window for Chandler strawberries in subtropical climates. Harvesting was performed manually on multiple dates as fruits reached full ripeness (entirely red). Only marketable-quality fruits free from deformities, pest damage, or disease symptoms were collected and included in the yield analysis. Treatments VC3 and VC3+VCT50 were excluded from the yield data due to insufficient production of marketable fruits.

The fresh weight of harvested strawberries was recorded immediately after harvest using a digital balance (precision ± 0.01 g). This approach reflects commercial practices, where strawberries are sold fresh. Brix sugar content was also measured immediately upon harvest using a hand-held Brix digital pocket refractometer (Atago Co., Ltd., Bellevue, WA, USA) to ensure accuracy, and the sugar content was compared with organic strawberries purchased from the grocery store (Publix, Miami, FL, USA). The total yield per treatment was calculated by summing the cumulative fresh fruit weight collected across all harvests. The yield was then expressed per acre (kg/acre) by scaling the total fruit mass to the area of the raised beds used in each treatment using the following equation:

$$Yield\ in\ \mathrm{kg\ per\ acre} = \frac{Total\ yield\ from\ plot\ in\ \mathrm{kg}}{Area\ of\ the\ plot\ in\ \mathrm{m}^2} \times 4046.86$$

2.4. *Plant and Soil Sampling and Chemical Analysis*

At the termination of the experiment (end of April 2023), despite ongoing fruiting, plants were uprooted, and biomass (shoots and roots) was harvested for final biomass determination. Strawberry plants were at the late reproductive stage, characterized by ongoing fruit production and senescence of early leaves. Fresh biomass per bed was weighed immediately after uprooting the plants. From this, representative subsamples were collected for laboratory analysis. These subsamples were carefully washed with clean water to remove soil and debris, particularly from the roots. The cleaned subsamples were oven-dried at 70 °C until constant weight to determine dry biomass. Once dried, the roots were manually separated from the shoot tissue. Both roots and leaves were then ground separately into a fine powder using a Mixer/Mill (SPEX Sample Prep, Metuchen, NJ, USA). The resulting ground root and leaf samples were analyzed for nutrient content following the procedure previously described in [25]. Briefly, the total carbon and total nitrogen content in the leaves and root samples were determined using dry combustion with a C/N elemental analyzer (LECO TruSpec, LECO Corporation, St. Joseph, MI, USA).

The elemental concentrations of Ca, K, Mg, P, S, B, Fe, Mn, Zn, and Cu were determined using Inductively Coupled Plasma Mass Spectroscopy (ICP-MS; Agilent 7900, Santa Clara, CA, USA). The powdered plant samples were digested using a standard acid digestion procedure, using concentrated nitric acid, to solubilize macro- and micronutrients. Quantification was performed using multi-element calibration standards, and quality control was maintained using reagent blanks and certified reference materials. Elemental concentrations were expressed in mg/kg or ppm (parts per million) dry weight.

Furthermore, the soil samples from all 12 plants were collected from two depths, that is, 0–10 cm (surface soils) and 10–20 cm (subsurface soils), and pooled to make three soil samples per bed per depth. The soil samples and background soil collected just before planting strawberries were air-dried and sieved through a 2 mm mesh and analyzed for nutrient and elemental concentrations [25]. The soil pH was determined in $CaCl_2$ solution, CEC was determined using unbuffered 1 M NH_4Cl, and organic matter was determined by the loss-on-ignition method. Total nitrogen in the soil was measured using the same C/N elemental analyzer. At the same time, P, K, Ca, Mg, Zn, and Na concentrations were determined using the same ICP-MS procedure for plant samples.

2.5. Statistical Analysis

All data collected in the study were analyzed using Minitab (version 21, Minitab LLC, State College, PA, USA) at a 5% significance level. A one-way analysis of variance (ANOVA) was conducted to evaluate the effect of different treatments on soil and plant parameters. For each parameter, ANOVA was first used to test whether there were significant differences among treatment means ($p < 0.05$). A post hoc test (Tukey's Honestly Significant Difference (HSD)) was performed to separate treatment means only when ANOVA indicated a significant difference. All assumptions for ANOVA, including normality and homogeneity of variances, were checked before conducting the analyses.

3. Results

3.1. Soil Chemical Characteristics

We assessed soil nutrients in the surface (0–10 cm) and subsurface (10–20 cm) layers after corn harvest and strawberry planting to understand residual fertility. The data (see Table 1, adapted from [25]) showed significant treatment effects, with VC3+VCT50 showing significantly higher N, P, Mg, organic matter (OM), Na, and cation exchange capacity (CEC) in surface soils compared to most of the treatments ($p < 0.05$). Treatments VC1 and VC3 also showed enriched nutrient profiles compared to the control, although to a lesser extent. Nutrient levels in the subsurface layers were generally lower, but VC3 maintained higher N and Ca levels. Furthermore, nutrient levels were generally lower in the subsurface layer (10–20 cm). However, VC3 and VC3+VCT50 exhibited significantly greater %OM and %C than VC1. Despite these differences, CEC values in subsurface soils did not differ significantly among treatments. These nutrient patterns showed that VC and VCT applications from the prior corn cycle left enough nutrient residues to influence subsequent strawberry growth; this nutrient carryover effect was explored in the current strawberry-focused results. Moreover, the high CEC observed under VC3+VCT50 treatment suggests enhanced soil capacity to retain water and nutrients, which is especially advantageous in organic systems with limited fertilizer inputs.

Table 1. Post-harvest soil nutrient concentrations at two depths (0–10 cm and 10–20 cm) following corn cultivation under different vermicompost (VC) and vermicompost tea (VCT) treatments. Data were extracted and adapted from [25] to illustrate residual soil fertility before strawberry planting.

Treatment	pH	N * (%)	P * (ppm)	K * (ppm)	Ca (ppm)	Mg * (ppm)	Zn (ppm)	Na * (ppm)	OM * (%)	CEC (cmol/kg)	C * (%)
Surface soils (0–10 cm)											
V0	7.9	0.79 a	286 a	181 a	2717	742 ab	15.1	50.6 b	25.9	20.3	14.6
VC1	8.0	0.63 b	247 c	161 b	2640	716 ab	14.9	48.0 bc	20.8	19.6	11.7
VCT100	8.0	0.63 b	209 d	165 b	2735	692 b	13.9	50.5 b	22.3	19.9	12.5
VC1+VCT50	8.0	0.63 b	205 d	118 c	2661	620 c	13.0	45.7 c	22.5	18.8	12.6
VC3	8.0	0.65 b	268 b	123 c	2549	744 ab	14.1	48.3 bc	21.4	19.3	12.0
VC3+VCT50	8.0	0.72 ab	289 a	185 a	2637	757 a	15.3	75.5 a	23.6	20.0	13.2
Subsurface soils (10–20 cm)											
V0	8.1	0.48 bc	117 b	279 a	2345	545 ab	10.9	41.9 b	18.9 ab	17.0	10.6 ab
VC1	8.1	0.47 c	132 a	219 b	2346	543 ab	11.3	43.4 b	17.6 b	16.8	9.9 b
VCT100	8.1	0.41 c	74.8 d	196 b	2314	403 c	10.4	44.6 b	18.7 ab	15.4	10.5 ab
VC1+VCT50	8.1	0.49 bc	99.4 c	140 c	2345	486 bc	10.7	40.9 b	19.4 ab	16.1	10.9 ab
VC3	8.1	0.54 ab	119 b	87.1 d	2372	589 a	8.2	45.7 b	23.0 a	17.0	12.9 a
VC3+VCT50	8.1	0.60 a	114 b	152 c	2326	578 ab	10.0	154 a	22.8 a	16.8	12.8 a

Nutrient concentrations are presented on a dry weight basis. The elements with an asterisk (*) are statistically significant. Means that do not share a letter differ significantly ($p < 0.05$).

Table 2 presents the nutrient content in the post-harvest soils of strawberries at both surface and subsurface depths across the different treatment groups. Significant increases in soil concentrations of N, P, Ca, Mg, Na, organic matter, and CEC were observed among all the treatments, with increases in VC and VCT concentrations in both surface and subsurface soils (Table 2). The pH, K, and Zn concentrations in the surface and subsurface soils were unaffected by the increased VC and VCT concentrations. The high VC application rates (VC3 and VC3+VCT50) generally exhibited higher soil nutrient content, while VC1+VCT50 exhibited the lowest nutrient concentration (Table 2). Furthermore, the N levels were significantly higher across all treatments compared to P and K in the surface and subsurface soils.

In the surface depth (0–10 cm), pH was relatively consistent across all treatments, ranging from 7.6 to 8.0. The VC3+VCT50 treatment exhibited the highest N content (0.81%), which was more significant than the other treatments, except VC3. It was observed that high VC rate (VC3 and VC3+VCT50) treatments contained more N compared to those with 1 ton/acre (VC1). Similar findings were made for P and Mg, where P levels were significantly higher in the VC3 treatment, with 216 ppm, significantly higher than VC1+VCT50 (94 ppm). VC3+VCT50 had significantly higher Mg content (648 ppm) than VC1+VCT50 (435 ppm). The highest Na content was found in the VC3+VCT50 treatment (67.8 ppm), which was significantly higher than the control group V0 (61.9 ppm) (Table 2). VC3+VCT50 resulted in 29% and 13% higher OM content and CEC in surface soils compared to the control (Table 2).

In the subsurface soils (10–20 cm), the highest N content was observed in the VC3 treatment (0.60%), which was significantly higher than the VC1+VCT50 treatment (0.44%). All treatments except for the control (V0) had significantly higher Ca content and CEC than VC1+VCT50. A similar trend was observed in Mg content, where VC1+VCT50 had significantly lower Mg content than the VC3 and VC3+VCT50 treatments. Furthermore, except for VC1+VCT50, the OM content was significantly higher in VC3 and VC3+VCT50 compared to the rest of the treatments. Moreover, the results show that the nutrient levels were generally higher in the surface soil samples than in the subsurface soil samples (Table 2). In addition, VC3+VCT50 had significantly higher N content in the surface soil samples than in the subsurface soil samples.

Table 2. Post-harvest soil nutrient content at two depths (surface and subsurface) after strawberry cultivation.

Treatment	pH	N * (%)	P * (%)	K (%)	Ca * (%)	Mg * (%)	Zn (ppm)	Na * (ppm)	OM * (%)	CEC *
Surface soils (0–10 cm)										
V0	8.0	0.59 bc	0.013 ab	0.009	0.30	0.05 abc	10.2	61.9 b	22.6 b	19.5 ab
VC1	7.6	0.61 bc	0.014 ab	0.009	0.34	0.05 bc	9.8	63.0 ab	24.9 ab	20.8 ab
VCT100	7.6	0.57 bc	0.013 ab	0.007	0.31	0.04 bc	9.5	65.2 ab	23.5 b	19.7 ab
VC1+VCT50	7.9	0.47 c	0.0094 b	0.007	0.29	0.04 c	11.6	66.9 ab	22.2 b	18.2 b
VC3	7.6	0.71 ab	0.022 a	0.009	0.34	0.06 ab	12.2	67.7 ab	26.1 ab	22.3 a
VC3+VCT50	7.8	0.81 a	0.020 ab	0.009	0.32	0.06 a	11.4	67.8 a	29.2 a	22.1 a
Subsurface soils (10–20 cm)										
V0	7.8	0.49 abc	0.008	0.006	0.27 ab	0.05 ab	7.2	61.3	22.2 b	17.6 ab
VC1	7.9	0.49 abc	0.008	0.006	0.29 a	0.05 ab	7.5	63.4	22.2 b	18.6 ab
VCT100	8.1	0.48 bc	0.007	0.006	0.30 a	0.05 ab	8.3	63.6	21.4 b	18.8 a
VC1+VCT50	8.0	0.44 c	0.006	0.004	0.26 b	0.04 b	8.1	61.0	22.7 ab	16.6 b
VC3	8.0	0.60 a	0.100	0.006	0.29 a	0.05 a	7.7	62.7	25.0 a	19.2 a
VC3+VCT50	8.0	0.59 ab	0.010	0.0066	0.30 a	0.06 a	8.3	66.6	24.8 a	19.5 a

Nutrient concentrations are presented on a dry weight basis. The elements with an asterisk (*) are statistically significant. Means that do not share a letter are significantly different.

3.2. *Plant Chemical Concentrations*

The findings show interesting variations of nutrient content in both leaves and roots (Table 3), with significant differences observed across elemental concentrations except for Ca, S, B, and Zn. The N content in leaf samples was not statistically significant across treatments; however, a slight trend was observed where the VC3+VCT50-treated plants had the highest N content (1.89%) compared to the rest. The control-treated (V0) and VC3+VCT50-treated plants had higher P levels than the other treatments. Furthermore, K content was significantly higher in VC1 (1.96%) than in the VC1+VCT50-, VC3-, and VC3+VCT50-treated plants, showing an inverse relationship between K content and VC application rate (Table 3). Furthermore, VC3 had significantly higher Fe content than the control-treated (V0), VC1-treated, and VCT100-treated plants (Table 3).

Table 3. Nutrient content in strawberry leaves and roots.

Treatment	N * (%)	P * (%)	K * (%)	Ca (%)	Mg * (%)	S (%)	B (ppm)	Fe * (ppm)	Mn * (ppm)	Zn (ppm)	Cu * (ppm)
Leaves											
V0	1.83	0.28	1.72 ab	2.26	0.46	0.08	27.5	96 b	16.6	15.2	6.0
VC1	1.75	0.26	1.96 a	2.31	0.42	0.08	28.5	88 b	16.6	14.7	5.9
VCT100	1.67	0.24	1.32 ab	2.08	0.36	0.07	26.2	114 b	17.6	15.0	5.7
VC1+VCT50	1.60	0.26	1.20 b	2.28	0.37	0.07	25.5	138 ab	17.7	16.4	5.6
VC3	1.82	0.22	0.96 b	1.87	0.29	0.06	21.3	248 a	19.4	20.2	5.6
VC3+VCT50	1.89	0.28	0.93 b	1.99	0.39	0.06	22.1	160 ab	20.6	20.3	5.6
Roots											
V0	1.05 ab	0.22 ab	1.68 a	4.39	0.49 a	0.10	19.1	618 bc	34.8 bc	49.5	11.1 ab
VC1	1.24 a	0.23 a	1.70 a	4.58	0.36 ab	0.09	14.1	389 c	26.7 c	45.2	7.9 b
VCT100	0.90 b	0.12 c	0.60 b	5.68	0.21 b	0.09	16.0	804 ab	38.0 bc	51.8	11.5 ab
VC1+VCT50	0.91 b	0.14 c	0.71 b	5.65	0.26 b	0.10	16.4	997 a	43.4 b	47.2	12.1 ab
VC3	0.90 b	0.16 abc	0.52 ab	3.71	0.24 b	0.09	19.6	797 abc	41.6 abc	50.6	15.2 ab
VC3+VCT50	0.97 ab	0.12 bc	0.28 b	3.65	0.17 b	0.09	18.2	948 ab	62.0 a	55.3	20.7 a

The elements with an asterisk (*) are statistically significant. Means that do not share a letter are significant differences.

In the root samples, VC1 showed significantly higher N content (1.24%) than the VCT100-, VC1+VCT50-, and VC3-treated plants. The P content in root samples varied significantly, with VC1 showing the highest concentration level (0.23%), whereas the VCT100- and VC1+VCT50-treated plants showed significantly lower P levels (Table 3). The K content in root samples was significantly higher in the control-treated (V0) and VC1-treated plants than in the VCT100- and VC1+VCT50-treated plants (Table 3). The Mg content was highest in the V0 treatment (0.49%), while treatments with higher vermicompost rates, such as VC3 and VC3+VCT50, had significantly lower Mg levels, consistent with the findings in leaves. The VC1+VCT50-treated plants exhibited significantly higher Fe content in the roots (997 ppm) than VC1. Furthermore, the VC3- and VC3+VCT50-treated plants had significantly higher Mn levels than all treatments. The Cu content in root samples increased significantly with increased VC concentration along the treatments (Table 3).

When comparing the nutrient contents in leaves and roots, the results showed that the leaves contained higher levels of N, P, K, Mg, and B concentrations, while the roots contained higher levels of Ca, S, Fe, Mn, Zn, and Cu concentrations. Furthermore, when comparing the relative nutrient content in the roots (Table 3) and subsurface soils (Table 2), the results showed that the roots contained higher N, P, K, Ca, Mg, and Zn concentrations than those in surface soils.

3.3. Strawberry Biomass, Yield, and Brix Sugar Content

The VC and VCT applications substantially impacted the biomass, yield, and sugar content of the strawberries, as shown in Figure 1. The VC1 treatment resulted in the highest dry biomass production (38,364 kg/acre), followed by the control (V0), VCT100, and VC1+VCT50 treatments, with a biomass of 11,015 kg/acre. There was statistical significance in strawberry biomass yield across all treatments. The VC3 and VC3+VCT50 treatments faced growth challenges, and most were not as healthy as the rest; therefore, their biomass was not determined.

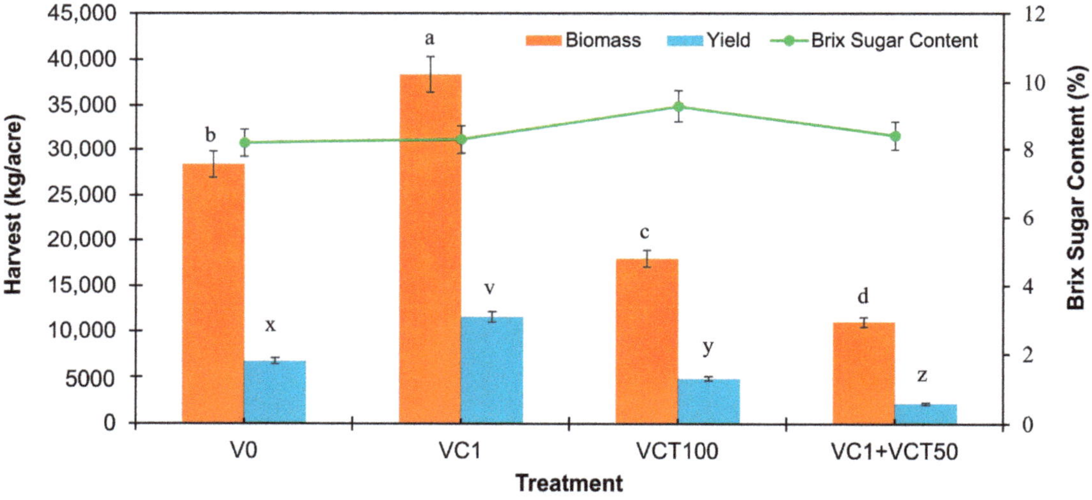

Figure 1. Effects of residual nutrients of vermicompost (VC) and vermicompost tea (VCT) treatments on strawberry biomass, yield, and Brix sugar content. Bars not sharing a common letter (a–d for biomass and v–z for yield) are significantly different at $p < 0.05$. No statistically significant differences were observed in Brix sugar content across treatments at $p < 0.05$, except between VCT100 and market-purchased strawberries. Brix sugar content error bars represent standard error at a 5% significance level ($p < 0.05$).

In terms of yield, the highest yield was observed in the VC1 treatment (11,573 kg/acre), which was higher than that of the control (V0). The VCT100 and VC1+VCT50 treatments yielded only 2203 kg/acre (Figure 1), both of which were lower than the control. A statistical significance in strawberry yield was observed across all treatments. However, VC3 and VC3+VCT50 did not yield enough marketable yield.

Brix sugar content, a measure of sugar concentration in fruit, showed some variation across treatments. The sugar content was determined from 12 fruits per treatment for V0, VC1, VCT100, and VC1+VCT50 treatments, as shown in Figure 1, with the highest sugar content observed in the VCT100 treatment. Although not many fruits were obtained from the VC3 and VC3+VCT50 treatments, strawberries had sugar content values of 8.7%

and 8.5%, respectively. Furthermore, when the sugar content in these treatments was compared to organic strawberries on the market, the results showed that the store-bought strawberries had a Brix sugar content of 6.8%, which was lower than all the treatments. A statistical difference was observed between VCT100 strawberries and organic strawberries on the market.

3.4. Relative Plant Chlorophyll Content

SPAD values, a measurement of relative chlorophyll content and an indicator of plant health and N status, were analyzed across different treatments over several weeks after planting. The trend and variations in SPAD values are illustrated in Figure 2. At the early stages of growth (weeks 10 to 17), all treatments show relatively higher SPAD values, but as the weeks progressed, there was a decline in SPAD values across all treatments. VC3+VCT50 consistently had lower SPAD values in comparison to other treatments.

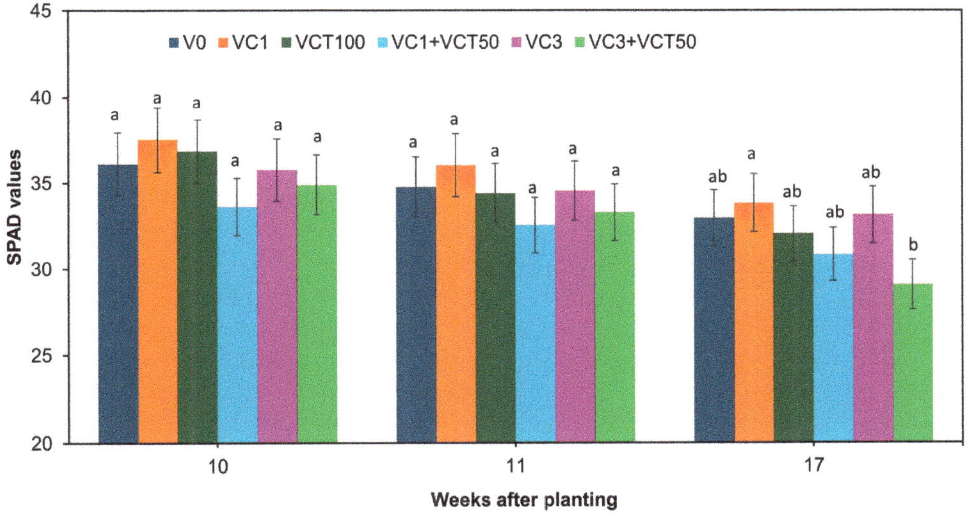

Figure 2. Variation of SPAD values of strawberry plants over weeks after planting, in response to residual nutrients from vermicompost (VC) and vermicompost tea (VCT) applied during the previous corn cycle. Error bars indicate the range of variation at the 5% significance level ($p < 0.05$). Bars not sharing a common letter are significantly different.

3.5. Plant Stomatal Conductance

Stomatal conductance is a critical physiological parameter that indicates how open the stomata are, affecting transpiration and photosynthesis. Most treatments showed an overall increase in stomatal conductance from week 10 to week 17 (Figure 3). The high VC rate applications, VC3 and VC3+VCT50, showed the lowest stomatal conductance. In week 17, treatments diverged significantly, with stomatal conductance in VC1 and VC3+VCT50 increasing more than in the other treatments. A one-way ANOVA followed by Tukey's HSD test ($p < 0.05$) confirmed that V0 and VC1 had significantly higher stomatal conductance than VC1+VCT50-treated plants.

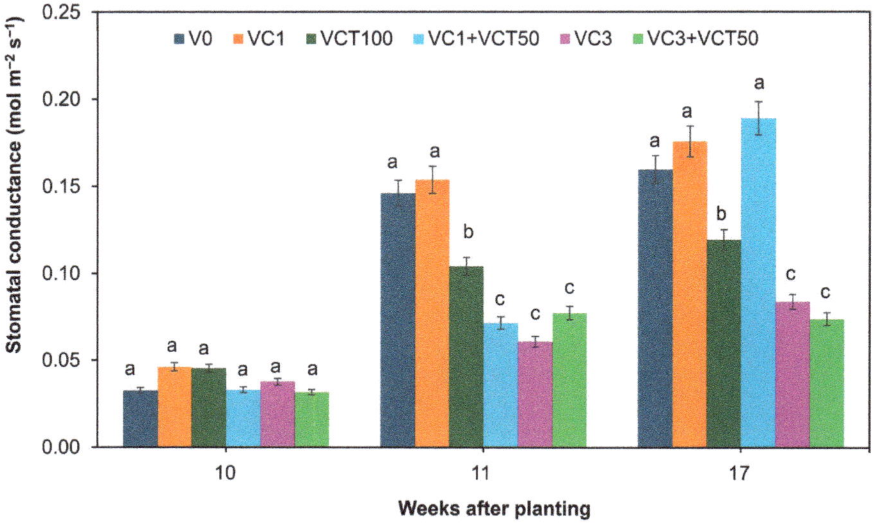

Figure 3. Stomatal conductance variation in strawberry plants over weeks after planting, in response to residual nutrients from vermicompost (VC) and vermicompost tea (VCT) applied during the previous corn cycle. Error bars indicate differences at the 5% significance level ($p < 0.05$). Bars not sharing a common letter are significantly different.

3.6. Plant- and Foliar-Level Spectral Reflectance and Vegetative Indices

The spectral reflectance values of the strawberries taken at leaf level and canopy level are shown in Figures 3 and 4, respectively. The visible region (400–700 nm) has a distinct chlorophyll peak centered at 550 nm, which is proportional to the chlorophyll content or greenness of the leaf. The leaf-level spectra (Figure 4) represent the reflectance of the younger, fully matured single leaf. In contrast, the canopy-level spectra (Figure 5) represent the plant's canopy level. The leaf-level spectra of VC3+VCT50 have lower chlorophyll peaks than other treatments. Also, there is low spectral absorption in the blue (400–500 nm) and red (600–700 nm) regions of the spectra. The near infrared (NIR) plateau (800 nm region) has lower leaf-level reflectance for VC3+VCT50 than other treatments. The spectral reflectance values in the water absorption bands at 1400 nm and 1900 nm remain higher in both leaf-level and canopy-level reflectance for VC3+VCT50. The low reflectance of VC3+VCT50 is consistent with the SPAD values, yield, Brix sugar content, and stomatal conductance observations for this treatment.

The vegetative indices NDVI, GNDVI, and NDMI are presented in Table 4 for leaf-level and canopy-level spectral reflectance values across different treatments. The V0 and VCT100 treatments showed the highest leaf-level NDVI values (0.83), significantly higher than those of VC3+VCT50 and VC3 ($p < 0.05$). The GNDVI did not show significant differences in reflectance at the leaf and canopy levels. The leaf-level NDMI values were low, although VC3 and VC3+VCT50 had significantly lower values than the rest of the treatments (Table 4). At the canopy level, only NDMI showed significant differences with higher VC application rates, with VC3 and VC3+VCT50 having the lowest values compared to VC1. Also, the reflectance was lower at the canopy level than at the leaf level in the water bands.

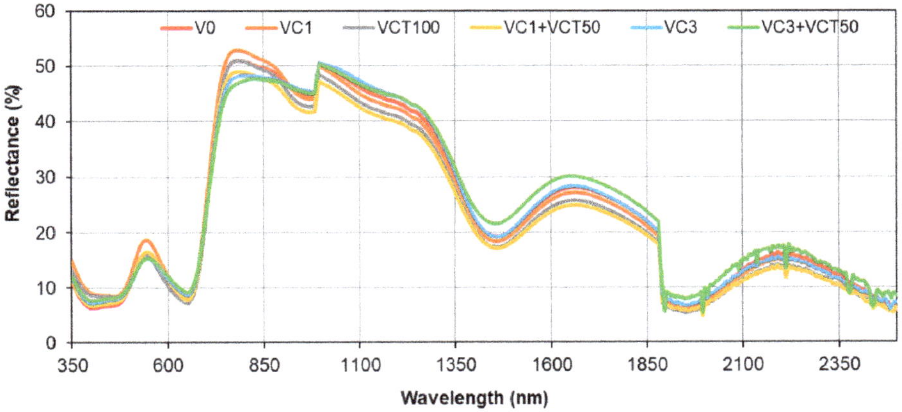

Figure 4. Average leaf-level spectral reflectance of fully expanded mature strawberry leaves located at the mid-height of the plant. Values represent the mean reflectance across five leaves per plant and four plants per replicate, averaged over three replicates, measured at 350–2500 nm using a hand-held spectroradiometer.

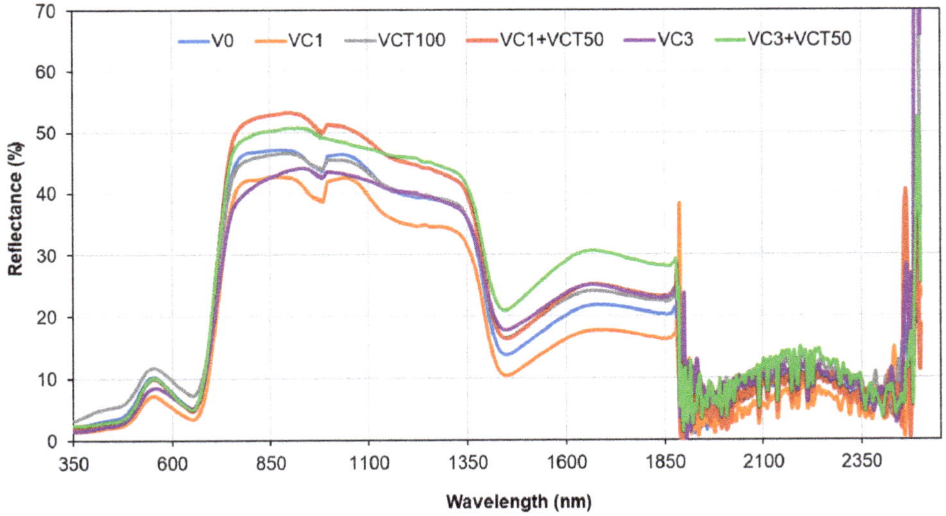

Figure 5. Average canopy-level spectral reflectance of strawberry plants. Values represent the mean reflectance across three plants per replicate, averaged over three replicates, measured at 350–2500 nm using a hand-held spectroradiometer.

Table 4. Vegetative indices for leaf-level and canopy-level spectral reflectance.

	Leaf Level			Canopy Level		
Treatment	**NDVI ***	**GNDVI**	**NDMI ***	**NDVI**	**GNDVI**	**NDMI ***
V0	0.83 a	0.69	0.13 ab	0.87	0.77	0.22 ab
VC1	0.82 ab *	0.66 *	0.16 a	0.90 *	0.82 *	0.27 a
VCT100	0.83 a	0.70	0.17 a	0.83	0.74	0.16 ab
VC1+VCT50	0.81 a–c	0.67 *	0.17 a	0.88	0.80 *	0.21 ab
VC3	0.80 bc	0.69 *	0.10 bc	0.85	0.78 *	0.09 b
VC3+VCT50	0.79 c *	0.68 *	0.06 c	0.87 *	0.79 *	0.08 b

Asterisk (*) indicates indices that showed a statistically significant treatment effect ($p < 0.05$) based on one-way ANOVA. Different letters within a row indicate significant differences among treatments based on Tukey's HSD test ($p < 0.05$).

Comparing the vegetative indices at the leaf and canopy levels, only statistical significance was observed for NDVI and GNDVI. For NDVI at the leaf level, VC1 and VC3+VCT50 values were significantly lower than their corresponding canopy-level NDVI values. GNDVI values at the leaf level for all treatments, except for the control (V0) and VCT100, were significantly lower than their corresponding canopy-level GNDVI values. There was no significant difference between leaf-level and canopy-level NDMI.

4. Discussion

This study explored the residual effects of VC and VCT treatments on strawberry growth and physiology in a rotation system following corn cultivation. The findings show the significance of VC applications for immediate crop enhancement and their prolonged impact on subsequent crop rotations.

4.1. Soil Chemical Analysis

Soil nutrient content data (Table 1) from the preceding corn experiment [25] confirmed that VC and VCT treatments significantly influenced post-harvest soil fertility across surface and subsurface layers. These residual nutrients provided a fertility baseline for the subsequent strawberry trial. However, while subsurface %OM and %C increased under these treatments (Table 1), CEC remained statistically unchanged. This is due to the short-term nature of the amendment, limited biological activity at this depth, and the insensitivity of the unbuffered 1 M NH_4Cl method to modest organic matter contributions in lower horizons [26].

After strawberry cultivation, post-harvest soil nutrient content analysis showed significant differences among the various treatment groups for all the elemental concentrations, especially in the surface soils, except for pH, K, and Zn. High vermicompost application rates, specifically VC3 and VC3+VCT50, were associated with elevated N, P, Mg, Na, OM, and CEC (Table 2). This demonstrates vermicompost's potential to significantly enhance soil fertility by increasing the availability of macro- and micronutrients and improving soil structure, aligning with previous research findings [27]. It also shows the slow-release nature of vermicompost, which provides a steady nutrient supply essential for chlorophyll synthesis and plant growth [28]. Additionally, the increased P availability in all treatments illustrates VC's ability to solubilize P, enhancing energy transfer and root development [13].

The elevated Na concentrations in the soils of all treatments could be attributed to using vermicompost with high Na content [25], which affected the VC and VCT applications. The high soil Na concentration might have contributed to high plant stress and poor growth in the VC3+VCT50-treated plants since strawberries are highly salinity-sensitive [29]. Stress in the VC3+VCT50-treated plants was also evident with low SPAD readings (Figure 2), low conductance (Figure 3), and lower chlorophyll reflectance peak (Figures 3 and 4). These stress indicators corresponded to the reduced strawberry yield and slightly lower Brix sugar content observed in the VC3+VCT50 treatment.

In subsurface soils (10–20 cm), while nutrient content was generally lower than in surface soils, significant improvements were still observed in treatments with higher VC applications. The VC3 treatment, for instance, showed the highest N content (0.60%) in subsurface soils (Table 2), illustrating VC's potential to promote nutrient availability beyond the immediate root zone and support deeper nutrient uptake [13–15]. Furthermore, higher Ca and Mg levels, as well as increased CEC, were observed in the VC3 and VC3+VCT50 treatments, improving soil structure and plant health [30]. The higher OM content in these treatments indicates enhanced soil aggregation and microbial activity, promoting long-term soil fertility.

The study findings demonstrate that nutrient levels are generally higher in surface soils than in subsurface soils across all treatments. The limited vertical movement of nutrients from the surface to the subsurface soil depths emphasizes the effectiveness of surface application techniques, which maximize nutrient availability in the critical root zone. This pattern can be attributed to the direct application of vermicompost to the soil surface, where it undergoes decomposition and nutrient release. The higher nutrient content in surface soils is crucial for crop establishment and early growth, providing readily available nutrients for young plants. The findings from Table 2 affirm that the observed strawberry responses were due to inherent soil properties and a direct consequence of nutrient carryover from the previous corn cropping season.

4.2. Plant Chemical Analysis

The distinct partitioning of nutrients between leaves and roots shows the different roles these plant parts play in nutrient dynamics. Leaves had higher concentrations of macronutrients, which are essential for photosynthesis and growth, while roots exhibited higher levels of micronutrients, reflecting their role in absorbing nutrients from the soil.

The leaf N concentrations were highest in the VC3+VCT50 treatment, indicating that a combination of vermicompost and VCT can potentially enhance N uptake (Table 3). In contrast, roots showed the highest N content in the VC1 treatment, suggesting that moderate vermicompost applications might be more effective for N uptake in roots. This differentiation in nutrient uptake patterns between leaves and roots could be attributed to how residual vermicompost influences soil nutrient availability, potentially enhancing microbial activity and N mineralization [31]. The P content in leaves was relatively higher in the control (V0) and VC3+VCT50 treatments, indicating that residual vermicompost might improve P availability even when applied in previous crop cycles. The inverse relationship between K content and the VC application rate in leaves suggests that higher vermicompost rates might inhibit K uptake or redistribution. This could be due to competitive absorption dynamics with other nutrients, such as Ca and Mg.

Iron and Mn showed enhanced availability in treatments with higher vermicompost rates, particularly VC3 and VC3+VCT50. This indicates that residual vermicompost may improve the solubility and uptake of certain micronutrients by producing chelating agents and improving soil structure. Interestingly, the VC1+VCT50 treatment exhibited the highest Fe content (Table 3) in roots, emphasizing the role of specific treatment combinations in optimizing micronutrient uptake. The differential effects observed across treatments highlight the importance of optimizing vermicompost application rates to achieve balanced nutrient availability. These findings align with previous research, such as Refs. [18,32], which reported positive impacts of vermicompost on plant growth and soil health.

4.3. Strawberry Biomass, Brix Sugar Content, and Yield

4.3.1. Biomass Production

The VC1 treatment resulted in the highest dry biomass at 38,364 kg/acre, followed by the control (V0) with 28,364 kg/acre. This suggests that moderate vermicompost applications can significantly enhance biomass production due to improved nutrient availability and soil structure, a consequence of the slow-release nature of vermicompost. This finding is consistent with studies indicating that moderate organic amendment rates optimize biomass production by improving soil health without overwhelming the plant's nutrient uptake systems [30].

The statistical analysis and Figure 1 showed that the superior performance of VC1 likely stems from its optimal nutrient profile, which promoted plant vigor without inducing salt stress. In contrast, treatments such as VC3 and VC3+VCT50 were excluded from biomass measurement due to poor plant establishment and health, consistent with their high sodium content (Table 1), which can lead to osmotic stress and growth inhibition.

In contrast, excessive VC applications likely introduced increased salinity, as shown by the elevated Na concentrations in the post-corn soils under VC3 and VC3+VCT50 treatments (Table 1). This salinity stress may have persisted into the strawberry season (Table 2), contributing to inhibited plant growth and reduced biomass accumulation. This finding aligns with previous studies indicating that strawberries are susceptible to soil salinity. High salinity levels can lead to leaf edge burn, necrosis, nutrient imbalances, specific ion toxicity, reduced fruit quality and yield, and even potential plant death if salinity stress persists or worsens [33].

4.3.2. Brix Sugar Content

This measures sugar concentration in fruit and is an essential quality indicator, showing variability across treatments but less pronounced differences than biomass and yield. The VCT100 treatment exhibited the highest Brix sugar content (9.3%), although it was not significantly different from other treatments, suggesting that VC and VCT applications enhance fruit quality. Notably, the lack of statistical significance across treatments (except between VCT100 and store-bought strawberries) indicates that while nutrient amendments may enhance sweetness, other factors such as genetics, maturity, and water availability may also play critical roles.

Interestingly, when comparing the Brix sugar content of strawberries from these treatments to organic strawberries available on the market, the market-bought strawberries had a lower Brix sugar content of 6.8%. This finding aligns with other studies that show organic amendments can enhance fruit quality by improving sugar accumulation compared to conventional practices [34].

4.3.3. Strawberry Yield

The impact of VC and VCT on strawberry yield was notable. The VC1 treatment produced the highest yield, comparable to the average of 11,300 kg/acre reported for South Florida [35]. This suggests that moderate vermicompost applications can optimize strawberry production by providing balanced nutrient availability and improving soil structure. In contrast, yields declined sharply in treatments with higher VC and VCT rates (e.g., VC3 and VC3+VCT50), likely due to increased soil Na levels causing salinity stress [18]. The low yield observed in VC1+VCT50 also suggests that even moderate rates, when combined, may affect nutrient dynamics or exacerbate salinity-related effects if not well-calibrated.

4.4. Relative Chlorophyll Content in Plants

The analysis of chlorophyll content, as indicated by SPAD values, shows varying trends across different treatments over several weeks after planting. The chlorophyll content is a critical indicator of plant health and N status, and its variation provides insights into how different treatments influence plant physiology and growth.

The decrease in chlorophyll content (Figure 2) from weeks 10 to 17 after planting is likely due to the plants reallocating energy resources from chlorophyll production to fruit development as the strawberry plants matured. As the plant shifts its focus to developing flowers, fruits, and seeds, less energy is available to maintain chlorophyll levels in the leaves. This is consistent with the understanding that plants often redirect energy to support reproductive structures as they mature, which can result in reduced leaf chlorophyll levels [36,37]. All treatments showed relatively high SPAD values in weeks 10 to 17, indicating that the residual effects of vermicompost and the inherent soil nutrients provided adequate N and other nutrients to support early growth and chlorophyll synthesis.

The reduced chlorophyll content in VC3+VCT50-treated plants can be attributed to the higher sodium levels observed in these treatments (Table 2), which are known to disrupt chlorophyll biosynthesis and accelerate leaf senescence [33]. Vermicompost and VCT applications influence chlorophyll indirectly by altering soil nutrient availability and salinity balance. Moderate VC rates (e.g., VC1) likely provided sufficient nitrogen and micronutrients to support healthy chlorophyll formation, while excessive applications led to ion toxicity and reduced photosynthetic pigment retention.

4.5. Stomatal Conductance in Plants

Stomatal conductance is a physiological parameter that reflects the openness of plant stomata, affecting both transpiration and photosynthesis. It is an essential indicator of plant response to environmental conditions and nutrient availability. In this study, stomatal conductance showed an overall increase from week 10 to week 17, suggesting an adaptation of the plants to their environment and treatments.

At the initial stage (week 10), all treatments exhibited relatively similar stomatal conductance, ranging from 0.10 to 0.15 (Figure 3), except for high VC rates (VC3 and VC3+VCT50), which showed the lowest initial conductance. This initial low conductance could be due to the higher organic matter content in these treatments, potentially leading to increased soil moisture retention and reduced stomatal opening, as plants may not need to open their stomata as broadly or as often because there is less need for water uptake from the soil [38–40]. The steady conductance observed in VC3 up to week 17 indicates that plants in this treatment maintained a relatively constant water and nutrient uptake strategy. Furthermore, VC3+VCT50 consistently has lower stomatal conductance in weeks 10 to 17, consistent with lower SPAD values. The persistently low stomatal conductance in VC3+VCT50 plants could also be attributed to residual salinity stress from the elevated sodium levels observed in these soils following the corn experiment (Table 2), which may have limited water uptake and physiological activity.

Treatment-related changes similarly influenced stomatal conductance in soil chemistry. In treatments with moderate nutrient levels (e.g., VC1), plants maintained optimal stomatal function, likely due to balanced water and nutrient uptake. In contrast, VC3+VCT50 showed persistently low conductance, which can be explained by salinity-induced osmotic stress, limiting water availability at the root–soil interface, and prompting partial stomatal closure to conserve moisture [39]. This physiological response is consistent with lower chlorophyll content, reduced SPAD values, and spectral reflectance patterns observed in the same treatment.

4.6. Plant- and Leaf-Level Spectral Reflectance and Vegetative Indices

The spectral reflectance data show differences between leaf-level and canopy-level measurements, particularly in the visible and near-infrared (NIR) regions. The leaf-level spectra of VC3+VCT50 have lower chlorophyll peaks than other treatments, indicating lower chlorophyll content. Also, low spectral absorption in the blue (400–500 nm) and red (600–700 nm) regions of the VC3+VCT50 spectra indicates a lower rate of photosynthesis, which is consistent with SPAD and stomatal conductance values. The high-water absorption bands at 1400 nm and 1900 nm in both leaf-level and canopy-level reflectance for VC3+VCT50 indicate water stress in plants. These spectral patterns correspond with the observed reductions in biomass, yield, and Brix sugar content in the same treatment, highlighting a clear link between physiological stress signals and diminished agronomic performance.

The higher reflectance in the green band at the leaf level suggests that individual leaves reflect more green light, which is typical for healthy, chlorophyll-rich leaves. The absorption troughs observed in the blue (400–500 nm) and red (600–700 nm) regions further indicate active photosynthesis and chlorophyll absorption [41]. At the canopy level, the reduced reflectance in the green band and more pronounced absorption in the blue and red regions can be attributed to the increased complexity of the canopy structure, including overlapping leaves, shadows, and varying leaf orientations [42,43]. In the NIR region (760–900 nm), the distinct separation of spectral curves at the canopy level, compared to the indistinguishable curves at the leaf level, reflects differences in canopy architecture, biomass, and leaf area index (LAI). The internal leaf structure and canopy geometry influence the NIR reflectance, with higher NIR reflectance indicating healthier and denser canopies [44]. These findings suggest that canopy-level measurements provide a more integrated view of plant health and structure, whereas leaf-level measurements offer detailed physiological insights.

The Normalized Difference Vegetation Index (NDVI) is a widely used indicator of plant health, reflecting chlorophyll content and photosynthetic activity. The lower NDVI values (Table 4) observed in high VC application rates (VC3 and VC3+VCT50) may indicate nutrient imbalances affecting chlorophyll content and photosynthetic efficiency. At the canopy level, the NDVI values were generally higher than at the leaf level, reflecting the integrated health and vigor of the entire plant canopy. The differences between leaf and canopy NDVI highlight the complexity of scaling leaf-level measurements to whole-canopy assessments, as canopy-level indices capture the combined effects of the leaf area, density, and structure [44,45]. The comparison of vegetative indices at leaf and canopy levels showed that leaf-level NDVI values were lower for VC1 and VC3+VCT50 compared to their respective canopy level values. Leaf-level GNDVI values were lower for all treatments except V0 and VCT100 compared to their canopy-level counterparts, suggesting that canopy measurements integrate more comprehensive plant responses, capturing variations in leaf area and distribution.

The Normalized Difference Moisture Index (NDMI) assesses plant water content and stress. NDMI values were generally lower than 0.4 for all treatments at both the leaf and canopy levels, indicating potential water stress [46,47]. NDMI is particularly useful for assessing relative plant water content and detecting moisture-related stress before visual symptoms appear. The consistently low NDMI values observed in the VC3 and VC3+VCT50 treatments likely reflect reduced water uptake due to elevated soil salinity, as indicated by high Na concentrations in Table 2. This is on par with the lower stomatal conductance (Figure 3) and reduced SPAD values (Figure 2), suggesting that the high Na content in the VC used may have disrupted plant water balance in this study. It is, therefore, essential to characterize vermicompost composition, as salinity-related stress effects may vary depending on feedstock sources. In contrast, higher NDMI in VC1 reflects better water retention and nutrient availability, contributing to improved physiological performance and yield.

Therefore, the corn–strawberry rotation used in this study was intended to simulate conditions where organic inputs are applied to one crop, and their residual effects are observed on a subsequent high-value crop. While this rotation is not standard practice in many production systems, it allowed for a controlled evaluation of nutrient carryover, soil salinity, and plant physiological responses. We recognize that outcomes might differ under alternative rotations, soil types, or amendment application timings. Therefore, caution should be taken when generalizing these results beyond the experimental setup.

5. Conclusions

This study showed the importance of the residual effects of vermicompost (VC) and vermicompost tea (VCT) on strawberry production in a corn–strawberry rotation system. A moderate vermicompost application (1 ton/acre, VC1) significantly enhanced strawberry yield, biomass accumulation, and fruit quality, as evidenced by increased Brix sugar content. These improvements reflect the role of VC in maintaining residual soil fertility and supporting plant physiological performance without additional fertilizer inputs.

However, high-rate applications (VC3 and VC3+VCT50) reduced plant vigor and yield, coinciding with elevated soil sodium concentrations and stress responses in plant physiology. In addition, VCT100, VC1+VCT50, and VC3+VCT50 treatments also resulted in diminished physiological performance and yields compared to VC1. It is important to note that these outcomes were specific to the vermicompost source used in this study, which contained atypically high sodium levels. Such salinity-induced effects may not be observed in vermicompost derived from low-sodium feedstocks. We recommend routine analysis of vermicompost sodium content, especially when applying high rates or using composts derived from salt-accumulating species, to prevent potential salinity stress in salt-sensitive crops like strawberries.

This study was limited to a single growing season under subtropical field conditions, and findings may not fully represent long-term or broader climatic responses. Additionally, the experiment was based on a specific corn–strawberry rotation designed to assess residual amendment effects. Therefore, outcomes may vary depending on alternative cropping systems or environmental conditions. Nevertheless, vermicompost and its derivatives offer a sustainable, rotation-compatible, cost-effective alternative to synthetic fertilizers; it is one that enhances strawberry productivity while reducing chemical input dependence, as long as salinity risks are managed through appropriate feedstock selection and pre-application testing.

Author Contributions: Conceptualization, I.O.; methodology, I.O. and M.S.B.B.; software, I.O.; validation, I.O.; formal analysis, I.O.; investigation, I.O.; resources, M.S.B.B.; data curation, I.O.; writing—original draft preparation, I.O.; writing—review and editing, I.O. and M.S.B.B.; visualization, I.O. and M.S.B.B.; supervision, M.S.B.B.; project administration, M.S.B.B.; funding acquisition, M.S.B.B. All authors have read and agreed to the published version of the manuscript.

Funding: This research was partially funded by the Fulbright Foreign Students' program and the USDA-NRCS, USDA-NIFA grants under the award numbers NR224209XXXXG001; AWD13273; NR233A750011G026; AWD15131; 2023-70001-40999, 2023-77040-41154.

Data Availability Statement: The data presented in this study are available on request from the corresponding author. The data is not publicly available due to privacy or ethical restrictions.

Acknowledgments: The authors are grateful for the help from Fiona Nakabuye, Samarah Martin, Moses Kiwanuka, Priyanka Belbase, and Peter Fedele of the Lions Fruit Farm, Miami, Florida.

Conflicts of Interest: The authors declare no conflicts of interest.

References

1. Vondracek, K.; Altpeter, F.; Liu, T.; Lee, S. Advances in Genomics and Genome Editing for Improving Strawberry (*Fragaria* × *ananassa*). *Front. Genet.* **2024**, *15*, 1382445. [CrossRef] [PubMed]
2. Yeh, D.A.; Kramer, J.; Calvin, L.; Weber, C. *The Changing Landscape of U.S. Strawberry and Blueberry Markets: Production, Trade, and Challenges from 2000 to 2020*; Report No. Eib-257; U.S. Department of Agriculture, Economic Research Service: Washington, DC, USA, 2023.
3. Samtani, J.B.; Rom, C.R.; Friedrich, H.; Fennimore, S.A.; Finn, C.E.; Petran, A.; Wallace, R.W.; Pritts, M.P.; Fernandez, G.; Chase, C.A.; et al. The Status and Future of the Strawberry Industry in the United States. *HortTechnology* **2019**, *29*, 11–24. [CrossRef]

4. Brad, B. Peak Strawberry Season Means More Florida-Grown, Flavorful Fruit at the Market. Available online: https://blogs.ifas.ufl.edu/news/2024/01/18/peak-strawberry-season-means-more-florida-grown-flavorful-fruit-at-the-market/ (accessed on 8 August 2024).
5. Giampieri, F.; Tulipani, S.; Alvarez-Suarez, J.M.; Quiles, J.L.; Mezzetti, B.; Battino, M. The Strawberry: Composition, Nutritional Quality, and Impact on Human Health. *Nutrition* **2012**, *28*, 9–19. [CrossRef]
6. Agehara, S.; Hochmuth, G. Fertilization of Strawberries in Florida. In *EDIS*; IFAS Communications: Gainesville, FL, USA, 2023.
7. Oscar, L.; Elena, R. Management of Strawberry Insect and Mite Pests in Greenhouse and Field Crops. In *Strawberry*; Asao, T., Asaduzzaman, M., Eds.; IntechOpen: Rijeka, Croatia, 2019; Chapter 6.
8. Rondon, S.I.; Price, J.F.; Cantliffe, D.J. Strawberries: Main Pests and Beneficials in Florida: Hs1018/Hs265, 7/2005. In *EDIS*; IFAS Communications: Gainesville, FL, USA, 2005.
9. Rosskopf, E.; Gioia, F.D.; Vincent, I.; Hong, J.; Zhao, X. Impacts of the Ban on the Soil-Applied Fumigant Methyl Bromide. *Phytopathology* **2024**, *114*, 1161–1175. [CrossRef] [PubMed]
10. Evans, B.; Krey, K.; Renkema, J. Susceptibility of *Phytoseiulus persimilis* and *Neoseiulus californicus* (Acari: Phytoseiidae) to Commonly-Used Insecticides Approved for Managing Arthropod Pests in Florida Strawberries. In *EDIS*; IFAS Communications: Gainesville, FL, USA, 2018.
11. Nguyen, T.T.; Rosello, C.; Bélanger, R.; Ratti, C. Fate of Residual Pesticides in Fruit and Vegetable Waste (Fvw) Processing. *Foods* **2020**, *9*, 1468. [CrossRef]
12. Environmental Working Group. The 2024 Dirty Dozen. Available online: https://www.ewg.org/foodnews/dirty-dozen.php (accessed on 22 April 2024).
13. Oyege, I.; Bhaskar, M.S.B. Effects of Vermicompost on Soil and Plant Health and Promoting Sustainable Agriculture. *Soil Syst.* **2023**, *7*, 101. [CrossRef]
14. Souffront, D.K.S.; Salazar-Amoretti, D.; Jayachandran, K. Influence of Vermicompost Tea on Secondary Metabolite Production in Tomato Crop. *Sci. Hortic.* **2022**, *301*, 111135. [CrossRef]
15. Yatoo, A.M.; Ali, M.N.; Baba, Z.A.; Hassan, B. Sustainable Management of Diseases and Pests in Crops by Vermicompost and Vermicompost Tea. A Review. *Agron. Sustain. Dev.* **2021**, *41*, 7. [CrossRef]
16. Stavridou, E.; Giannakis, I.; Karamichali, I.; Kamou, N.N.; Lagiotis, G.; Madesis, P.; Emmanouil, C.; Kungolos, A.; Nianiou-Obeidat, I.; Lagopodi, A.L. Biosolid-Amended Soil Enhances Defense Responses in Tomato Based on Metagenomic Profile and Expression of Pathogenesis-Related Genes. *Plants* **2021**, *10*, 2789. [CrossRef]
17. Oyege, I.; Bhaskar, M.S.B. Residual Effects of Vermicompost Derived Soil Nutrients on Growth and Physiology of Subsequent Strawberry Rotation after Corn. In Proceedings of the ASA, CSSA, SSSA International Annual Meeting 2023, St. Louis, MO, USA, 29 October–1 November 2023.
18. Arancon, N.Q.; Edwards, C.A.; Bierman, P.; Welch, C.; Metzger, J.D. Influences of Vermicomposts on Field Strawberries: 1. Effects on Growth and Yields. *Bioresour. Technol.* **2004**, *93*, 145–153. [CrossRef]
19. Chatterjee, R.; Debnath, A.; Mishra, S. Vermicompost and Soil Health. *Soil Health* **2020**, *59*, 69–88.
20. USDA. Corn, Soybeans Accounted for Half of U.S. Crop Cash Receipts in 2023. Available online: https://www.ers.usda.gov/data-products/chart-gallery/gallery/chart-detail?chartId=76946 (accessed on 10 January 2025).
21. USDA. Production—Soybeans. Available online: https://www.fas.usda.gov/data/production/commodity/2222000 (accessed on 10 January 2025).
22. McWhirt, A. Long-Term Study of the Impacts of Crop Rotation on Southeastern Plasticulture Strawberry Production Principal Investigator (S). In *2023–2024 SRSFC Research Report*; Sugar Research Australia Ltd.: Brisbane, Australia, 2024.
23. Thompson, S. Crop Rotation: What Can Strawberries Follow in Home Garden? 2014. Available online: https://storables.com/garden/crop-rotation-what-can-strawberries-follow-in-home-garden/ (accessed on 10 April 2025).
24. Silva, A.M. *The First Usda People's Garden in South Florida*; U.S. Department of Agriculture: Washington, DC, USA, 2010. Available online: https://www.usda.gov/about-usda/news/blog/first-usda-peoples-garden-south-florida (accessed on 24 October 2024).
25. Oyege, I.; Bhaskar, M.S.B. Evaluation of Vermicompost and Vermicompost Tea Application on Corn (*Zea Mays*) Growth and Physiology Using Optical Plant Sensors. *Plant Nutr.* **2024**, *48*, 1275–1293. [CrossRef]
26. Ross, D.S.; Ketterings, Q. Recommended Methods for Determining Soil Cation Exchange Capacity. *Recomm. Soil Test. Proced. Northeast. United States* **1995**, *493*, 62.
27. Rehman, S.U.; De Castro, F.; Aprile, A.; Benedetti, M.; Fanizzi, F.P. Vermicompost: Enhancing Plant Growth and Combating Abiotic and Biotic Stress. *Agronomy* **2023**, *13*, 1134. [CrossRef]
28. Azizi Yeganeh, M.; Shahabi, A.A.; Ebadi, A.; Abdossi, V. Vermicompost as an Alternative Substrate to Peat Moss for Strawberry (*Fragaria ananassa*) in Soilles Culture. *BMC Plant Biol.* **2024**, *24*, 149. [CrossRef]
29. Denaxa, N.-K.; Nomikou, A.; Malamos, N.; Liveri, E.; Roussos, P.A.; Papasotiropoulos, V. Salinity Effect on Plant Growth Parameters and Fruit Bioactive Compounds of Two Strawberry Cultivars, Coupled with Environmental Conditions Monitoring. *Agronomy* **2022**, *12*, 2279. [CrossRef]

30. Gutiérrez-Miceli, F.A.; Santiago-Borraz, J.; Molina, J.A.M.; Nafate, C.C.; Abud-Archila, M.; Llaven, M.A.O.; Rincón-Rosales, R.; Dendooven, L. Vermicompost as a Soil Supplement to Improve Growth, Yield and Fruit Quality of Tomato (*Lycopersicum esculentum*). *Bioresour. Technol.* **2007**, *98*, 2781–2786. [CrossRef]
31. Wang, G.; Ren, Y.; Bai, X.; Su, Y.; Han, J. Contributions of Beneficial Microorganisms in Soil Remediation and Quality Improvement of Medicinal Plants. *Plants* **2022**, *11*, 3200. [CrossRef]
32. Laugale, V.; Dane, S.; Strautiņa, S.; Kalniņa, I. Influence of Vermicompost on Strawberry Plant Growth and Dehydrogenase Activity in Soil. *Agron. Res.* **2020**, *18* (Suppl. S4), 2742–2751.
33. Sun, Y.; Niu, G.; Wallace, R.; Masabni, J.; Gu, M. Relative Salt Tolerance of Seven Strawberry Cultivars. *Horticulturae* **2015**, *1*, 27–43. [CrossRef]
34. Khalid, S.; Qureshi, K.M.; Hafiz, I.A.; Khan, K.S.; Qureshi, U.S. Effect of Organic Amendments on Vegetative Growth, Fruit and Yield Quality of Strawberry. *Pak. J. Agric. Res.* **2013**, *26*, 104–112.
35. Guan, Z.; Suh, D.H.; Khachatryan, H.; Wu, F. Import Growth and the Impact on the Florida Strawberry Industry: Fe1022, 7/2017. In *EDIS*; IFAS Communications: Gainesville, FL, USA, 2018.
36. Yavari, N.; Tripathi, R.; Wu, B.-S.; MacPherson, S.; Singh, J.; Lefsrud, M. The Effect of Light Quality on Plant Physiology, Photosynthetic, and Stress Response in Arabidopsis Thaliana Leaves. *PLoS ONE* **2021**, *16*, e0247380. [CrossRef]
37. Zhao, W.; Zhao, H.; Wang, H.; He, Y. Research Progress on the Relationship between Leaf Senescence and Quality, Yield and Stress Resistance in Horticultural Plants. *Front. Plant Sci.* **2022**, *13*, 1044500. [CrossRef]
38. Buckley, T.N. Modeling Stomatal Conductance. *Plant Physiol.* **2017**, *174*, 572–582. [CrossRef] [PubMed]
39. Chadha, A.; Florentine, S.K.; Chauhan, B.S.; Long, B.; Jayasundera, M. Influence of Soil Moisture Regimes on Growth, Photosynthetic Capacity, Leaf Biochemistry and Reproductive Capabilities of the Invasive Agronomic Weed; *Lactuca serriola*. *PLoS ONE* **2019**, *14*, e0218191. [CrossRef] [PubMed]
40. Xu, S.; Li, P.; Sayer, E.J.; Zhang, B.; Wang, J.; Qiao, C.; Peng, Z.; Diao, L.; Chi, Y.; Liu, W.; et al. Initial Soil Organic Matter Content Influences the Storage and Turnover of Litter, Root and Soil Carbon in Grasslands. *Ecosystems* **2018**, *21*, 1377–1389. [CrossRef]
41. Liu, J.; van Iersel, M.W. Photosynthetic Physiology of Blue, Green, and Red Light: Light Intensity Effects and Underlying Mechanisms. *Front. Plant Sci.* **2021**, *12*, 619987. [CrossRef]
42. Terashima, I.; Fujita, T.; Inoue, T.; Chow, W.S.; Oguchi, R. Green Light Drives Leaf Photosynthesis More Efficiently Than Red Light in Strong White Light: Revisiting the Enigmatic Question of Why Leaves Are Green. *Plant Cell Physiol.* **2009**, *50*, 684–697. [CrossRef]
43. Ustin, S.L.; Jacquemoud, S. How the Optical Properties of Leaves Modify the Absorption and Scattering of Energy and Enhance Leaf Functionality. In *Remote Sensing of Plant Biodiversity*; Cavender-Bares, J., Gamon, J.A., Townsend, P.A., Eds.; Springer International Publishing: Cham, Switzerland, 2020; pp. 349–384.
44. Ollinger, S.V. Sources of Variability in Canopy Reflectance and the Convergent Properties of Plants. *New Phytol.* **2011**, *189*, 375–394. [CrossRef]
45. Imran, H.A.; Gianelle, D.; Rocchini, D.; Dalponte, M.; Martín, M.P.; Sakowska, K.; Wohlfahrt, G.; Vescovo, L. Vis-Nir, Red-Edge and Nir-Shoulder Based Normalized Vegetation Indices Response to Co-Varying Leaf and Canopy Structural Traits in Heterogeneous Grasslands. *Remote Sens.* **2020**, *12*, 2254. [CrossRef]
46. Berca, M.; Horoias, R. Ndmi Use in Recognition of Water Stress Issues, Related to Winter Wheat Yields in Southern Romania. *Sci. Pap. Ser. Manag. Econ. Eng. Agric. Rural. Dev.* **2022**, *22*, 105–112.
47. Ghazaryan, G.; Dubovyk, O.; Graw, V.; Kussul, N.; Schellberg, J. Local-Scale Agricultural Drought Monitoring with Satellite-Based Multi-Sensor Time-Series. *GISci. Remote Sens.* **2020**, *57*, 704–718. [CrossRef]

environments

MDPI

Review

Biochar Amendments for Soil Restoration: Impacts on Nutrient Dynamics and Microbial Activity

Kuok Ho Daniel Tang

Department of Environmental Science, The University of Arizona, Tucson, AZ 85721, USA; danielkhtang@arizona.edu

Abstract: Biochar is increasingly recognized as a multifunctional soil amendment that improves soil fertility, nutrient cycling, and crop productivity. Studies across field, greenhouse, and incubation settings show that biochar enhances nutrient retention, reduces leaching, and regulates carbon, nitrogen, and phosphorus cycling. Its effects are shaped by intrinsic physicochemical properties and interactions with soil minerals, microbial communities, and enzymatic processes. Short-term benefits of biochar applications often include improved nutrient adsorption and water regulation, while long-term applications support stable soil organic matter formation, root development, and fertilizer use efficiency. Biochar also reshapes soil microbial diversity and activity. Beneficial bacterial groups such as Proteobacteria and Actinobacteria, along with fungi such as *Mortierella*, respond positively, enhancing nitrogen fixation, phosphorus solubilization, and organic matter decomposition. Meanwhile, biochar applications could suppress pathogens. Enzyme activities, including urease and phosphatase, are typically stimulated, driving nutrient mobilization. Yet outcomes remain context-dependent, with biochar feedstock, application rate, soil conditions, and crop type influencing results; excessive use may suppress enzymatic activity, reduce nutrient availability, or shift microbial communities unfavorably. Practically, biochar can improve fertilizer efficiency, restore degraded soils, and reduce greenhouse gas emissions, contributing to climate-smart agriculture. Future work should prioritize long-term, multi-site trials and advanced analytical tools to refine sustainable application strategies.

Keywords: biochar; enzyme activities; fertility; microbial diversity; nutrient cycling; soil amendment

1. Introduction

The excessive application of chemical fertilizers has triggered multiple environmental challenges, including soil degradation caused by acidification, nutrient leaching, deterioration of soil structure, decline in beneficial microbial populations, and an increase in greenhouse gas emissions [1]. Biochar, a pyrolyzed carbon-rich solid produced under oxygen-limited conditions from diverse biomass sources, has emerged as a sustainable amendment due to its capacity to improve soil properties and foster beneficial microbial activity [2,3]. Because of its high carbon content, porous structure, negatively charged surface, and alkaline properties, biochar can greatly change soil physicochemical properties, such as lowering bulk density, increasing pH, and improving nutrient retention and availability, water holding capacity, cation exchange capacity, and aggregate stability [4,5]. Nonetheless, these effects are influenced by multiple factors, including feedstock type, pyrolysis conditions, biochar application rate, soil type, and climate [6,7].

A growing body of evidence highlights its effectiveness in lowering greenhouse gas emissions [8–10]. A critical review by Lyu et al. [8] demonstrated a general inhibitory effect of biochar across various environments, with the impact on greenhouse gases decreasing in the order: $N_2O > CH_4 > CO_2$. This suggests that biochar modifies the physicochemical characteristics of soil and compost under diverse conditions, thereby influencing gas emissions. While relatively few, studies have also shown that biochar has no significant effect on or may even increase greenhouse gas emissions. A smaller review by Atilano-Camino et al. [11] on total greenhouse gas emissions from soils with and without biochar amendment found that the average CO_2 (7.7 t/ha) and CH_4 (4.9 t/ha) outputs were higher in soils amended with biochar than those from soils without biochar, whereas N_2O emissions (1.6 t/ha) were lower than those in the untreated control. Additionally, biochar has been reported to promote long-term crop productivity by fostering synergistic interactions between biochar, soil, and microbial communities [12,13]. Soil is a highly intricate ecosystem that sustains diverse microorganisms essential to ecosystem functioning. The incorporation of biochar has been shown to influence these functions by enhancing enzyme activity and restructuring microbial community composition [14]. Soil enzymes are critical for mobilizing recalcitrant nutrients, thereby ensuring a continuous energy supply for soil microbiota [15]. Moreover, biochar-induced increases in the pH of acidic soil promote crop growth by releasing phosphorus (P) bound to iron (Fe) and alleviating aluminum (Al) toxicity, ultimately regulating key aspects of soil nutrient dynamics [16].

The nutrient composition of biochar is largely influenced by the type of feedstock and the pyrolysis conditions. For instance, biochar produced from manure and biosolid feedstocks generally contains higher levels of nitrogen (N) and P compared to that generated from straw or wood materials [17,18]. During pyrolysis, an increase in ash content (inorganic residues) elevates the concentrations of P, calcium (Ca), magnesium (Mg), silicon (Si), and potassium (K), whereas the N content declines at higher pyrolysis temperatures due to gaseous losses [19,20].

Biochar also functions as a nutrient reservoir, effectively retaining macronutrients such as N, P, and K, as it can store and hold these nutrients within its structure and on its surface and slowly release them, keeping them available in soil longer [21,22]. However, excessive application of biochar may raise soil pH above the optimal level for crop growth, which can alter nutrient availability and ultimately diminish soil fertility and crop productivity [23]. A study reported that potato yields dropped by over 6% with mineral and 10% with organic fertilization, likely due to biochar-induced alkalinity that raised soil pH and reduced nutrient availability [24]. Although less reported, nutrient losses, particularly N, may occur through leaching and nitrous oxide emissions. For instance, at a high application rate of 30 t/ha, biochar increased nitrate and dissolved organic nitrogen leaching by 38% and 170%, respectively, relative to the control. This treatment resulted in the highest N losses through both nitrous oxide emissions and leaching, indicating that application rate plays a crucial role in N loss [25]. Nonetheless, a meta-analysis has shown that, at application rates of up to 120 t/ha, biochar generally reduces nitrous oxide emissions, as larger quantities can limit N availability for denitrification through the adsorption of nitrate. This could raise N use efficiency [26]. Yu et al. [27] noted that the nutrient retention capacity of biochar is primarily governed by its porosity and surface charge, which determine its ability to adsorb both cations and anions.

In recent years, biochar has attracted attention as a strategy for mitigating soil pollution and enhancing soil restoration [28–30]. Reviews of current studies indicate that biochar contributes to the immobilization of heavy metals and persistent organic pollutants in soils, thereby lowering their bioavailability. This remediation effect is largely attributed to processes such as precipitation, electrostatic attraction, surface adsorption, structural entrapment, and promoted degradation. However, the efficiency of contaminant removal is influenced by factors including the type of biochar, the application rate, the characteristics of the soil, and the nature of the pollutants [31–33]. By acting as a nutrient reservoir and altering soil microbial composition, biochar contributes to both contaminant removal from soil and alterations in soil nutrient content. However, few reviews focus specifically on the latter.

Dai et al. [34] reviewed how biochar properties influence N and P cycles in soil and the abundance of soil microbial communities, but provided limited explanation on how biochar applications affect the levels and transformation of various nutrients in soil and the microbial communities for cycling of nutrients beyond N and P. Tsolis and Barouchas [35] reviewed the influence of adding biochar on soil properties and its biological interactions in soil, but did not discuss the influence of biochar on the retention, alteration, and leaching of various soil nutrients in detail. Similarly to the review of Dai et al. [34], the review of Anyebe et al. [36] highlights how biochar properties, shaped by feedstock and pyrolysis temperature, affect soil ecosystem services, nutrient cycling, and crop productivity, with the emphasis placed on biochar synthesis and properties. Singh et al. [37] reviewed biochar production, key properties, and its role in restoring degraded soils, with emphasis on its effectiveness in remediating saline and acidic soils, its interactions with soil properties, and its ecological functions influencing enzymes and crop performance. Their review did not focus on the underlying molecular and microbial mechanisms that alter soil nutrient levels.

It is important to review the effects of biochar on nutrient dynamics and microbial activity in soil because both are central to soil fertility, ecosystem functioning, and sustainable crop production. Biochar can alter nutrient availability by influencing soil pH, cation exchange capacity, and sorption processes, which in turn affect the efficiency with which plants and microbes access essential elements. The two are closely connected: microbial processes (e.g., mineralization, nitrogen fixation, and decomposition) drive nutrient cycling, while nutrient availability regulates microbial growth and metabolism. Thus, understanding how biochar simultaneously influences nutrient dynamics and microbial activity is critical for predicting its long-term impacts on soil health and agricultural productivity. Therefore, this review aims to examine how biochar influences soil nutrient levels, particularly nitrogen, phosphorus, carbon, and cations. Specifically, it aims to elucidate the physicochemical mechanisms by which biochar alters nutrient retention, transformation, and leaching. Additionally, it presents recent progress in understanding how biochar influences soil microbial communities (bacteria and fungi). It seeks to elucidate the mechanisms by which biochar–microbe interactions enhance nutrient availability. In doing so, it provides deeper mechanistic insights into how it regulates soil health.

2. Review Methodology

This narrative review was conducted to ensure a comprehensive and balanced synthesis of current knowledge on biochar amendments in soil, with a particular focus on nutrient dynamics and microbial activity. A broad search was undertaken in major scientific databases, including Web of Science, Scopus, and Google Scholar, to identify peer-reviewed journal articles, reviews, and relevant book chapters published between 2015 and 2025. Search terms were combined using Boolean operators and included variations of the following keywords: (1) "biochar" AND "soil restoration", (2) "biochar" AND "nutrient dynamics" OR "nutrient cycling" OR "fertility", (3) "biochar" AND "microbial activity" OR "microbial community" OR "soil microbiota", and (4) "biochar" AND "soil health" OR "ecosystem functioning".

Studies were included if they: (1) reported experimental, field, or modeling data on the impacts of biochar on nutrient availability, nutrient cycling, or microbial processes in soil; (2) provided mechanistic insights into biochar–soil interactions relevant to soil fertility, ecosystem functioning, or crop productivity focusing on the nutrient and microbial aspects; and (3) were published in English and in peer-reviewed outlets.

Studies were excluded if they: (1) focused solely on biochar production techniques without discussing impacts on nutrient dynamics or microbial activity; (2) addressed only physical soil properties (e.g., bulk density, porosity) without linking them to nutrient or microbial processes; (3) mixed biochar with other soil amendments or treatments except fertilization as the simplest form of soil improvement; (4) involved functionalized biochar due to the myriad of functionalizing materials and techniques; and (5) were non-peer-reviewed sources such as conference abstracts, opinion articles, or policy briefs.

Key information from each study was extracted, including biochar characteristics, soil type and experimental setting (laboratory, greenhouse, or field), reported impacts on nutrient dynamics (e.g., N, P, K availability, pH changes, cation exchange capacity), as well as reported impacts on microbial activity (e.g., microbial biomass, community composition, enzymatic activity, nutrient mineralization).

Extracted data were synthesized thematically to highlight the mechanisms by which biochar influences nutrient dynamics, the variations in microbial activity and community structure due to biochar addition, and the interlinkages between nutrient dynamics and microbial processes under biochar amendment. Thematic synthesis combined both a priori and emergent coding. Priori themes, namely nutrient dynamics and microbial community composition, were identified based on the review's objectives. Additional subthemes, including variations by soil type, biochar feedstock, and environmental conditions, emerged inductively during literature analysis. No formal effect size calculations were performed, as this review aimed for qualitative integration rather than meta-analysis; however, directional trends were noted (e.g., the proportion of studies reporting increased nutrient availability or microbial biomass). Contradictory results were reconciled by considering contextual factors, including biochar properties, soil characteristics, and experimental duration. Study robustness was evaluated qualitatively, with greater weight given to peer-reviewed experimental data with clear methodological detail.

3. Impacts of Biochar on Nutrient Dynamics

Biochar amendments are known to significantly influence soil nutrient dynamics by altering chemical properties, nutrient availability, and nutrient cycling processes, thereby shaping soil fertility and plant productivity.

3.1. Phosphorus and Nitrogen Dynamics

Biochar plays a pivotal role in regulating soil P and N dynamics. A field study in western Montana examined the effects of wood biochar (0, 20, and 40 tons per ha), with or without short-term cattle trampling, on soil properties, nutrient dynamics, and grass nutrient uptake [38]. Trampling alone compacted surface soils and reduced nitrogen availability, whereas biochar application improved infiltration, enhanced net nitrification, and promoted biochar incorporation into soil. Biochar also increased organic P while lowering inorganic forms, with stronger effects under trampling, and these changes were mirrored in grass P concentrations (Figure 1). Cattle trampling drove biochar deeper into the soil, increasing its contact with soil matrices and enhancing P sorption, which lowered $CaCl_2$-P and citrate-P levels. These P shifts may also reflect higher phosphatase activity from plants or microbes responding to greater P demand under adequate N and better moisture conditions [38].

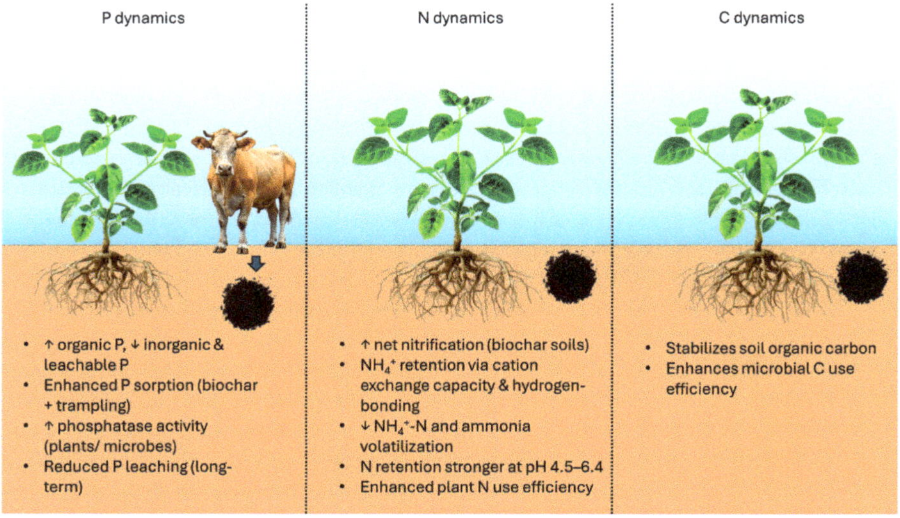

Figure 1. Impacts of biochar on P, N, and C dynamics in soil. (Note: ↑ represents increase and ↓ represents decrease).

Long-term trials (7 years) confirm biochar's ability to limit nutrient losses. On vegetated roofs in Finland, biochar consistently reduced P leaching in both seeded plantings and pre-grown mats, while N leaching decreased only in plantings, coinciding with improved plant cover [39]. Nutrient losses declined over time as roofs aged; however, fertilization six years post-establishment caused elevated P leaching that biochar did not mitigate. Runoff from plantings contained more total P than mats, despite mats having 1.4 times higher P in the substrate. P dynamics were governed mainly by substrate texture, chemistry, and pH, whereas N dynamics were driven by biological processes. Substrate pH declined from 7.2–7.9 (2016) to 6.3–7.0 (2020) over the study duration, a range that keeps P plant-available rather than particle-bound.

Dubrovina [40] examined how wood biochar affects phosphate forms in two Karelian soils with different phosphorus availability, namely a sandy Umbric Podzol and a heavy loamy Umbric Retisol. Biochar fractions ($\leq$2 mm and 3–5 mm) were applied at 2% and 5% rates. Biochar, especially the $\leq$2 mm fraction at 5%, raised soil pH, available phosphorus (by 20–40%), and the proportions of Ca-, Al-, and loosely bound P, while enhancing phosphatase activity in the Umbric Podzol. In pot tests with spring barley, biochar further increased P_2O_5 levels. In the Umbric Retisol, biochar had a smaller effect, slightly increasing available and Ca-bound P but not phosphatase activity; significant P_2O_5 gains occurred only with the combined biochar–fertilizer treatment. Overall, fine biochar ($\leq$2 mm) at 5% showed the strongest positive impact across parameters.

Soil incubation experiments provide further insight into N cycling. Gupta et al. [41] tested rice straw (RSB) (pH = 9.1) and acacia wood biochar (ACB) (pH = 7.3) (at 0–1% in clay loam (pH = 7.6) and loamy sand soils (pH = 7.0). Ammonium-N declined faster in biochar-treated soils, while RSB consistently raised nitrate-N more than ACB. In clay loam, nitrate levels increased with higher biochar application rates up to 15 days. However, in loamy sand, the 1% amendment reduced nitrate levels after 30–60 days compared to the 0.5% amendment. With ACB, nitrate remained higher at 1% than 0.5% for longer periods in both soils. Phosphorus stayed elevated across all biochar treatments throughout the experiment. Biochar application markedly reduced NH_4^+-N in both soils compared to controls throughout the 60-day incubation. This effect is linked to carboxyl and ketone groups in biochar, which enhance NH_4^+ retention via hydrogen bonding and electrostatic forces (Figure 1). By improving soil cation exchange capacity, biochar also suppressed ammonia volatilization, thereby lowering cumulative NH_4^+-N losses in a dose-dependent manner [42].

In an acidic citrus orchard soil, pH was adjusted to five levels (4.0–7.2) with CaO. The soil was subsequently amended with 0 or 1% biochar [43]. Using 15N-tracing, the study showed that biochar generally increased gross mineralization (TM) and NH_4^+ immobilization (TI), except for TI at pH 4.0. It also promoted heterotrophic nitrification across a pH range of 4.0–7.4, while autotrophic nitrification occurred above a pH of 6.4. Overall, biochar improved nitrogen retention within pH 4.5–6.4, with stronger effects on TI than TM, and reduced autotrophic nitrification, suggesting enhanced N use efficiency and lower NO_3^--N losses in subtropical pomelo orchards (Figure 1). Soil pH strongly influenced how biochar affected nitrogen cycling in subtropical citrus soils. Biochar enhanced N retention mainly at pH 4.5–6.4, with greater effects on NH_4^+ immobilization than on mineralization, while suppressing autotrophic nitrification as a key mechanism for improved retention. This aligns with the findings of [41].

Ammonia nitrogen in farmland soils is easily lost due to environmental and human factors. Lu et al. [44] assessed how *Gracilaria lemaneiformis* biochar influences ammonium behavior in soils. Results showed that adding 10% biochar increased ammonium in the 0–10 cm layer by 1.57–2.36 times while reducing leaching losses by 44.83–72.27%. Adsorption and retention were mainly governed by electrostatic attraction and ion exchange, with additional effects from molecular interactions and system energy. Low-energy sites near $C_6H_{12}O_6$ favored ammonium binding, while structures like $C_5H_{10}O_5$, C-S-H, C-SO_3, and $C_4H_7NO_4$ contributed via physical or chemical adsorption mechanisms. The negative adsorption free energy (−394,590.84 kcal/mol) confirmed stable binding of ammonium to biochar surfaces.

Additionally, a study using pot experiments (farmland soil) reported that biochar improved soil N availability when applying corn straw biochar at rates of 0%, 0.5%, 1.0%, 2.0%, and 4.0%. It also evaluated soil nutrients and bacterial diversity [45]. Total and ammonium nitrogen rose by 4.7–32.3% and 8.3–101.5%, respectively, while microbial biomass nitrogen increased with higher biochar levels. This is primarily attributed to biochar's highly porous structure and large surface area, which enable it to adsorb and retain nitrogen, minimize leaching losses, and enhance soil nitrogen levels [46].

Overall, biochar significantly regulates soil P and N dynamics across diverse systems by enhancing nutrient retention, limiting leaching, and improving plant uptake. Biochar increases organic P while reducing inorganic and leachable forms. It also promotes nitrification and enhances ammonium retention through electrostatic and chemical interactions. Soil type, pH, and biochar feedstock strongly mediate these effects, with optimal N retention often occurring under moderately acidic conditions. Trampling and substrate properties can further influence nutrient shifts, while crop- and system-specific responses highlight both benefits and potential constraints at high application rates.

3.2. Carbon Dynamics

Biochar influences soil carbon dynamics by stabilizing organic matter, enhancing carbon sequestration, and moderating microbial activity. Biochar stabilizes organic matter through several mechanisms. Its highly aromatic and condensed carbon structure resists microbial decomposition, thereby slowing down organic carbon mineralization. Additionally, biochar provides microsites that physically protect labile organic matter within its porous matrix and through organo-mineral associations. The surface functional groups (e.g., carboxyl, hydroxyl) can also adsorb soluble organic compounds, reducing their accessibility to decomposers. These physicochemical interactions contribute to long-term carbon stabilization and enhanced soil organic carbon persistence. The details are presented in the following reviews.

Wu et al. [47] conducted an experiment that included a no-amendment control and a treatment with peanut shell biochar. After 70 days of incubation, peanut shell biochar application increased soil organic carbon by 44.8% and particulate organic carbon by 27.4% compared to the control. Mantel analysis and structural equation modeling revealed strong links among *cbbL*, *cbbM*, Bacteroidota, total organic carbon, and particulate organic carbon. The biochar treatment enriched soil carbon fractions, enhanced Bacteroidetes abundance, and reduced Acidobacteria richness, influencing key carbon cycle genes (*cbbL* and *cbbM*) (Figure 1). Overall, peanut shell biochar markedly improved soil carbon content and fertility.

Furthermore, Abdul-Aziz et al. [48] assessed how biochar type and application rate affect soil fertility and maize yield over two field seasons (2022–2023). Biochar materials from rice husk, groundnut husk, and sawdust were applied at 0–8 t ha^{-1}. Groundnut husk biochar proved to be the most effective, with 8 t ha^{-1} increasing grain yield by up to 218% in 2022 and 106% in 2023. Soil organic matter rose by 90–343%, and nitrogen availability by up to 220% (2022) and 70% (2023). Biochar's large surface area and reactive groups (carboxyl, hydroxyl) enable it to hold NH_4^+ and limit NO_3^- leaching. Maize yield strongly correlated with soil fertility indicators, except harvest index. Regarding the increase in soil organic matter, two main mechanisms may be involved: biochar directly adds stable carbon to the soil, and it suppresses the mineralization of soil organic matter by immobilizing microbes and adsorbing easily degradable organic substances.

Liu et al. [49] compared how biochar (1.0% *w*/*w*) with and without N addition influences the fate of photosynthetically fixed carbon in rice plant–soil systems using $^{13}CO_2$ pulse labeling at key growth stages. Biochar did not alter the distribution of ^{13}C among plant–soil components but increased its accumulation in shoots, roots, and soil, particularly when combined with N addition. During the growth period, biochar raised total ^{13}C in shoots, roots, and soil by 23%, 14%, and 20%, respectively, and improved nitrogen use efficiency by about 23% at the heading and ripening stages. This was likely due to two factors, namely its porous structure adsorbs labile compounds and enhances microbial carbon use efficiency, thereby stabilizing more photo-assimilated C [50], and greater root biomass under biochar plus N addition contributed additional ^{13}C inputs to the soil, even though allocation patterns of ^{13}C within the plant–soil system remained unchanged [51]. Overall, biochar promotes higher plant productivity and soil carbon sequestration through enhanced N use efficiency.

A field trial applied biochar at 0% (control), 1% (LB), 2% (MB), and 4% (HB) of topsoil dry weight before planting winter wheat [52]. Soil organic carbon and particulate organic carbon rose with higher biochar rates. After 8 months, readily oxidizable carbon was reduced but later returned to control levels. Biochar increased dissolved organic carbon without affecting water-soluble organic carbon, though both dissolved organic carbon and water-soluble organic carbon declined over time. Soil organic carbon rose with higher biochar rates due to its high stable carbon content (60–80%) [53]. While biochar did not affect water-soluble organic carbon, it was positively linked to dissolved organic carbon, likely from labile carbon release, leaching of dissolved organic matter, and selective adsorption of small dissolved organic molecules. Both water-soluble organic carbon and dissolved organic carbon declined over time, reflecting microbial carbon utilization [53].

Mesocosm experiments with *Festuca arundinacea* were conducted to investigate the impact of cornstalk biochar (100 g/kg soil) on soil organic carbon under simulated grazing conditions in two different soil types [54]. Biochar boosted soil organic carbon by 595% and mineral-associated organic carbon by 39%, mainly by stimulating microbes that promoted stable organo-metal complexes with Fe/Al in red soils and Ca in calcareous soils. Effects were stronger in red soils due to pH and mineral differences, while in calcareous soils, biochar also offered long-term protection against acidification. Biochar enriches soil carbon pools by directly adding carbon and stabilizing mineral-associated organic carbon through its high carbon content, aromatic structure, and resistance to decomposition [55]. It also raises soil pH, causing Fe and Al to precipitate, which allows high-molecular-weight organics to bind to these oxides and form stable Fe- or Ca-associated complexes. This process enhances the incorporation of plant inputs into stable soil organic carbon, particularly in acidic to neutral soils like red soils [56].

In Russia, an incubation study was conducted on three Podzol Antric soils differing in initial total organic carbon content for 90 days, using birch- and aspen-derived biochar (produced by fast pyrolysis at 550 °C) at application rates of 0.1% and 1.0%. The 0.1% dose showed no effect on soil organic matter mineralization, while 1.0% biochar increased it by 15–18%, accompanied by shifts in humic substance composition [57]. The higher proportion of aromatic humic acid fragments suggests greater stability. In humus-rich soils with abundant insoluble material, mineralization and humic acid formation occurred concurrently, likely due to enhanced transformation of resistant organic fractions and partial humification of the biochar itself.

Collectively, biochar enhances soil carbon dynamics by stabilizing organic matter, improving carbon sequestration, and boosting nutrient use efficiency. It increases soil organic and particulate carbon, stimulates carbon-fixing microbes, and reduces nitrogen losses. Higher application rates promote stable mineral-associated carbon and long-term C retention, especially in acidic and red soils. Even at moderate levels, biochar improves humic composition and aromatic carbon stability. Empirical evidence indicates that biochar carbon may persist for 100–1000 years depending on feedstock, production temperature, and environmental conditions [58], underscoring its potential as a long-term carbon sink and soil fertility enhancer. Overall, biochar enhances soil fertility and carbon storage through the activation of microbes and the chemical stabilization of soil.

3.3. Nutrient Dynamics in General

Numerous studies have examined the effects of biochar on various soil nutrients, including C, N, P, and minerals. For instance, peanut shell biochar (0–10%) was tested in acidified soils using experiments and simulations. The highest adsorption was for K^+ (820.38 mg/g), followed by PO_4^{3-} (270.51 mg/g), NO_3^- (235.65 mg/g), and NH_4^+ (130.93 mg/g) [59]. At 10% biochar, outlet volume decreased by 65.32%, nutrient leaching dropped by 48.40–68.28%, and surface soil nutrient retention increased by 437.80–913.87%. Nutrient concentrations declined with depth, confirming biochar's potential for improving nutrient efficiency in acidic soils.

Similarly, Lewoyehu et al. [60] conducted an incubation study in Ethiopia, testing water hyacinth biochar (WHB) against lime and fertilizer in acidic Nitisol. WHB slowed nitrification, lowered H^+ release, and improved N and P availability. Soil pH rose by 0.30–0.35 (1% WHB) and 0.72–0.86 (2% WHB), while exchangeable acidity dropped by 26.5–28.8% and 58.4–63%, respectively. The 2% WHB + fertilizer treatment further raised the pH by 0.71–0.90 and cut acidity by 49.8–64.7% compared to lime + fertilizer or fertilizer alone, showing WHB outperformed lime in resisting acidification. Biochar reduced soil acidity by retaining NH_4^+ and neutralizing Al^{3+}, H^+, and Fe^{3+} via its surface functional groups. The more porous, phenolic- and carboxyl-rich pyrolyzed WHB raised the soil pH more than the ground WHB [60]. Overall, WHB mitigated acidification by slowing nitrification and enhancing soil pH buffering capacity [61]. The improvement in soil P availability was attributed to higher soil pH dissolving Al^{3+}/Fe^{3+}-bound phosphates, functional groups on biochar releasing P bound to metal cations, and the direct release of soluble P from biochar.

Long-term effects were explored by Burgeon et al. [62], who studied century-old kiln charcoal (~200 years) as a proxy for historical biochar in three land covers. They compared soils with old biochar (CoBC), fresh oak biochar (YBC, 80 t ha^{-1}), and no biochar. Results showed that the age of biochar strongly influenced nutrient dynamics. YBC reduced NO_3^- and K^+ leaching but increased PO_4^{3-} in topsoil pore water, while CoBC raised subsoil K^+ and Mg^{2+}, enhanced total N, K^+, and Ca^{2+}, but decreased available P. YBC reduced NO_3^- leaching via ion diffusion into its porous structure and temporarily raised topsoil PO_4^{3-}, though alkaline pH limited its availability. In contrast, CoBC immobilized P through cationic bridging on oxidized surfaces but improved Ca^{2+} availability through functional groups. While yields were unaffected, biochar reduced N, K, and Ca uptake but increased Mg uptake, showing that YBC affects short-term pore water dynamics, whereas CoBC drives long-term nutrient cycling [62].

Long-term field experiments have further highlighted biochar's role in crop systems. In a seven-year rice study in northwest China, annual biochar application (9 t ha^{-1} yr^{-1}) improved shoot and root growth, enhanced root traits, and delayed root senescence during grain filling [63]. Biochar alone increased root biomass and length, while its combination with N fertilizer altered root morphology and growth patterns. Beyond its effects on plants, biochar reduced soil bulk density by 10–12%, increased total organic carbon, and raised available N, P, and K, underscoring its contributions to soil quality and fertility.

Additionally, dos Santos et al. [64] examined how biochar materials from sugarcane bagasse (SB), orange bagasse (OB), and corncobs (CB) affect ion leaching in clay soil, with untreated soil (CT) and gypsum-amended soil (CTG) as controls. Biochar raised water infiltration and flow by 80% and 71%, respectively. SB and CB enhanced ion removal by 48%, while OB achieved 19%, likely due to its lower surface area and water retention. Biochar reduced exchangeable Ca^{2+} and Mg^{2+} by 44% and 76% but increased K^+ by 233%. Notably, SB and CB cut exchangeable Na^+ by 90% and OB by 50%, indicating strong potential for improving soil quality and reclaiming salt-affected soils [64]. Biochar improved anion removal from soil, but the effect differed by type, likely due to variations in elemental composition, high electronegativity that repels similar ions, and enhanced soil water distribution and retention.

Nutrient release from animal waste biochar has also been studied. Hadroug et al. [65] tested poultry manure biochar (RPM-B, 600 °C) under static and dynamic conditions. In a 10-day leaching test, P and K were gradually but substantially released (93.6 and 17.1 mg g^{-1}, about 95% and 43% of initial contents), outperforming many other biochar materials. Column leaching showed RPM-B (5–8% w/w) retained NO_3^--N and NH_4^+-N in alkaline sandy soil but promoted leaching of PO_4^{3-}-P, K^+, Mg^{2+}, and Ca^{2+} at low, sustained rates (<10 mg kg^{-1} d^{-1} over 40 days) [65]. This indicates that RPM-B could be a sustainable amendment for nutrient-poor semi-arid soils compared with synthetic fertilizers.

A six-week pot experiment assessed how wood biochar influences boreal soil chemistry and spring barley growth in sandy loam Umbric Podzol and clay loam Umbric Retisol [66]. Using two biochar fractions (3–5 mm and ≤2 mm) at 5% (w/w), soil pH, available P and K, and mineral N were measured, along with barley growth and protein content. In the Umbric Podzol, biochar raised pH by 0.5 units and increased K (17 mg/kg), P (7 mg/kg), and nitrate (4 mg/kg) levels, promoting plant growth, biomass yield, and protein accumulation, especially with fine biochar (3–5 mm) plus NPK fertilizer. In contrast, the same treatments in the Umbric Retisol slowed barley growth and lowered nitrate content by 4–7 mg/kg, likely due to enhanced denitrification.

In summary, biochar significantly improves soil nutrient dynamics by enhancing nutrient retention, reducing leaching, and alleviating acidity. Its porous structure and functional groups adsorb essential ions, buffer soil pH, and increase the availability of N, P, and K across diverse soil types. Long-term applications further stabilize nutrient cycling and improve root growth, soil structure, and fertility. Performance varies with feedstock and soil conditions, with materials like sugarcane bagasse and poultry manure biochar showing strong potential for nutrient recovery and soil reclamation. Overall, biochar enhances soil productivity by improving nutrient efficiency and providing long-term chemical stabilization. A summary of the effects of biochar on P, N, C, and nutrient dynamics, in general, is presented in Table 1.

Table 1. Impacts of biochar on P, N, C, and general nutrient dynamics.

Reference	Location/System	Biochar Treatment	Main Findings	Implications
Effects on P and N dynamics				
[38]	Field study on grassland, Montana	Wood biochar (0, 20, 40 t ha^{-1}) with/without trampling	• Trampling compacted soils, reduced N availability • Biochar improved infiltration (no trampling: 70.5% at 20 t ha^{-1} and 26.5% at 40 t ha^{-1}), nitrification, and nutrient incorporation • Increased organic P, reduced inorganic P. • Trampling enhanced P sorption (lower $CaCl_2$-P, citrate-P)	Biochar offsets negative trampling effects, enhances P availability in plants, and stabilizes nutrients through deeper incorporation
[39]	Long-term (7 yr) green roof trial, Finland (plantings & mats)	Biochar in substrates	• Reduced P leaching consistently • Reduced N leaching only in plantings • Fertilization after 6 years caused high P leaching not mitigated by biochar • P dynamics governed by substrate chemistry/pH • N regulated by biological processes	Biochar enhances long-term P retention, but effectiveness depends on fertilization and substrate properties
[40]	Karelia, Russia–Sandy Umbric Podzol & heavy loamy Umbric Retisol	Wood biochar ($\leq$2 mm and 3–5 mm) applied at 2% and 5% (*w*/*w*); with and without NPK fertilizer	• Fine biochar ($\leq$2 mm, 5%) increased soil pH and available P (by 20–40%), and enhanced Ca-, Al-, and loosely bound P and phosphatase activity in Umbric Podzol • Minor effects in Retisol; notable P_2O_5 gains only with biochar–fertilizer combination	Fine biochar at higher application rates improves P availability and soil enzymatic activity in low-P sandy soils but is less effective in clayey soils

Table 1. *Cont.*

Reference	Location/System	Biochar Treatment	Main Findings	Implications
[41]	Incubation in clay loam & loamy sand	Rice straw (RSB) & acacia wood biochar (ACB), 0–1%	• NH_4^+ declined faster with biochar • RSB raised NO_3^- more than ACB • Effects varied by soil type and rate • Biochar reduced NH_4^+ losses via retention (H-bonding, electrostatic)	Feedstock and soil type strongly affect N cycling; biochar improves NH_4^+ retention and reduces volatilization
[43]	Subtropical pomelo orchards with acidic soil	0–1% biochar at different pH (4.0–7.2)	• Biochar increased mineralization & NH_4^+ immobilization (except TI at pH 4.0) • Enhanced heterotrophic nitrification (pH 4–7.4) • Suppressed autotrophic nitrification (pH > 6.4)	Biochar improves N retention mainly at pH 4.5–6.4, enhancing N use efficiency and lowering NO_3^- losses
[44]	Farmland soil	10% amendment with *Gracilaria lemaneiformis* biochar	• Increased NH_4^+ by 1.57–2.36× in 0–10 cm layer • Reduced leaching losses by 45–72% • Retention via electrostatic attraction, ion exchange, stable molecular binding	Strong potential to reduce N losses and stabilize ammonium in soils
[45]	Farmland soil (pot experiment)	Corn straw biochar (0–4% peat)	• Total and ammonium nitrogen rose by 4.7–32.3% and 8.3–101.5% • Microbial biomass nitrogen increased with higher biochar levels	Biochar enhanced soil nitrogen availability and crop uptake by increasing ammonium and nitrate levels. Low doses suppressed ammonia oxidation, while higher application rates improved N use efficiency by stimulating nitrification and denitrification
Effects on C dynamics				
[47]	Peanut shell biochar (treatment vs. no-amendment control)	70-day incubation experiment	• Soil organic carbon ↑ 44.8% and particulate organic carbon ↑ 27.4% compared to control • Enhanced Bacteroidetes abundance and reduced Acidobacteria richness • Strong associations among *cbbL*, *cbbM*, Bacteroidota, total organic carbon, and particulate organic carbon	Biochar enriched soil carbon fractions, stimulated beneficial microbes, and improved soil fertility

Table 1. *Cont.*

Reference	Location/System	Biochar Treatment	Main Findings	Implications
[48]	Field maize trials, two seasons (2022–2023)	Biochar from different feedstocks (rice husk, groundnut husk, sawdust) at rates up to 8 t/ha	• Groundnut husk biochar highest effect • Soil organic matter increased by ~90–343% • Nitrogen availability also increased • Maize yield strongly correlated with increased soil fertility indicators • Biochar's reactive groups helped hold NH_4^+ and limit NO_3^- leaching	Biochar can significantly enhance soil C pools and fertility in real field settings; feedstock matters for maximizing effect
[49]	Rice plant–soil systems; $^{13}CO_2$ pulse labeling	1.0% biochar with/without N fertilization	• Biochar + N enhanced accumulation of newly fixed C in shoots (↑23%), roots (↑14%), and soil (↑20%) • Improved nitrogen use efficiency (~23%) at heading and ripening stages • Porous structure adsorbs labile compounds; more root biomass leads to more carbon inputs	Biochar + N fertilization can increase carbon retention from recent photosynthesis, boosting both plant growth and soil C storage
[52]	Field winter wheat trial, topsoil amendment	Biochar at control, low (LB-~1%), medium (MB-~2%), high (HB-~4%) rates (dry weight basis)	• Soil organic carbon & particulate organic carbon increased with higher biochar rate • Readily oxidizable carbon initially reduced, but later returned to control level • Dissolved organic carbon increased; water-soluble organic carbon not strongly changed • Over time, both dissolved organic carbon and water-soluble organic carbon declined (microbial use)	Higher rates of biochar build more stable carbon pools; labile C dynamics shift over time (initial flush then decline)
[54]	Mesocosm experiments with *Festuca arundinacea* planted on red soils and calcareous soils; simulated grazing	Cornstalk biochar at 100 g biochar per kg soil	• Huge boost in soil organic carbon (~+595%) and mineral-associated organic carbon (~+39%) in red soils • Stronger effects in red (acidic) soils than in calcareous soils • Biochar helped stabilize organic matter via organo-metal complexes (with Fe, Al in red soils; with Ca in calcareous soils) • Also mitigated soil acidification in calcareous soils	Biochar is especially effective in acidic soils for stabilizing carbon; mineral interactions are key (Fe/Al/Ca binding)

Table 1. *Cont.*

Reference	Location/System	Biochar Treatment	Main Findings	Implications
[57]	Northwestern Russia–Three Podzol Antric soils with varying organic C	Birch and aspen biochar (fast pyrolysis, 550 °C) at 0.1% and 1.0% (*w*/*w*); 90-day incubation	• 1.0% biochar increased soil organic matter mineralization by 15–18% and altered humic substances • Higher aromatic humic acid fraction indicated greater stability • In humus-rich soils, mineralization and humic acid formation occurred simultaneously	Moderate biochar addition stimulates organic matter turnover and humic substance formation, enhancing soil carbon dynamics
Nutrient dynamics in general				
[59]	Acidified soils (experiments + simulations)	Peanut shell biochar (0–10%)	• High adsorption: K^+ (820.38 mg/g) > PO_4^{3-} (270.51 mg/g) > NO_3^- (235.65 mg/g) > NH_4^+ (130.93 mg/g) • At 10% biochar: outlet volume ↓65.32%, nutrient leaching ↓48.40–68.28%, retention ↑437.80–913.87%	Improves nutrient efficiency and reduces leaching in acid soils
[60]	Ethiopia, acidic Nitisol (incubation study)	WHB (1–2%), with lime/fertilizer comparisons	• Slowed nitrification, lowered H^+ release, improved N and P availability • pH ↑0.30–0.35 (1% WHB) and ↑0.72–0.86 (2% WHB) • Exchangeable acidity ↓26.5–28.8% (1%) and ↓58.4–63% (2%) • WHB + fertilizer outperformed lime + fertilizer • Functional groups neutralized Al^{3+}/H^+/Fe^{3+}	WHB is more effective than lime in resisting acidification, enhancing soil buffering and nutrient availability
[62]	Historical kiln sites, various land covers	Century-old charcoal (CoBC), fresh oak biochar (YBC, 80 t/ha), no biochar	• YBC: reduced NO_3^- and K^+ leaching, increased PO_4^{3-} in topsoil pore water • CoBC: raised subsoil K^+, Mg^{2+}, total N, Ca^{2+}, but reduced P • YBC improved short-term nutrient retention; CoBC influenced long-term cycling	Biochar age shapes nutrient dynamics: YBC affects short-term processes, CoBC sustains long-term soil fertility
[63]	7-year rice field	Annual biochar (9 t/ha/yr) with/without N fertilizer	• Improved root biomass, morphology, delayed senescence, boosted shoot/root growth • Soil bulk density ↓10–12%, total organic carbon and available N, P, K ↑	Enhances soil fertility, crop growth, and long-term soil quality

Table 1. *Cont.*

Reference	Location/System	Biochar Treatment	Main Findings	Implications
[64]	Clay soil (Brazil), with gypsum control	Biochar from sugarcane bagasse (SB), orange bagasse (OB), corncobs (CB)	• Biochar ↑ infiltration (80%) and flow (71%) • SB & CB enhanced ion removal (48%), OB less effective (19%) • Reduced exchangeable Ca^{2+}, Mg^{2+}; ↑K^+ (233%) • Strong Na^+ reduction: SB & CB (90%), OB (50%)	Effective for salt-affected soil remediation and improving soil–water distribution; performance varies by biochar type
[65]	Alkaline sandy soil	Poultry manure biochar (RPM-B, 600 °C, 5–8% *w*/*w*)	• Released P (93.6 mg/g) and K (17.1 mg/g) gradually (10-day test) • Retained NO_3^--N, NH_4^+-N, but promoted slow leaching of PO_4^{3-}-P, K^+, Mg^{2+}, Ca^{2+} at <10 mg/kg/day (40 days)	Sustainable alternative to synthetic fertilizers for nutrient-poor semi-arid soils
[66]	Karelia, Russia–Boreal soils (Sandy loam Umbric Podzol; clay loam Umbric Retisol)	Wood biochar (≤2 mm and 3–5 mm) at 5% (*w*/*w*); 6-week pot study with spring barley	• In Umbric Podzol, biochar increased pH (+0.5), K (+17 mg/kg), P (+7 mg/kg), NO_3^- (+4 mg/kg), and barley growth and protein yield • In Retisol, growth slowed and NO_3^- decreased by 4–7 mg/kg, likely from denitrification	Biochar improves fertility and crop performance in acidic sandy soils but may hinder N dynamics in dense, clayey systems.

Note: The reported changes are statistically significant ($p < 0.05$) in the studies reviewed, unless otherwise specified with words like 'slight' and 'not strongly'; ↑ represents increase and ↓ represents decrease.

3.4. Nutrient Dynamics in Fertilized Soils

A two-year field experiment evaluated how rice-straw biochar influences the nutrient supply of organic fertilizer (OF) under rice–rapeseed rotation [67]. Results showed that using OF alone reduced rice yields by 2% and rapeseed yields by 6–10% compared with chemical fertilizer (CF). However, applying biochar at 15 t ha^{-1} with OF (OF + B15) boosted rice yields by 10–17% and rapeseed yields by 20–25% relative to OF, and still outperformed CF with yield increases of 7–14% for rice and 12–13% for rapeseed. The OF + B15 treatment also improved soil quality: during rice and rapeseed seasons, it raised soil organic carbon by 57–90% and enhanced enzyme activities, i.e., catalase (14–19%), invertase (14–20%), urease (17–24%), and phosphatase (13–17%) [67]. Biochar improved the nutrient-supplying ability of organic manure by raising soil pH and organic carbon. This pH increase was linked to the high ash content (19.43%) and alkalinity (pH 9.42) of the biochar, which released cations that neutralized soil acidity [68]. Overall, biochar addition enhanced soil properties, enzymatic activity, and nutrient availability from OF.

Lu et al. [69] tested how biochar type (fine pine wood powder, 0.1 mm, vs. coarse fescue grass, 0.5–2 mm) and application method (surface vs. incorporated at 1% *w*/*w*) influenced N and P leaching from soils with inorganic fertilizer or biosolids. Soil columns were leached for six months, yielding ~16 pore volumes of drainage. Biochar showed limited retention of ammonium (NH_4^+) and no sorption of nitrate (NO_3^-). However, incorporating powder biochar reduced cumulative leaching of NH_4-N, NO_3-N, and total N by about 25% compared with controls. Although biochar's negatively charged surface was expected to strongly adsorb NH_4^+-N [70], powder biochar reduced NH_4^+–N in leachate during only 6 of 16 events, mainly in biosolids and control soils, with effects declining after day 50. Still, incorporating powder biochar lowered cumulative NH_4^+-N, NO_3^--N, and total N losses, especially in the early leaching phase (Days 1–36) when N release was greatest. Incorporation proved more effective than surface application, likely due to greater contact between biochar and soluble N [69].

In a greenhouse pot trial, Llovet et al. [71] compared three commercial organo-mineral fertilizers with their biochar-amended versions (NPK + B, NP + B, K + B) using barley as the test crop. Results showed that NPK + B slowed nutrient leaching, particularly nitrate (−28%) and potassium (−22%) in the first two weeks, though this effect reversed after week three, likely due to microbial nutrient immobilization supported by biochar. NPK + B also boosted barley straw biomass by 23.4% compared to NPK, while all biochar-based fertilizers improved plant nutrient content and export (K, S, Ca, Mn), suggesting biochar contributed as an additional nutrient source.

Furthermore, a lab incubation study tested biochar from spent mushroom substrate (2% *w*/*w*) and inorganic P addition (62 kg ha^{-1}) on soil P availability, pH, acid phosphatase activity, and bacterial abundance in acidic paddy and red soils from southern China at 15 °C and 25 °C [72]. Both amendments significantly raised soil pH by up to ~39–41% in paddy soil and ~43–45% in red soil. Inorganic P addition produced the highest available P, reaching 111.47 mg/kg at 15 °C and 100.17 mg/kg at 25 °C in paddy soil. The biochar contained labile P that directly released phosphate into the soil. Along with inorganic P, it reduced adsorption and complex formation, mobilizing native soil P beyond critical levels. Changes in soil pH, porosity, and microbial activity further enhanced available P.

Together, these studies demonstrate that biochar can significantly enhance fertilizer efficiency by improving soil fertility, reducing nutrient losses, and increasing crop yields. Its effects vary with feedstock, application method, and soil type, but overall, biochar strengthens nutrient retention, mobilizes otherwise unavailable pools, and supports long-term soil health, making it a promising amendment for sustainable crop production.

3.5. Effects of Nutrient-Enriched Biochar on Nutrient Dynamics

Sewage sludge biochar generally contains little K because much of it is lost with water during treatment. Supplementing sewage sludge biochar with mineral K can produce organo-mineral fertilizers that release K slowly, but their release patterns in different soils are not well understood. Ndoung et al. [73] examined K release from biochar-based fertilizers in pellet and granule forms in clayey soil, sandy soil, and silica over 60 days at an application rate of 4 Mg ha^{-1}. Compared to KCl, biochar-based fertilizers released K more slowly, with pellets showing the slowest release. K availability was greatest in clay soil, followed by sandy soil and silica. K release patterns from biochar-based fertilizers and KCl were similar across soils and silica, and particle size had little effect. However, release rates differed with soil moisture at 10 kPa tension: clay soil showed the fastest K release, followed by sandy soil and silica. The higher water content in clay likely enhanced fertilizer dissolution by maintaining closer contact with soil water [73]. Overall, enriched biochar-based fertilizers act as slow-release fertilizers, suggesting potential for improved plant K uptake during growth.

Yin et al. [74] assessed the impacts of N-enriched biochar (4 and 8 t ha^{-1} y^{-1}) on C emissions, storage, rice yield, and Fe dynamics in subtropical paddy fields. Both application rates raised C emissions, but 4 t ha^{-1} significantly improved yields. Field and culture experiments showed increased soil Fe^{2+} with biochar, which correlated positively with CO_2, CH_4, total C, and dissolved organic carbon, but negatively with Fe^{3+}. In culture experiments without plants, biochar reduced cumulative CO_2 and CH_4 emissions. Both N-enriched biochar rates increased soil Fe^{2+}, likely due to higher pH, lower redox potential, and enhanced Fe^{3+} reduction. Biochar's porous structure and surface area also favor Fe-reducing microbes [75,76]. Additionally, biochar may serve as an electron donor or shuttle for Fe^{3+} reduction, roles considered more important than its potential function as an electron acceptor for Fe^{2+} oxidation [77]. Overall, moderate application (4 t ha^{-1}) enhanced yield, biomass, and dissolved organic carbon while helping regulate C emissions through effects on Fe dynamics.

Javeed et al. [78] tested the application of date palm biochar (at 2%) enriched with buffalo slurry (SEB) on soil C–N dynamics, nutrient leaching, and wheat growth. Enrichment raised biochar C and N contents by up to 310% and 286%, respectively. Both standard and enriched biochar reduced nitrification (81–94%), ammonification (48–74%), and CO_2 emissions (50–92%) compared to the control. SEB was especially effective, cutting leaching losses of C, N, P, K, Na, Ca, and Mg by 9–125%. Dissolved organic carbon supplied electrons for denitrification, releasing CO_2 as oxygen was consumed. In SEB treatments, higher water-filled pore space and organic carbon created anaerobic conditions that accelerated denitrification while reducing microbial respiration [78].

Additionally, a study developed two enriched biochar (EB) fertilizers with a 6–6–4 N-P_2O_5-K_2O grade: EB-1 (NPK intercalated into biochar) and EB-2 (EB-1 supplemented with humic acid and seaweed extract) [79]. Nutrient release tests showed EB fertilizers slowed the release of NH_4^+, P, and K^+ compared to conventional fertilizers, while NO_3^- release remained similar. Although total N release was comparable, EB fertilizers released considerably less P and K within 36 h. In field trials with direct-seeded rice, EB fertilizers improved apparent nutrient use efficiency, achieving 11.7–29.5% higher N, 32.9–64.0% higher P, and 31.4–38.0% higher K efficiency than conventional fertilizers at full application rates. These results highlight the potential of EB fertilizers for sustainable rice production by reducing nutrient losses and enhancing nutrient use efficiency.

Enriched biochar materials demonstrate strong potential as sustainable slow-release fertilizers that enhance nutrient efficiency and soil health. Sewage sludge- and organo-mineral-based biochar materials released K gradually, particularly in clay soils, improving nutrient availability over time. N-enriched biochar optimized rice yield and carbon regulation by promoting Fe reduction and moderating greenhouse gas emissions. Similarly, slurry- and nutrient-enriched biochar materials improved C–N cycling, reduced nutrient leaching, and increased fertilizer use efficiency in cereals. Overall, moderate application of enriched biochar supports higher productivity while minimizing nutrient losses and environmental impacts.

3.6. *Implications*

Biochar plays a multifaceted role in soil nutrient and carbon dynamics by combining physical, chemical, and biological mechanisms. Its porous structure and stable carbon backbone enhance soil organic carbon storage while also influencing CO_2 and CH_4 fluxes through changes in redox potential and microbial respiration [54,78,80]. Acting as an electron shuttle or donor, biochar supports Fe-reducing microbes, which further regulate carbon turnover and greenhouse gas emissions [74,75].

Nutrient cycling is similarly affected, as biochar improves N and P retention while reducing losses through leaching, runoff, and volatilization [38,39,41]. Functional groups such as carboxyl and ketone bond with NH_4^+, while its high cation exchange capacity stabilizes mineral N [42]. Phosphorus is retained in more stable pools but is also made more plant-available through enhanced phosphatase activity and P-solubilizing microbes [38,72]. These processes increase nutrient-use efficiency and promote sustained plant uptake.

The magnitude and direction of these effects depend on soil type, feedstock, and environmental context. Acidic soils show greater ammonium immobilization and suppress nitrification, while clay soils enhance K and P availability through better water–fertilizer interactions [41,73]. Feedstocks such as rice straw enriched with Mg increase P solubility, whereas sewage sludge biochar requires mineral supplementation to supply K [73,81]. Moisture, pH buffering, and redox conditions further shape how biochar interacts with soil nutrients and microbial communities.

Mechanistically, biochar mediates nutrient dynamics not only through sorption and ion exchange but also by reshaping microbial networks. Its porous surface provides habitats for microbes, promoting groups such as Proteobacteria and Actinobacteria that drive organic matter decomposition, nitrification, and P solubilization [81]. In this way, biochar integrates physical stabilization with microbial mediation, supporting both soil fertility and environmental sustainability. Overall, its potential as a tool for nutrient management and climate mitigation is clear, though context-specific application strategies remain essential.

A zonal comparison of soils reveals that biochar's influence on nutrient dynamics varies strongly with soil type, texture, and pH. In sandy Umbric Podzols, biochar (especially fine fractions $\leq$ 2 mm at 5%) significantly elevated soil pH, available P, and phosphatase activity, enhancing plant P uptake and growth [40]. In contrast, in heavy loamy Umbric Retisols, biochar produced smaller improvements in P availability and negligible effects on enzyme activity, with major benefits only when combined with fertilizer [40]. Similarly, studies on clay loam and loamy sand soils showed that biochar improved N retention and reduced NH_4^+ losses more effectively in fine-textured soils, while in coarse soils, the benefits declined over time due to lower cation exchange capacity [66]. In acidic soils, biochar buffered pH and enhanced N and P availability, whereas in neutral to calcareous soils, its main role was improving C stabilization and reducing nutrient leaching [43,54,72,82]. Collectively, these zonal variations demonstrate that biochar's efficiency depends on the inherent physicochemical properties of soils, where fine-textured and moderately acidic soils show the most consistent gains in nutrient retention, biological activity, and fertility improvement.

4. Impacts of Biochar on Microbial Activity

Biochar has gained significant attention as a soil amendment due to its potential to enhance soil fertility, sequester carbon, and mitigate environmental pollution. Beyond its physical and chemical contributions to soil systems, biochar exerts profound effects on soil microbial communities, which are central to nutrient cycling, organic matter decomposition, and overall ecosystem functioning. By modifying soil properties such as pH, porosity, moisture retention, and nutrient availability, biochar can create favorable or, in some cases, inhibitory conditions for microbial growth and activity. These shifts in microbial dynamics not only influence soil health and productivity but also determine the efficacy of biochar in delivering long-term agronomic and environmental benefits.

4.1. Recent Findings on the Influence of Biochar on Microbial Activity

Zhang et al. [83] examined the impacts of straw-based carbon substrate (100 kg), a new form of biochar, on soil fertility, microbial communities, and maize yield compared with traditional straw amendments. Both crop stage and treatment significantly shaped bacterial and fungal diversity. Key taxa, including Rhizobiales, Saccharimonadales, and Eurotiales, responded positively to the biochar substrate by enhancing nutrient cycling. Rhizobiales abundance was linked to higher soil nitrogen during early maize growth, while Saccharimonadales and Eurotiales correlated with elevated carbon and phosphorus at later stages (Figure 2). These findings indicate that straw-based carbon substrate fosters stage-specific recruitment of beneficial microbes, thereby improving nutrient supply and boosting crop productivity.

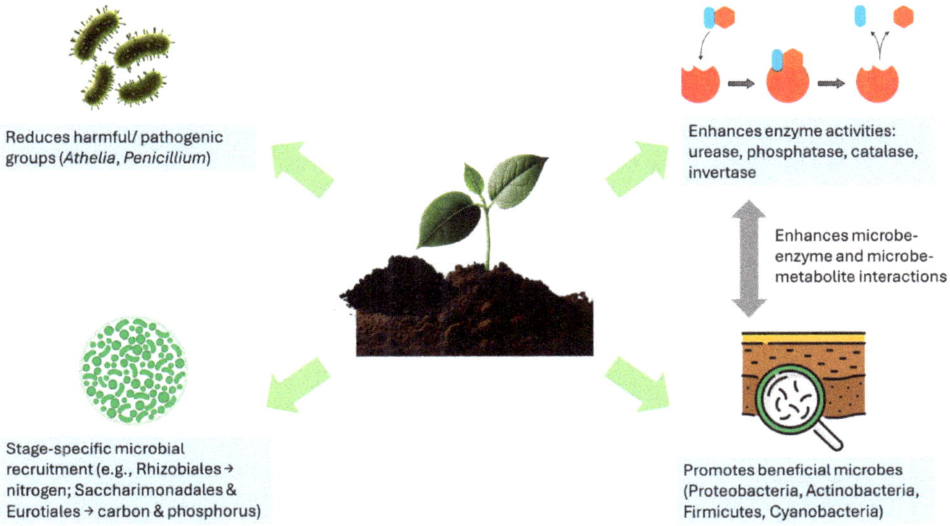

Figure 2. Impacts of biochar on soil microbial dynamics.

A two-year pot experiment (2020–2021) investigated how biochar (0–60 t ha^{-1}) and nitrogen fertilizer (0–450 kg ha^{-1}) influence soil properties, enzyme activity, nitrogen use efficiency, and rapeseed yield [84]. Compared with the control, biochar with high nitrogen markedly enhanced soil enzymes: B30 + N450 increased urease by 73–75%, B60 + N450 raised catalase by 16–17%, and B60 + N225 improved alkaline phosphatase by 17–19% (Figure 2). Although excessive nitrogen alone reduced plant nitrogen metabolism enzymes, adding 30 t ha^{-1} biochar alleviated this stress and stimulated nitrate reductase, nitrite reductase, glutamate synthase, and glutamine synthetase at both seedling and flowering stages. The B30 + N450 treatment also increased plant height (11–18%), pods per plant (32–39%), and grain yield (54–64%), while B15 + N225 achieved the highest nitrogen use efficiency (29%) [84]. These results demonstrate that biochar enhances enzyme activity, offsets nitrogen stress, and improves nutrient efficiency and rapeseed productivity. Biochar addition enhanced key soil enzymes by improving nutrient cycling and microbial activity. Urease activity rose due to greater nitrogen mineralization and retention, while higher phosphatase activity was linked to biochar's porous structure, ash content, and organic carbon, which increased soil pH and moisture [85]. Catalase activity improved as biochar boosted microbial oxidative metabolism through better soil aeration and moisture, and invertase activity increased with higher soil organic carbon and nutrient availability. However, excessive biochar (60 t ha^{-1}) reduced urease and invertase, likely due to elevated pH and adsorption effects [86].

Application of an enriched rice straw biochar at 4.24 g per treatment altered soil microbial communities by promoting beneficial groups like Proteobacteria and Actinobacteria, which support phosphorus solubilization and organic matter decomposition, while reducing taxa such as Methylomirabilota and Desulfobacterota (Figure 2) [81]. Biochar treatments increased the number of unique microbial operational taxonomic units, indicating community shifts that support nutrient cycling, plant–microbe interactions, and overall soil health. Biochar-based phosphate fertilizers further enriched rhizosphere diversity, enhancing nutrient solubilization, organic matter breakdown, and pathogen suppression [17].

In a seven-year maize-field trial testing biochar applications (0–126 Mg ha^{-1}) and their effects on root traits and soil microbes, biochar was found to significantly enhance root length, surface area, volume, and biomass. It also increased bacterial diversity and altered community structure, though fungal communities showed little change [87]. Root development was strongly associated with bacterial, rather than fungal, communities, with microbial network analysis linking key bacterial taxa to root growth parameters. Generally, fungi can utilize the stable carbon in biochar more efficiently than bacteria, colonizing its pores and exploiting available resources [88]. However, in this study, the limited fungal response may be due to the long interval since application (7 years), the low proportion of stable carbon, and the gradual collapse of biochar pores over time. Consequently, biochar had a stronger influence on bacterial communities than on fungal ones. Biochar application (15.75–126 Mg ha^{-1}) was observed to increase Firmicutes while reducing Actinobacteria and Acidobacteria in rhizosphere soils. Biochar can reshape rhizosphere microbial communities by stimulating root exudation and supplying nutrients and energy that support microbial activity, potentially favoring Firmicutes [89].

A study separating corn-straw biochar into aqueous extractable (AE), organic extractable (OE), and residual solid fractions (EBC) and incubating it with purple soil for 30 days revealed that biochar increased available P, K, and organic C but lowered pH and available N [90]. Sequencing showed fungal richness and diversity increased, while bacterial richness remained unchanged. Community shifts differed by group: AE mainly shaped bacterial composition through nutrient inputs, whereas EBC influenced fungal communities via its structure and stable compounds. The limited impact of biochar on fungi observed by Yan et al. [87] may be due to the seven-year gap since application, the low proportion of stable carbon in the biochar, and the gradual collapse of its pore structure over time.

In apple root zone soils, apple wood biochar (1% *w*/*w*) was found to reshape bacterial, fungal, and archaeal communities, raising bacterial α-diversity in the rhizoplane but lowering fungal and archaeal diversity in the rhizosphere [91]. It boosted nitrogen-fixing microbes and N_2O-reducing activity in root-associated soils, cutting N_2O emissions by up to 35%, while showing little effect on *nifH* and *amoA* genes in non-rhizosphere soils. Network analysis further linked nitrogen-cycling genes to diverse microbial groups.

Shang et al. [92] examined enzyme activity, microbial growth, and bacterial community dynamics in maize rhizospheres following straw and straw-derived biochar addition. Straw returned to soil boosted microbial biomass and specific growth rates (1.2–1.6 and 1.7–2.0 times higher, respectively), favoring fast-growing Firmicutes, which carried more β-1,4-glucosidase (BG) and β-N-acetylglucosaminidase (NAG) genes. Correspondingly, BG activity and affinity rose 2.2- and 1.8-fold, while NAG rose 4.0- and 2.0-fold. By contrast, biochar favored slower-growing Actinobacteria (K-strategists), resulting in weaker enzymatic responses but enhanced root biomass (+31%) and expanded rhizosphere hotspots for BG (+26%) and NAG (+47%). Biochar also increased network stability, whereas straw mainly accelerated nutrient cycling by promoting rapid microbial proliferation and enzyme production. Biochar enhanced nutrient mobilization in the rhizosphere by enlarging hotspot areas, allowing nutrient release from a broader soil volume [92].

Akumuntu et al. [93] assessed the impact of rice husk biochar (0–1.5%) on soil properties, lettuce growth, and earthworm toxicity. Rice husk biochar did not significantly alter soil pH but enhanced enzyme activities (dehydrogenase, alkaline phosphatase, and beta-glucosidase) after 28 days and increased beneficial growth-promoting microbes, such as *Massilia*, *Bacillus*, and *Trichocladium* (Figure 2). Lettuce root weight and leaf number rose with higher rice husk biochar levels, with growth improvements linked to enhanced nutrient availability driven by microbial and enzymatic shifts. Liu et al. [94] reported that the porous nature of biochar supports microbial diversity and protects microbes against stresses such as freeze–thaw events. Consequently, biochar contributes to creating favorable habitats for soil microorganisms and promotes the stability of microbial networks.

A six-year field study (2013–2018) examined how different biochar application rates (0, 15.75, 31.5, and 47.25 t ha^{-1}) with fertilizers influence soil fertility, organic N fractions, and microbial communities [95]. Biochar raised soil organic matter, C/N ratio, and pH, and significantly altered acid-hydrolyzable N and its fractions. Amino acid N was associated with enriched Verrucomicrobia, Cyanobacteria, Glucosylbacteria, and Nitrospira; ammonium N with Verrucomicrobia, Bacteroidetes, Gemmatimonadetes, Acidobacteria, and Proteobacteria; amino sugar N with Acidobacteria; and unknown N with Gemmatimonadetes (Figure 2). These shifts, along with changes in soil enzyme activities, indicate that biochar drives N cycling through specific microbial–enzyme interactions.

Moreover, Yin et al. [96] investigated the impact of rice and corn straw biochar materials on microbial communities in the rhizosphere of rice grown in albic soil. In a pot experiment, the application of biochar increased soil nutrient levels and enhanced bacterial and fungal abundance and diversity. Compared with the control, rice straw biochar increased the bacterial abundance-based coverage estimate (ACE) by 3.1%, and the fungal ACE and Chao indices by 7.9% and 14.2%, respectively. Both biochar materials increased the operational taxonomic units. Biochar also strengthened bacteria–fungi interactions, with shifts in community structure closely tied to changes in soil nutrients. Overall, rice straw biochar had the strongest positive effect on microbial diversity and soil quality. Biochar boosts microbial abundance by supplying labile C and minor nutrients, while in nutrient-poor, compact albic soil, it also alters community composition indirectly through improvements in pH, structure, and related properties. Biochar raised the relative abundance of Proteobacteria, Verrucomicrobia, and Bacteroidetes. Proteobacteria, the dominant phylum, increased by 13.95% with rice straw biochar and 9.94% with corn straw biochar compared to the control. Their enrichment reflects their eutrophic nature, as biochar improves nutrient conditions in albic soils, consistent with findings of Di et al. [95].

A three-month field trial tested peanut shell biochar (0, 5, and 10% *v*/*v*) in landfill cover soil to assess its impacts on plants and microbes [97]. Biochar raised soil pH, nutrients, and organic matter up to two times, boosting plant growth. It also increased bacterial richness, diversity, and community heterogeneity. However, plant diversity declined, and dominant species shifted. Specifically, it promoted *Cyperus rotundus* L. dominance, likely through its allelopathic traits in nutrient-rich conditions. Biochar significantly increased bacterial richness and diversity at all levels, while only 5% biochar boosted fungal richness and diversity. Functional genes linked to amino acid metabolism were abundant, likely from root growth, and biochar also raised genes for cofactor and vitamin metabolism. Improved soil chemistry from biochar also reinforced interactions between fungi and *Cyperus rotundus* L. [97].

In addition, a field experiment tested three treatments: no biochar, 10 t/m^3, and 20 t/m^3, with soil microbial diversity analyzed using high-throughput sequencing [98]. Biochar improved soil properties, and alpha diversity was highest at 20 t/m^3. The dominant phyla included Proteobacteria, Cyanobacteria, and Actinobacteria, with *Skermanella*, *Nostoc*, *Frankia*, and unclassified Proteobacteria as key genera (Figure 2). Actinobacteria varied significantly across treatments. Microbial community structure was mainly shaped by soil porosity, moisture, and nitrogen and potassium fertilizers, rather than phosphate or organic matter.

Zhang et al. [99] tested four wheat straw biochar levels (0, 5, 10, 15 kg/plant) across two soil depths (0–20 cm, 20–40 cm). Biochar improved citrus fruit quality and soil properties (pH, organic matter, nutrients). It enhanced bacterial richness, evenness, and diversity but reduced fungal evenness, while enriching beneficial bacteria and saprophytic (Basidiomycota) fungi involved in nutrient cycling. Furthermore, Yang et al. [100] examined *Panax quinquefolium* L. under different rice straw biochar levels (0–2.4%) to assess plant performance, active compounds, soil enzymes, and microbial communities. Biochar enhanced plant growth, photosynthesis, chlorophyll b, and ginsenoside content, while also increasing sucrase, urease, and acid phosphatase activities (Figure 2). In contrast, catalase activity and microbial diversity declined. Notably, bacterial phyla (e.g., Proteobacteria) and fungal genera (e.g., *Chalara*, *Sistotrema*) varied between the control and 1.8% biochar. Linear discriminant analysis further indicated significant shifts in rhizosphere microbial taxa under 1.8% and 2.4% treatments. Biochar can enhance microbial growth by supplying organic carbon and nutrients [100]. In this study, the rhizosphere fungal community of *P. quinquefolium* was more affected by biochar, differing from Zhang et al. [98], who reported increased bacterial diversity but reduced fungal evenness.

In a separate study, Ren et al. [101] assessed biochar (20 kg/plant) effects on bayberry decline by examining tree growth, fruit quality, soil properties, microbial communities, and metabolites. Biochar improved tree vigor, fruit quality, and rhizosphere microbial diversity. It increased *Mycobacterium*, *Crossiella*, *Geminibasidium*, and *Fusarium*, while reducing *Acidothermus*, *Bryobacter*, *Acidibacter*, *Cladophialophora*, *Mycena*, and *Rickenella*. Redundancy analysis showed that soil pH, organic matter, nutrients, and exchangeable cations strongly shaped microbial composition, with a greater influence on fungi than bacteria at the genus level, in line with the findings of Yang et al. [100].

A pot experiment with cotton seedlings was conducted to test rice straw biochar at three levels (0, 1, and 2%, *w*/*w*) for saline–alkaline soil treatment [102]. At 2%, biochar lowered soil pH, EC, and soluble salts, while increasing organic matter, nutrients, and enzyme activity. It reduced salinity in both soil and plant tissues and altered microbial community composition, enhancing salt-tolerant phyla like Proteobacteria, Bacteroidota, Acidobacteriota, and Actinobacteriota. Biochar markedly enhanced extracellular enzyme activity, with nitrogen- and phosphorus-cycling enzymes rising sharply by 32.79% and 34.25%, respectively. Biochar also strengthened microbe–metabolite interactions, mitigating salt stress on microorganisms [102].

A field study in a tropical ecosystem tested biochar from corn cobs at 15 and 30 t/ha, with or without phosphate fertilizer, on soil microbes and enzymes [103]. Biochar boosted microbial biomass C (4.5–8.2×), N (1.4–2.7×), and mineralized C (1.2–1.7×), while 30 t/ha amendments reduced specific maintenance respiration rates by 66–73%, indicating higher efficiency. It also increased dehydrogenase and urease activity, as well as the abundance of arbuscular mycorrhiza and soil fungi. Only 30 t/ha and 30 t/ha + P raised Gram-positive (1.7–1.9×) and Gram-negative (1.5–1.6×) bacteria, and total phospholipid fatty acid was highest under 30 t/ha treatments [103]. Specific maintenance respiration reflects microbial oxygen use and stress levels. Biochar improves soil porosity and gas diffusion [104], raising O_2 availability and stimulating microbial activity. This explains the sharp decline in specific maintenance respiration with biochar, while higher specific maintenance respiration in controls indicates carbon stress. The lower stress in biochar-amended soils also favors greater fungal abundance, as fungi typically thrive under stable conditions [105].

In a double rice system, the effects of wheat straw biochar on microbes and soil fertility were assessed 3–4 years after a single application (0, 24, 48 t/ha) [106]. Biochar increased bacterial and fungal abundances by up to 102% and 178%, linked to higher total organic carbon, nitrogen, and rice biomass, while archaea rose only slightly. The bacteria/fungi ratio fell by up to 61.4%, as fungi dominated in decomposing recalcitrant carbon. Biochar also promoted Acidobacteria, *Mortierella*, and *Westerdykella*, which support carbon cycling and plant growth, while reducing phytopathogens like *Athelia* and *Penicillium*. A summary of these recent findings is shown in Table 2.

Table 2. Biochar effects on soil microbial communities.

Ref.	Biochar Treatment	Crop/System	Duration	Key Findings	Microbial & Enzymatic Effects
[83]	Straw-based carbon substrate (100 kg)	Maize/Field	5 months	Enhanced soil fertility & yield; stage-specific nutrient effects (N early, C & P later)	↑ Rhizobiales, Saccharimonadales, Eurotiales; improved N, C, P cycling
[84]	Wood biochar (0–60 t/ha) + N (0–450 kg/ha)	Rapeseed	2 years	↑ Soil enzymes, plant height, pods, yield; alleviated N stress	↑ Urease, catalase, phosphatase, nitrate reductase, etc.; excess biochar (60 t/ha) inhibited urease/invertase
[81]	Enriched rice straw biochar (4.24 g/treatment)	General soils	60 days	Altered microbial structure; ↑ nutrient solubilization	↑ Proteobacteria, Actinobacteria; ↓ Methylomirabilota, Desulfobacterota; ↑ rhizosphere diversity
[87]	Biochar (0–126 Mg/ha)	Maize/Field	7 years	↑ Root traits, bacterial diversity (5–9%); minor fungal response	↑ Firmicutes; ↓ Actinobacteria, Acidobacteria; bacterial communities linked to root traits
[90]	Corn-straw biochar fractions (AE, OE, EBC)	Purple soil	30 days	↑ Available P, K, organic C; ↓ pH & N	AE affected bacteria via nutrients; EBC affected fungi via structure; ↑ fungal diversity
[91]	Apple wood biochar (1% *w/w*)	Apple orchard	1 year	↓ N_2O emissions (−35%); altered rhizosphere diversity	↑ N-fixing microbes; altered bacterial/fungal/archaeal structure
[92]	Straw vs. straw-biochar	Maize rhizosphere	21 days	Straw ↑ microbial biomass; biochar ↑ root biomass & stability	Straw favored Firmicutes (r-strategists, ↑ enzyme genes); biochar favored Actinobacteria (K-strategists)
[93]	Rice husk biochar (0–1.5%)	Lettuce	28 days	↑ Enzyme activity, root & leaf growth	↑ *Massilia*, *Bacillus*, *Trichocladium*; biochar stabilized microbial networks
[95]	Biochar (0–47.25 t/ha) + fertilizer	Field	6 years	↑ Soil organic matter, pH, C/N; altered N fractions	N fractions correlated with Verrucomicrobia, Acidobacteria, Proteobacteria; ↑ N-cycling enzymes
[96]	Rice & corn straw biochar	Rice (albic soil)/Pot	Till rice maturity	↑ Nutrients, diversity, ACE & Chao indices (2–3% for bacteria; 7–14% for fungi)	↑ Proteobacteria, Verrucomicrobia, Bacteroidetes; stronger bacteria–fungi interactions
[97]	Peanut shell biochar (0–10%)	Landfill soil	3 months	↑ pH, nutrients, organic matter; ↑ plant growth, ↓ plant diversity	↑ Bacterial richness; 5% biochar ↑ fungal diversity; enriched amino acid metabolism genes
[98]	Biochar (0, 10, 20 t/m^3)	Field soil	Not specified	↑ α-diversity (20 t/m^3); improved soil properties	↑ Proteobacteria, Cyanobacteria, Actinobacteria; shaped by porosity & moisture

Table 2. *Cont.*

Ref.	Biochar Treatment	Crop/System	Duration	Key Findings	Microbial & Enzymatic Effects
[99]	Wheat straw biochar (0–15 kg/plant)	Citrus	7 months	↑ Soil fertility & fruit quality	↑ Bacterial richness & diversity; ↑ Basidiomycota; ↓ fungal evenness
[100]	Rice straw biochar (0–2.4%)	*Panax quinquefolium*/Pot	7 months	↑ Growth, photosynthesis, ginsenosides (23–101%); ↓ catalase	↑ Proteobacteria, *Chalara*, *Sistotrema*; altered rhizosphere fungi strongly
[101]	Biochar (20 kg/plant)	Bayberry	11 months	↑ Growth (27–57% for various parameters), fruit quality, soil nutrients (14–58% for organic matter, alkali hydrolyzed N, exchangeable Ca, and Mg)	↑ *Mycobacterium*, *Crossiella*, *Geminibasidium*, *Fusarium*; ↓ *Acidothermus*, *Mycena*; fungi > bacteria
[102]	Rice straw biochar (0–2%)	Cotton, saline–alkaline soil/Pot	6 months	↓ pH (1%), electrical conductivity (19%), salts (25–28%); ↑ nutrients, enzyme activity	↑ Proteobacteria, Bacteroidota, Actinobacteriota; ↑ N, P cycling enzymes (32–34%)
[103]	Corn cob biochar (15–30 t/ha) ± P	Tropical soil/Field	2015 to Jan 2017	↑ Microbial biomass C (4.5–8.2 times), N (1.4–2.7 times), and enzymes; ↓ stress respiration	↑ *Gram+* (1.7–1.9 times) & Gram− (1.5–1.6 times) bacteria; ↑ fungi & arbuscular mycorrhizal fungi; improved porosity, O_2 diffusion
[106]	Wheat straw biochar (0–48 t/ha)	Double rice system	3–4 yrs post-application	↑ Bacteria (+102%), fungi (+178%); ↑ total organic carbon & N	↑ Acidobacteria, *Mortierella*, *Westerdykella*; ↓ *Athelia*, *Penicillium* (pathogens)

Note: The reported changes are statistically significant ($p < 0.05$) in the studies reviewed, unless otherwise specified with words like 'slight' and 'not strongly'; ↑ represents increase and ↓ represents decrease.

4.2. Variations in Bacterial and Fungal Communities

Biochar generally exerts positive effects on bacterial communities. It increases bacterial richness, diversity, and network stability by improving soil conditions, such as pH, porosity, nutrient availability, and aeration. Beneficial bacterial phyla, such as Proteobacteria, Actinobacteria, Rhizobiales, and Firmicutes, as well as genera like *Massilia*, often proliferate under biochar amendment, thereby supporting nutrient cycling, nitrogen fixation, and root development. Biochar's porous structure provides microhabitats and labile carbon that stimulate bacterial activity, while its ability to retain nutrients enhances bacterial-mediated processes such as phosphorus solubilization, nitrogen mineralization, and organic matter decomposition [81,92,93,95]. However, at very high application rates, biochar can reduce bacterial enzyme activity or shift community composition in ways that are less favorable, often due to excessive pH increases or adsorption of available nutrients [86].

The effects of biochar on fungi are more variable and context-dependent. In many cases, biochar increases fungal abundance and richness, particularly of beneficial saprophytic fungi (e.g., Basidiomycota, *Mortierella*, *Westerdykella*) that drive carbon cycling and plant–soil symbioses [99,106]. Fungal communities often show stronger responses to biochar's structural fractions (e.g., stable carbon compounds and pore architecture) than bacteria do [96,100]. Positive impacts include enhanced phosphorus mobilization, pathogen suppression, and increased resilience under stressors such as salinity.

However, biochar can also reduce fungal evenness or suppress certain groups, especially under high application rates, long-term use, or when biochar significantly alters soil pH [98]. Some studies reported little change in fungi compared to bacteria, while others observed declines in diversity or abundance of specific fungal taxa (e.g., *Mycena*, *Rickenella*) [100].

Biochar generally plays a suppressive role against soil-borne pathogens by modifying the soil environment in ways that reduce their competitiveness. Many studies show that biochar lowers the abundance of harmful fungi such as *Athelia* and *Penicillium* in rice systems, as well as *Cladophialophora*, *Mycena*, and *Rickenella* in bayberry soils [100,106]. These reductions are linked to improvements in soil fertility, organic matter, pH, and redox conditions, which create less favorable niches for pathogens to thrive. By improving overall soil quality, biochar makes conditions more conducive to beneficial microbes while simultaneously weakening pathogen survival.

An important mechanism for pathogen suppression is competitive exclusion by beneficial microbial taxa. Biochar stimulates the proliferation of groups like Proteobacteria, Actinobacteria, *Mortierella*, and *Westerdykella*, which are known for their roles in nutrient cycling, organic matter decomposition, and plant growth promotion [106]. These beneficial organisms occupy ecological niches and utilize resources that might otherwise support pathogens, thereby reducing their abundance and influence.

Enhanced soil enzyme activity also contributes to pathogen suppression. Biochar addition increases activities of urease, phosphatase, catalase, invertase, and other enzymes that strengthen nutrient cycling and microbial oxidative metabolism [67,84,103]. This enzymatic activity supports antagonistic microbial populations and biochemical processes that hinder pathogen growth. In particular, biochar's porous structure and high surface area provide habitats for beneficial microbes that can outcompete or inhibit pathogens.

However, biochar does not always act purely suppressively. Some evidence suggests that, under certain conditions, biochar may shift microbial communities in ways that favor particular taxa with pathogenic potential. For example, increases in *Fusarium* in bayberry soils following biochar application suggest that some pathogenic groups may benefit from the altered nutrient and pH environment [101].

4.3. Connections Between Microbial Activity and Nutrient Dynamics

Biochar alters nutrient dynamics largely through its influence on microbial activity. By improving soil conditions, such as pH, aeration, and organic matter content, biochar creates favorable habitats that stimulate microbial growth and activity. Beneficial groups such as Proteobacteria and Actinobacteria become more abundant, alongside fungi like *Mortierella* and *Westerdykella*, which contribute to organic matter decomposition and nutrient release [92,98,106]. These shifts in community composition strengthen soil biochemical cycling.

Nitrogen cycling is particularly sensitive to microbial responses under biochar amendment. Proteobacteria, many of which include nitrogen-fixing (*Rhizobium*) and denitrifying (*Pseudomonas*) genera, are stimulated by improved soil conditions [83,107]. Increased urease activity, often linked to bacterial and fungal groups, accelerates N mineralization, raising the pool of plant-available nitrogen. Biochar also promotes Fe-reducing microbes, such as members of the *Geobacter* genus within Deltaproteobacteria, which drive Fe^{3+} reduction [74,75]. This process lowers redox potential and indirectly regulates denitrification pathways, thereby affecting nitrogen losses and greenhouse gas emissions.

Phosphorus dynamics are also strongly microbially mediated. Enrichment of Proteobacteria and Actinobacteria enhances the activity of phosphatase-producing taxa, which liberate inorganic phosphate from organic matter [81]. Some groups within Actinobacteria are known for secreting organic acids that solubilize phosphate minerals, complementing biochar's direct effects on P sorption–desorption equilibria [108]. In this way, microbial turnover ensures greater and more sustained phosphorus availability.

For potassium, microbial activity plays a supporting role through mineral weathering and organic matter decomposition. Certain bacterial genera within Acidobacteria and Bacteroidetes contribute to the breakdown of plant residues, gradually releasing K^+. This complements the cation exchange capacity of biochar, which stabilizes K in the soil and prolongs its availability [65,109].

Overall, microbial activities under biochar amendment drive nutrient dynamics by accelerating organic matter decomposition, stimulating enzyme-mediated processes, and regulating redox-sensitive pathways. The enhanced abundance of phyla such as Proteobacteria, Actinobacteria, Bacteroidetes, and Acidobacteria, along with fungal genera like *Mortierella* and *Westerdykella*, underscores the biological mechanisms linking biochar's physicochemical properties with nutrient cycling. These microbial shifts underpin improvements in N, P, and K availability, thereby supporting soil fertility and plant nutrition.

5. Conclusions

Across a zonal series of soils, from acidic Podzols and Retisols to neutral loams and calcareous soils, biochar exerts soil-specific effects on nutrient dynamics, particularly phosphorus (P) and nitrogen (N) transformations. In coarse-textured, acidic soils such as Umbric Podzols, biochar markedly enhances nutrient availability by raising pH, stimulating phosphatase activity, and increasing exchangeable and Ca-bound P. Conversely, in fine-textured or neutral soils like Umbric Retisols and loamy sands, biochar's influence is more moderate, often improving nutrient retention but showing limited stimulation of microbial processes. High surface area and functional groups in biochar promote cation exchange and ammonium adsorption, thereby reducing volatilization and leaching losses. However, its efficacy depends strongly on soil reaction and mineralogy: optimal N retention typically occurs under mildly acidic to neutral conditions (pH 4.5–6.4), where biochar enhances NH_4^+ immobilization and suppresses nitrification. These zonal contrasts demonstrate that while biochar universally contributes to nutrient conservation and soil fertility enhancement, its mechanisms and magnitudes vary with soil type, texture, and pH. Hence, region-specific optimization of biochar formulation and application rate is essential for maximizing nutrient efficiency and sustaining soil productivity across diverse agroecosystems. For instance, higher application rates may be required for acidic soil to raise pH, increase available P, and improve nutrient retention, while lower rates are beneficial for calcareous or neutral to alkaline soils to avoid further alkalinization, which can immobilize P and hinder nutrient uptake.

Biochar has demonstrated consistent potential to improve soil fertility by enriching microbial diversity, stimulating enzyme activity, and supporting nutrient cycling, thereby boosting crop productivity. Its porous structure and nutrient content provide favorable habitats for beneficial bacterial and fungal taxa, which drive nitrogen fixation, phosphorus solubilization, and organic matter decomposition, while also suppressing certain pathogens. However, outcomes remain context-dependent, influenced by soil type, biochar feedstock, application rate, and crop species, with excessive use sometimes reducing enzyme activity, shifting microbial communities unfavorably, raising pH, and reducing nutrient availability.

Future research should prioritize long-term, multi-site field trials across varying climates, soils, and cropping systems to better understand biochar's role in microbial succession and nutrient-use efficiency. Combining molecular, isotopic, and metabolomic tools will help clarify the mechanisms underlying biochar–microbe interactions. In addition, integrating biochar with nanotechnology, organic amendments, and precision fertilization strategies may reveal synergies that optimize soil health, crop productivity, and climate resilience in sustainable agricultural systems.

Funding: This research received no external funding.

Data Availability Statement: This review paper does not report original data. All data presented or discussed are derived from previously published studies, which are cited throughout the manuscript. No new datasets were generated or analyzed for this study.

Acknowledgments: The author wishes to thank the University of Arizona for the administrative support provided.

Conflicts of Interest: The authors declare no conflicts of interest.

References

1. Hartmann, M.; Six, J. Soil structure and microbiome functions in agroecosystems. *Nat. Rev. Earth Environ.* **2023**, *4*, 4–18. [CrossRef]
2. Ren, T.; Feng, H.; Xu, C.; Xu, Q.; Fu, B.; Azwar, E.; Wei, Y.; Lam, S.S.; Liu, G. Exogenous application and interaction of biochar with environmental factors for improving functional diversity of rhizosphere's microbial community and health. *Chemosphere* **2022**, *294*, 133710. [CrossRef]
3. Qian, R.; Yu, K.; Chen, N.; Li, R.; Tang, K.H.D. Adsorptive immobilization of cadmium and lead using unmodified and modified biochar: A review of the advances, synthesis, efficiency and mechanisms. *Chemosphere* **2025**, *370*, 143988. [CrossRef]
4. Hossain, M.Z.; Bahar, M.M.; Sarkar, B.; Donne, S.W.; Ok, Y.S.; Palansooriya, K.N.; Kirkham, M.B.; Chowdhury, S.; Bolan, N. Biochar and its importance on nutrient dynamics in soil and plant. *Biochar* **2020**, *2*, 379–420. [CrossRef]
5. Kang, M.W.; Yibeltal, M.; Kim, Y.H.; Oh, S.J.; Lee, J.C.; Kwon, E.E.; Lee, S.S. Enhancement of soil physical properties and soil water retention with biochar-based soil amendments. *Sci. Total Environ.* **2022**, *836*, 155746. [CrossRef]
6. Kuok Ho Daniel, T. Valorization of organic waste as biosorbents for wastewater treatment. *Water Emerg. Contam. Nanoplastics* **2024**, *3*, 25. [CrossRef]
7. Liu, Y.; Pan, J.; Wang, J.; Yang, X.; Zhang, W.; Tang, K.H.D.; Wang, H.; Zhang, X.; Gao, R.; Yang, G.; et al. Insight into the humification and carbon balance of biogas residual biochar amended co-composting of hog slurry and wheat straw. *Environ. Sci. Pollut. Res.* **2025**, *32*, 21919–21930. [CrossRef]
8. Lyu, H.; Zhang, H.; Chu, M.; Zhang, C.; Tang, J.; Chang, S.X.; Mašek, O.; Ok, Y.S. Biochar affects greenhouse gas emissions in various environments: A critical review. *Land Degrad. Dev.* **2022**, *33*, 3327–3342. [CrossRef]
9. Xia, F.; Zhang, Z.; Zhang, Q.; Huang, H.; Zhao, X. Life cycle assessment of greenhouse gas emissions for various feedstocks-based biochars as soil amendment. *Sci. Total Environ.* **2024**, *911*, 168734. [CrossRef]
10. Tang, K.H.D. Urban Solutions to Climate Change: An Overview of Latest Progress. *Acad. Environ. Sci. Sustain.* **2024**, *1*. [CrossRef]
11. Atilano-Camino, M.M.; Canizales Laborin, A.P.; Ortega Juarez, A.M.; Valenzuela Cantú, A.K.; Pat-Espadas, A.M. Impact of Soil Amendment with Biochar on Greenhouse Gases Emissions, Metals Availability and Microbial Activity: A Meta-Analysis. *Sustainability* **2022**, *14*, 5648. [CrossRef]
12. Xie, Y.; Dong, C.; Chen, Z.; Liu, Y.; Zhang, Y.; Gou, P.; Zhao, X.; Ma, D.; Kang, G.; Wang, C.; et al. Successive biochar amendment affected crop yield by regulating soil nitrogen functional microbes in wheat-maize rotation farmland. *Environ. Res.* **2021**, *194*, 110671. [CrossRef]
13. Li, X.; Wu, D.; Liu, X.; Huang, Y.; Cai, A.; Xu, H.; Ran, J.; Xiao, J.; Zhang, W. A global dataset of biochar application effects on crop yield, soil properties, and greenhouse gas emissions. *Sci. Data* **2024**, *11*, 57. [CrossRef]
14. Das, S.K.; Ghosh, G.K.; Mishra, V.K.; Choudhury, B.U.; Dutta, S.K.; Hazarika, S.; Kalita, H.; Roy, A.; Singh, N.U.; Gopi, R.; et al. Utilizing dissimilar feedstocks derived biochar amendments to alter soil biological indicators in acidic soil of Northeast India. *Biomass Convers. Biorefinery* **2023**, *13*, 10203–10214. [CrossRef]
15. Pokharel, P.; Ma, Z.; Chang, S.X. Biochar increases soil microbial biomass with changes in extra- and intracellular enzyme activities: A global meta-analysis. *Biochar* **2020**, *2*, 65–79. [CrossRef]
16. Hajiboland, R.; Panda, C.K.; Lastochkina, O.; Gavassi, M.A.; Habermann, G.; Pereira, J.F. Aluminum Toxicity in Plants: Present and Future. *J. Plant Growth Regul.* **2023**, *42*, 3967–3999. [CrossRef]
17. El-Naggar, A.; Lee, S.S.; Rinklebe, J.; Farooq, M.; Song, H.; Sarmah, A.K.; Zimmerman, A.R.; Ahmad, M.; Shaheen, S.M.; Ok, Y.S. Biochar application to low fertility soils: A review of current status, and future prospects. *Geoderma* **2019**, *337*, 536–554. [CrossRef]

18. Purakayastha, T.J.; Bera, T.; Bhaduri, D.; Sarkar, B.; Mandal, S.; Wade, P.; Kumari, S.; Biswas, S.; Menon, M.; Pathak, H.; et al. A review on biochar modulated soil condition improvements and nutrient dynamics concerning crop yields: Pathways to climate change mitigation and global food security. *Chemosphere* **2019**, *227*, 345–365. [CrossRef]
19. Leng, L.; Xu, S.; Liu, R.; Yu, T.; Zhuo, X.; Leng, S.; Xiong, Q.; Huang, H. Nitrogen containing functional groups of biochar: An overview. *Bioresour. Technol.* **2020**, *298*, 122286. [CrossRef]
20. Christel, W.; Bruun, S.; Magid, J.; Kwapinski, W.; Jensen, L.S. Pig slurry acidification, separation technology and thermal conversion affect phosphorus availability in soil amended with the derived solid fractions, chars or ashes. *Plant Soil* **2016**, *401*, 93–107. [CrossRef]
21. Zhao, B.; O'Connor, D.; Zhang, J.; Peng, T.; Shen, Z.; Tsang, D.C.W.; Hou, D. Effect of pyrolysis temperature, heating rate, and residence time on rapeseed stem derived biochar. *J. Clean. Prod.* **2018**, *174*, 977–987. [CrossRef]
22. Tomczyk, A.; Sokołowska, Z.; Boguta, P. Biochar physicochemical properties: Pyrolysis temperature and feedstock kind effects. *Rev. Environ. Sci. Bio/Technol.* **2020**, *19*, 191–215. [CrossRef]
23. Domingues, R.R.; Sánchez-Monedero, M.A.; Spokas, K.A.; Melo, L.C.A.; Trugilho, P.F.; Valenciano, M.N.; Silva, C.A. Enhancing Cation Exchange Capacity of Weathered Soils Using Biochar: Feedstock, Pyrolysis Conditions and Addition Rate. *Agronomy* **2020**, *10*, 824. [CrossRef]
24. Salem, T.M.; Refaie, K.M.; Sherif, A.E.-H.E.-G.A.E.-L.; Eid, M.A.M. Biochar application in alkaline soil and its effect on soil and plant. *Acta Agric. Slov.* **2019**, *114*, 85–96. [CrossRef]
25. Zhou, R.; Chen, Z.; Ei-Naggar, A.; Tian, L.; Huang, C.; Zhang, Z.; Palansooriya, K.N.; Li, Y.; Yu, B.; Chang, S.X.; et al. Contrasting effects of rice husk and its biochar on N_2O emissions and nitrogen leaching from Lei bamboo soils under subtropical conditions. *Biol. Fertil. Soils* **2023**, *59*, 803–817. [CrossRef]
26. Kaur, N.; Kieffer, C.; Ren, W.; Hui, D. How much is soil nitrous oxide emission reduced with biochar application? An evaluation of meta-analyses. *GCB Bioenergy* **2023**, *15*, 24–37. [CrossRef]
27. Yu, H.; Zou, W.; Chen, J.; Chen, H.; Yu, Z.; Huang, J.; Tang, H.; Wei, X.; Gao, B. Biochar amendment improves crop production in problem soils: A review. *J. Environ. Manag.* **2019**, *232*, 8–21. [CrossRef]
28. Kołtowski, M.; Hilber, I.; Bucheli, T.D.; Oleszczuk, P. Effect of steam activated biochar application to industrially contaminated soils on bioavailability of polycyclic aromatic hydrocarbons and ecotoxicity of soils. *Sci. Total Environ.* **2016**, *566–567*, 1023–1031. [CrossRef]
29. Ippolito, J.A.; Berry, C.M.; Strawn, D.G.; Novak, J.M.; Levine, J.; Harley, A. Biochars Reduce Mine Land Soil Bioavailable Metals. *J. Environ. Qual.* **2017**, *46*, 411–419. [CrossRef]
30. Tang, K.H.D.; Kristanti, R.A. Bioremediation of perfluorochemicals: Current state and the way forward. *Bioprocess Biosyst. Eng.* **2022**, *45*, 1093–1109. [CrossRef]
31. O'Connor, D.; Peng, T.; Zhang, J.; Tsang, D.C.W.; Alessi, D.S.; Shen, Z.; Bolan, N.S.; Hou, D. Biochar application for the remediation of heavy metal polluted land: A review of in situ field trials. *Sci. Total Environ.* **2018**, *619–620*, 815–826. [CrossRef] [PubMed]
32. Zama, E.F.; Reid, B.J.; Arp, H.P.H.; Sun, G.-X.; Yuan, H.-Y.; Zhu, Y.-G. Advances in research on the use of biochar in soil for remediation: A review. *J. Soils Sediments* **2018**, *18*, 2433–2450. [CrossRef]
33. Kumpiene, J.; Antelo, J.; Brännvall, E.; Carabante, I.; Ek, K.; Komárek, M.; Söderberg, C.; Wårell, L. In situ chemical stabilization of trace element-contaminated soil—Field demonstrations and barriers to transition from laboratory to the field—A review. *Appl. Geochem.* **2019**, *100*, 335–351. [CrossRef]
34. Dai, Z.; Xiong, X.; Zhu, H.; Xu, H.; Leng, P.; Li, J.; Tang, C.; Xu, J. Association of biochar properties with changes in soil bacterial, fungal and fauna communities and nutrient cycling processes. *Biochar* **2021**, *3*, 239–254. [CrossRef]
35. Tsolis, V.; Barouchas, P. Biochar as Soil Amendment: The Effect of Biochar on Soil Properties Using VIS-NIR Diffuse Reflectance Spectroscopy, Biochar Aging and Soil Microbiology—A Review. *Land* **2023**, *12*, 1580. [CrossRef]
36. Anyebe, O.; Sadiq, F.K.; Manono, B.O.; Matsika, T.A. Biochar Characteristics and Application: Effects on Soil Ecosystem Services and Nutrient Dynamics for Enhanced Crop Yields. *Nitrogen* **2025**, *6*, 31. [CrossRef]
37. Singh, S.; Luthra, N.; Mandal, S.; Kushwaha, D.P.; Pathak, S.O.; Datta, D.; Sharma, R.; Pramanick, B. Distinct Behavior of Biochar Modulating Biogeochemistry of Salt-Affected and Acidic Soil: A Review. *J. Soil Sci. Plant Nutr.* **2023**, *23*, 2981–2997. [CrossRef]
38. Gao, S.; DeLuca, T.H. Rangeland application of biochar and rotational grazing interact to influence soil and plant nutrient dynamics. *Geoderma* **2022**, *408*, 115572. [CrossRef]
39. Kuoppamäki, K.; Setälä, H.; Hagner, M. Nutrient dynamics and development of soil fauna in vegetated roofs with the focus on biochar amendment. *Nat.-Based Solut.* **2021**, *1*, 100001. [CrossRef]
40. Dubrovina, I.A. Changes in the Phosphate Regime of Soils in the Middle Taiga under the Impact of Biochar. *Eurasian Soil Sci.* **2023**, *56*, 363–370. [CrossRef]
41. Gupta, R.K.; Vashisht, M.; Naresh, R.K.; Dhingra, N.; Sidhu, M.S.; Singh, P.K.; Rani, N.; Al-Ansari, N.; Alataway, A.; Dewidar, A.Z.; et al. Biochar influences nitrogen and phosphorus dynamics in two texturally different soils. *Sci. Rep.* **2024**, *14*, 6533. [CrossRef] [PubMed]

42. Amin, A.E.-E.A.Z. Carbon sequestration, kinetics of ammonia volatilization and nutrient availability in alkaline sandy soil as a function on applying calotropis biochar produced at different pyrolysis temperatures. *Sci. Total Environ.* **2020**, *726*, 138489. [CrossRef] [PubMed]
43. Qian, X.; Li, Q.; Chen, H.; Zhao, L.; Wang, F.; Zhang, Y.; Zhang, J.; Müller, C.; Yi, Z. Enhancing Soil Nitrogen Retention Capacity by Biochar Incorporation in the Acidic Soil of Pomelo Orchards: The Crucial Role of pH. *Agronomy* **2023**, *13*, 2110. [CrossRef]
44. Lu, J.; Li, Y.; Wang, B.; Hou, B.; Du, G.; Si, H. Analysis of the adsorption and fixation process of ammonium nitrogen in arable soil by biochar based on molecular dynamics simulation. *Sci. Total Environ.* **2024**, *930*, 172815. [CrossRef]
45. Hu, T.; Wei, J.; Du, L.; Chen, J.; Zhang, J. The effect of biochar on nitrogen availability and bacterial community in farmland. *Ann. Microbiol.* **2023**, *73*, 4. [CrossRef]
46. Zhang, M.; Song, G.; Gelardi, D.L.; Huang, L.; Khan, E.; Mašek, O.; Parikh, S.J.; Ok, Y.S. Evaluating biochar and its modifications for the removal of ammonium, nitrate, and phosphate in water. *Water Res.* **2020**, *186*, 116303. [CrossRef]
47. Wu, B.; Zhang, M.; Zhai, Z.; Dai, H.; Yang, M.; Zhang, Y.; Liang, T. Soil organic carbon, carbon fractions, and microbial community under various organic amendments. *Sci. Rep.* **2024**, *14*, 25431. [CrossRef]
48. Abdul-Aziz, A.-L.; Abukari, I.A.; Galadima, M.M.; Haruna, A.; Abubakari, M.; Abdulai, R. Biochar effects on soil properties and yield of maize in Northern region, Ghana. *Discov. Agric.* **2025**, *3*, 103. [CrossRef]
49. Liu, Z.; Wu, X.; Li, S.; Liu, W.; Bian, R.; Zhang, X.; Zheng, J.; Drosos, M.; Li, L.; Pan, G. Quantitative assessment of the effects of biochar amendment on photosynthetic carbon assimilation and dynamics in a rice–soil system. *New Phytol.* **2021**, *232*, 1250–1258. [CrossRef]
50. Ge, T.; Liu, C.; Yuan, H.; Zhao, Z.; Wu, X.; Zhu, Z.; Brookes, P.; Wu, J. Tracking the photosynthesized carbon input into soil organic carbon pools in a rice soil fertilized with nitrogen. *Plant Soil* **2015**, *392*, 17–25. [CrossRef]
51. Zang, H.; Xiao, M.; Wang, Y.; Ling, N.; Wu, J.; Ge, T.; Kuzyakov, Y. Allocation of assimilated carbon in paddies depending on rice age, chase period and N fertilization: Experiment with $^{13}CO_2$ labelling and literature synthesis. *Plant Soil* **2019**, *445*, 113–123. [CrossRef]
52. Qiu, H.; Liu, J.; Boorboori, M.R.; Li, D.; Chen, S.; Ma, X.; Cheng, P.; Zhang, H. Effect of biochar application rate on changes in soil labile organic carbon fractions and the association between bacterial community assembly and carbon metabolism with time. *Sci. Total Environ.* **2023**, *855*, 158876. [CrossRef]
53. Chagas, J.K.M.; Figueiredo, C.C.d.; Ramos, M.L.G. Biochar increases soil carbon pools: Evidence from a global meta-analysis. *J. Environ. Manag.* **2022**, *305*, 114403. [CrossRef]
54. Zhu, S.; Guo, Y.; Zhou, H.; Luo, W.; Yi, X.; Zhou, Y.; Wu, Y.; Petticord, D.F.; Song, X. Biochar efficacy in enhancing soil carbon fractions is mediated by parent soil type in grazing karst grassland. *Carbon Res.* **2025**, *4*, 52. [CrossRef]
55. Patel, M.R.; Panwar, N.L. Biochar from agricultural crop residues: Environmental, production, and life cycle assessment overview. *Resour. Conserv. Recycl. Adv.* **2023**, *19*, 200173. [CrossRef]
56. Chintala, R.; Mollinedo, J.; Schumacher, T.E.; Malo, D.D.; Julson, J.L. Effect of biochar on chemical properties of acidic soil. *Arch. Agron. Soil Sci.* **2014**, *60*, 393–404. [CrossRef]
57. Orlova, N.; Abakumov, E.; Orlova, E.; Yakkonen, K.; Shahnazarova, V. Soil organic matter alteration under biochar amendment: Study in the incubation experiment on the Podzol soils of the Leningrad region (Russia). *J. Soils Sediments* **2019**, *19*, 2708–2716. [CrossRef]
58. Spokas, K.A. Review of the stability of biochar in soils: Predictability of O:C molar ratios. *Carbon Manag.* **2010**, *1*, 289–303. [CrossRef]
59. Lu, J.; Luo, Y.; Huang, J.; Hou, B.; Wang, B.; Ogino, K.; Zhao, J.; Si, H. The effect of biochar on the migration theory of nutrient ions. *Sci. Total Environ.* **2022**, *845*, 157262. [CrossRef]
60. Lewoyehu, M.; Kohira, Y.; Fentie, D.; Addisu, S.; Sato, S. Water Hyacinth Biochar: A Sustainable Approach for Enhancing Soil Resistance to Acidification Stress and Nutrient Dynamics in an Acidic Nitisol of the Northwest Highlands of Ethiopia. *Sustainability* **2024**, *16*, 5537. [CrossRef]
61. Shi, R.-Y.; Ni, N.; Wang, R.-H.; Nkoh, J.N.; Pan, X.-Y.; Dong, G.; Xu, R.-K.; Cui, X.-M.; Li, J.-Y. Dissolved biochar fractions and solid biochar particles inhibit soil acidification induced by nitrification through different mechanisms. *Sci. Total Environ.* **2023**, *874*, 162464. [CrossRef]
62. Burgeon, V.; Fouché, J.; Garré, S.; Dehkordi, R.H.; Colinet, G.; Cornelis, J.-T. Young and century-old biochars strongly affect nutrient cycling in a temperate agroecosystem. *Agric. Ecosyst. Environ.* **2022**, *328*, 107847. [CrossRef]
63. Liu, B.; Li, H.; Li, H.; Zhang, A.; Rengel, Z. Long-term biochar application promotes rice productivity by regulating root dynamic development and reducing nitrogen leaching. *GCB Bioenergy* **2021**, *13*, 257–268. [CrossRef]
64. dos Santos, W.M.; Gonzaga, M.I.S.; da Silva, A.J.; de Almeida, A.Q. Improved water and ions dynamics in a clayey soil amended with different types of agro-industrial waste biochar. *Soil Tillage Res.* **2022**, *223*, 105482. [CrossRef]

65. Hadroug, S.; Jellali, S.; Jeguirim, M.; Kwapinska, M.; Hamdi, H.; Leahy, J.J.; Kwapinski, W. Static and Dynamic Investigations on Leaching/Retention of Nutrients from Raw Poultry Manure Biochars and Amended Agricultural Soil. *Sustainability* **2021**, *13*, 1212. [CrossRef]
66. Dubrovina, I. Biochar has Different Effects on Soil Fertility and Spring Barley Productivity in Two Boreal Soils with Contrasting Textures. *Agric. Res.* **2025**. [CrossRef]
67. Zhang, K.; Khan, Z.; Khan, M.N.; Luo, T.; Luo, L.; Bi, J.; Hu, L. The application of biochar improves the nutrient supply efficiency of organic fertilizer, sustains soil quality and promotes sustainable crop production. *Food Energy Secur.* **2024**, *13*, e520. [CrossRef]
68. Schulz, H.; Dunst, G.; Glaser, B. Positive effects of composted biochar on plant growth and soil fertility. *Agron. Sustain. Dev.* **2013**, *33*, 817–827. [CrossRef]
69. Lu, Y.; Silveira, M.L.; O'Connor, G.A.; Vendramini, J.M.B.; Li, Y.C. Biochar type and application methods affected nitrogen and phosphorus leaching from a sandy soil amended with inorganic fertilizers and biosolids. *Agrosyst. Geosci. Environ.* **2022**, *5*, e20236. [CrossRef]
70. Lian, F.; Xing, B. Black Carbon (Biochar) In Water/Soil Environments: Molecular Structure, Sorption, Stability, and Potential Risk. *Environ. Sci. Technol.* **2017**, *51*, 13517–13532. [CrossRef]
71. Llovet, A.; Vidal-Durà, A.; Alcañiz, J.M.; Ribas, A.; Domene, X. Biochar addition to organo-mineral fertilisers delays nutrient leaching and enhances barley nutrient content. *Arch. Agron. Soil Sci.* **2023**, *69*, 2537–2551. [CrossRef]
72. Sarfraz, R.; Nadeem, F.; Yang, W.; Tayyab, M.; Khan, M.I.; Mahmood, R.; Guo, X.; Xing, S.; Kim, G.W. Evaluation of Biochar and Inorganic Fertilizer on Soil Available Phosphorus and Bacterial Community Dynamics in Acidic Paddy Soils for Different Incubation Temperatures. *Agronomy* **2024**, *14*, 26. [CrossRef]
73. Ndoung, O.C.N.; Souza, L.R.d.; Fachini, J.; Leão, T.P.; Sandri, D.; Figueiredo, C.C.d. Dynamics of potassium released from sewage sludge biochar fertilizers in soil. *J. Environ. Manag.* **2023**, *346*, 119057. [CrossRef]
74. Yin, X.; Peñuelas, J.; Sardans, J.; Xu, X.; Chen, Y.; Fang, Y.; Wu, L.; Singh, B.P.; Tavakkoli, E.; Wang, W. Effects of nitrogen-enriched biochar on rice growth and yield, iron dynamics, and soil carbon storage and emissions: A tool to improve sustainable rice cultivation. *Environ. Pollut.* **2021**, *287*, 117565. [CrossRef]
75. Suksabye, P.; Pimthong, A.; Dhurakit, P.; Mekvichitsaeng, P.; Thiravetyan, P. Effect of biochars and microorganisms on cadmium accumulation in rice grains grown in Cd-contaminated soil. *Environ. Sci. Pollut. Res.* **2016**, *23*, 962–973. [CrossRef]
76. Wang, M.; Lan, X.; Xu, X.; Fang, Y.; Singh, B.P.; Sardans, J.; Romero, E.; Peñuelas, J.; Wang, W. Steel slag and biochar amendments decreased CO_2 emissions by altering soil chemical properties and bacterial community structure over two-year in a subtropical paddy field. *Sci. Total Environ.* **2020**, *740*, 140403. [CrossRef]
77. Joseph, S.; Husson, O.; Graber, E.R.; Van Zwieten, L.; Taherymoosavi, S.; Thomas, T.; Nielsen, S.; Ye, J.; Pan, G.; Chia, C.; et al. The Electrochemical Properties of Biochars and How They Affect Soil Redox Properties and Processes. *Agronomy* **2015**, *5*, 322–340. [CrossRef]
78. Javeed, H.M.; Ali, M.; Ahmed, I.; Wang, X.; Al-Ashkar, I.; Qamar, R.; Ibrahim, A.; Habib-Ur-Rahman, M.; Ditta, A.; El Sabagh, A. Biochar Enriched with Buffalo Slurry Improved Soil Nitrogen and Carbon Dynamics, Nutrient Uptake and Growth Attributes of Wheat by Reducing Leaching Losses of Nutrients. *Land* **2021**, *10*, 1392. [CrossRef]
79. Roy, A.; Chaturvedi, S.; Singh, S.V.; Kasivelu, G.; Dhyani, V.C.; Pyne, S. Preparation and evaluation of two enriched biochar-based fertilizers for nutrient release kinetics and agronomic effectiveness in direct-seeded rice. *Biomass Convers. Biorefinery* **2024**, *14*, 2007–2018. [CrossRef]
80. Lan, Y.; Chu, Q.; Liu, X.; Xu, S.; Li, D.; Zhang, C.; He, P.; Feng, X.; Zhang, H.; Sha, Z. Oxygen-nanobubble-loaded biochar increases soil carbon sequestration in rice paddies. *Soil Environ. Health* **2025**, *3*, 100174. [CrossRef]
81. Deng, L.; Tu, P.; Ahmed, N.; Zhang, G.; Cen, Y.; Huang, B.; Deng, L.; Yuan, H. Biochar-based phosphate fertilizer improve phosphorus bioavailability, microbial functioning, and citrus seedling growth. *Sci. Hortic.* **2024**, *338*, 113699. [CrossRef]
82. Mandal, S.; Donner, E.; Smith, E.; Sarkar, B.; Lombi, E. Biochar with near-neutral pH reduces ammonia volatilization and improves plant growth in a soil-plant system: A closed chamber experiment. *Sci. Total Environ.* **2019**, *697*, 134114. [CrossRef]
83. Zhang, J.; Ye, L.; Chang, J.; Wang, E.; Wang, C.; Zhang, H.; Pang, Y.; Tian, C. Straw Soil Conditioner Modulates Key Soil Microbes and Nutrient Dynamics across Different Maize Developmental Stages. *Microorganisms* **2024**, *12*, 295. [CrossRef]
84. Khan, Z.; Zhang, K.; Khan, M.N.; Bi, J.; Zhu, K.; Luo, L.; Hu, L. How Biochar Affects Nitrogen Assimilation and Dynamics by Interacting Soil and Plant Enzymatic Activities: Quantitative Assessment of 2 Years Potted Study in a Rapeseed-Soil System. *Front. Plant Sci.* **2022**, *13*, 853449. [CrossRef]
85. Gong, X.; Cai, L.; Li, S.; Chang, S.X.; Sun, X.; An, Z. Bamboo biochar amendment improves the growth and reproduction of Eisenia fetida and the quality of green waste vermicompost. *Ecotoxicol. Environ. Saf.* **2018**, *156*, 197–204. [CrossRef]
86. Elzobair, K.A.; Stromberger, M.E.; Ippolito, J.A.; Lentz, R.D. Contrasting effects of biochar versus manure on soil microbial communities and enzyme activities in an Aridisol. *Chemosphere* **2016**, *142*, 145–152. [CrossRef]
87. Yan, H.; Cong, M.; Hu, Y.; Qiu, C.; Yang, Z.; Tang, G.; Xu, W.; Zhu, X.; Sun, X.; Jia, H. Biochar-mediated changes in the microbial communities of rhizosphere soil alter the architecture of maize roots. *Front. Microbiol.* **2022**, *13*, 1023444. [CrossRef]

88. Lehmann, J.; Rillig, M.C.; Thies, J.; Masiello, C.A.; Hockaday, W.C.; Crowley, D. Biochar effects on soil biota—A review. *Soil Biol. Biochem.* **2011**, *43*, 1812–1836. [CrossRef]
89. Li, X.; Wu, M.; Xue, Y. Nickel-loaded shrimp shell biochar enhances batch anaerobic digestion of food waste. *Bioresour. Technol.* **2022**, *352*, 127092. [CrossRef]
90. Li, Y.; Yang, Y.; Shen, F.; Tian, D.; Zeng, Y.; Yang, G.; Zhang, Y.; Deng, S. Partitioning biochar properties to elucidate their contributions to bacterial and fungal community composition of purple soil. *Sci. Total Environ.* **2019**, *648*, 1333–1341. [CrossRef]
91. Cao, H.; Jia, M.; Xun, M.; Wang, X.; Chen, K.; Yang, H. Nitrogen transformation and microbial community structure varied in apple rhizosphere and rhizoplane soils under biochar amendment. *J. Soils Sediments* **2021**, *21*, 853–868. [CrossRef]
92. Shang, W.; Razavi, B.S.; Yao, S.; Hao, C.; Kuzyakov, Y.; Blagodatskaya, E.; Tian, J. Contrasting mechanisms of nutrient mobilization in rhizosphere hotspots driven by straw and biochar amendment. *Soil Biol. Biochem.* **2023**, *187*, 109212. [CrossRef]
93. Akumuntu, A.; Hong, J.-K.; Jho, E.H.; Omidoyin, K.C.; Park, S.-J.; Zhang, Q.; Zhao, X. Biochar derived from rice husk: Impact on soil enzyme and microbial dynamics, lettuce growth, and toxicity. *Chemosphere* **2024**, *349*, 140868. [CrossRef]
94. Liu, X.; Zhiming, Q.I.; Wang, Q.; Zhiwen, M.A.; Lanhai, L.I. Effects of biochar addition on CO_2 and CH_4 emissions from a cultivated sandy loam soil during freeze-thaw cycles. *Plant Soil Environ.* **2017**, *63*, 243–249. [CrossRef]
95. Di, W.; Lan, Y.; Chen, W.; Liu, Z.; Gao, J.; Cao, D.; Wang, Q.; Mazhang, C.; An, X. Response of bacterial communities, enzyme activities and dynamic changes of soil organic nitrogen fractions to six-year different application levels of biochar retention in Northeast China. *Soil Tillage Res.* **2024**, *240*, 106097. [CrossRef]
96. Yin, D.; Li, H.; Wang, H.; Guo, X.; Wang, Z.; Lv, Y.; Ding, G.; Jin, L.; Lan, Y. Impact of Different Biochars on Microbial Community Structure in the Rhizospheric Soil of Rice Grown in Albic Soil. *Molecules* **2021**, *26*, 4783. [CrossRef]
97. Liao, J.; Hao, G.; Guo, H.; Chen, H. Effects of biochar on plant and microbial communities in landfill soil. *Appl. Soil Ecol.* **2024**, *204*, 105749. [CrossRef]
98. Zhang, J.; Shen, J.-L. Effects of biochar on soil microbial diversity and community structure in clay soil. *Ann. Microbiol.* **2022**, *72*, 35. [CrossRef]
99. Zhang, M.; Zhang, L.; Riaz, M.; Xia, H.; Jiang, C. Biochar amendment improved fruit quality and soil properties and microbial communities at different depths in citrus production. *J. Clean. Prod.* **2021**, *292*, 126062. [CrossRef]
100. Yang, X.; Ran, Z.; Li, R.; Fang, L.; Zhou, J.; Guo, L. Effects of Biochar on the Growth, Ginsenoside Content, and Soil Microbial Community Composition of Panax quinquefolium L. *J. Soil Sci. Plant Nutr.* **2022**, *22*, 2670–2686. [CrossRef]
101. Ren, H.; Guo, H.; Shafiqul Islam, M.; Zaki, H.E.M.; Wang, Z.; Wang, H.; Qi, X.; Guo, J.; Sun, L.; Wang, Q.; et al. Improvement effect of biochar on soil microbial community structure and metabolites of decline disease bayberry. *Front. Microbiol.* **2023**, *14*, 1154886. [CrossRef]
102. Wang, X.; Riaz, M.; Babar, S.; Eldesouki, Z.; Liu, B.; Xia, H.; Li, Y.; Wang, J.; Xia, X.; Jiang, C. Alterations in the composition and metabolite profiles of the saline-alkali soil microbial community through biochar application. *J. Environ. Manag.* **2024**, *352*, 120033. [CrossRef]
103. Amoakwah, E.; Arthur, E.; Frimpong, K.A.; Lorenz, N.; Rahman, M.A.; Nziguheba, G.; Islam, K.R. Biochar amendment impacts on microbial community structures and biological and enzyme activities in a weathered tropical sandy loam. *Appl. Soil Ecol.* **2022**, *172*, 104364. [CrossRef]
104. Arthur, E.; Ahmed, F. Rice straw biochar affects water retention and air movement in a sand-textured tropical soil. *Arch. Agron. Soil Sci.* **2017**, *63*, 2035–2047. [CrossRef]
105. Rodriguez-Ramos, J.C.; Cale, J.A.; Cahill, J.F., Jr.; Simard, S.W.; Karst, J.; Erbilgin, N. Changes in soil fungal community composition depend on functional group and forest disturbance type. *New Phytol.* **2021**, *229*, 1105–1117. [CrossRef]
106. Wang, C.; Chen, D.; Shen, J.; Yuan, Q.; Fan, F.; Wei, W.; Li, Y.; Wu, J. Biochar alters soil microbial communities and potential functions 3–4 years after amendment in a double rice cropping system. *Agric. Ecosyst. Environ.* **2021**, *311*, 107291. [CrossRef]
107. Liu, Y.; Liu, S.; Yang, Z.; Xiao, L. Synergetic effects of biochars and denitrifier on nitrate removal. *Bioresour. Technol.* **2021**, *335*, 125245. [CrossRef] [PubMed]
108. Solans, M.; Messuti, M.I.; Reiner, G.; Boenel, M.; Vobis, G.; Wall, L.G.; Scervino, J.M. Exploring the response of Actinobacteria to the presence of phosphorus salts sources: Metabolic and co-metabolic processes. *J. Basic Microbiol.* **2019**, *59*, 487–495. [CrossRef]
109. Li, F.; Chen, L.; Zhang, J.; Yin, J.; Huang, S. Bacterial Community Structure after Long-term Organic and Inorganic Fertilization Reveals Important Associations between Soil Nutrients and Specific Taxa Involved in Nutrient Transformations. *Front. Microbiol.* **2017**, *8*, 187. [CrossRef]

MDPI AG
Grosspeteranlage 5
4052 Basel
Switzerland
Tel.: +41 61 683 77 34

Environments Editorial Office
E-mail: environments@mdpi.com
www.mdpi.com/journal/environments

www.ingramcontent.com/pod-product-compliance
Lightning Source LLC
LaVergne TN
LVHW070451061126
851858LV00011B/212
* 9 7 8 3 7 2 5 8 8 4 5 9 9 *